Fundamental Principles of Environmental Physics

Abel Rodrigues •
Raul Albuquerque Sardinha •
Gabriel Pita

Fundamental Principles of Environmental Physics

 Springer

Abel Rodrigues 🅿
UTI
Instituto Nacional de Investigação
Agrária e Veterinária
Oeiras, Portugal

Raul Albuquerque Sardinha
Carnaxide, Portugal

Gabriel Pita 🅿
IST—Mechanical Engineering
University of Lisbon
Lisbon, Portugal

ISBN 978-3-030-69027-4   ISBN 978-3-030-69025-0   (eBook)
https://doi.org/10.1007/978-3-030-69025-0

This Springer imprint is published by the registered company Springer Nature Switzerland AG
The registered company address is: Gewerbestrasse 11, 6330 Cham, Switzerland

Foreword I

The objective of this book is to promote an understanding of the qualitative characterization of heat and mass transfer processes in the atmospheric boundary layer. More specifically, it addresses the quantification of energy budgets and vertical gas fluxes over vegetated surfaces including evapotranspiration. The book abides by a classical definition of biophysical ecology in that "This understanding is enhanced by using mathematical formulations of physical processes and relating them to the unique properties of organisms".[1]

The topics covered by the book are instrumental to deal with issues of great societal relevance. Agriculture and forestry manage exchanges of light, water, and gases between vegetated land surfaces and the atmosphere to produce food, feed, fiber, wood, and fuels. From an environmental protection perspective, biophysical ecology plays equally significant roles in the study and development of solutions for the problems of climate change, desertification, and protection of biological diversity, which is critically dependent on healthy vegetated habitats.

The judicious choices of the scope and depth of the book contents and its clarity of organization are very well targeted for a public of undergraduate and master's level instructors, students, and practitioners in the environmental sciences, earth sciences, agriculture, forestry, and physical geography.

The main body of the text characterizes the atmospheric surface layer, including turbulent flow, flow over built-up surfaces, mass and energy flows over forests, essential concepts of heat and mass transfer, and fundamentals of the global carbon budget and climate change. The latter topic includes a component discussing the potential relevance of biochar as a global carbon sink, a theme of utmost importance nowadays. Worked-out exercises dealing with the concepts introduced in the previous chapters are well designed and will help consolidate understanding of the corresponding topics. A series of annexes effectively frame and complement some of the themes addressed in the main section of the book for the reader interested in additional information.

The three authors have academic backgrounds in Forestry (an engineering degree, in Portugal), and in Mechanical Engineering. Abel Rodrigues has an undergraduate degree in Forestry engineering obtained in 1983 from the School of

---

[1]Gates, D. M. (2012). *Biophysical ecology*. Courier Corporation.

Agriculture and Forestry (ISA, U. Lisbon), and from Instituto Superior Técnico (IST, U. Lisbon) an M.Sc. in Mechanical Engineering (1994), and a Ph.D. in Environmental Engineering (2002). He is a Senior Researcher at the Forestry Research Unit of the National Institute for Agricultural and Veterinary Research, where he developed research on atmospheric fluxes and on production and conversion of biomass from short rotation coppices.

Gabriel Pita has an undergraduate degree in Mechanical Engineering/Applied Thermodynamics (1975) and a Ph.D. in Mechanical Engineering (1988), both from IST. He is an Assistant Professor at the Department of Mechanical Engineering at IST, where he teaches Applied Thermodynamics, Physical Ecology, and Mass and Energy Transfer. Professor Pita's research has dealt with applied combustion, characterization of the impacts of droughts on Portuguese forests, and analysis of carbon uptake and productivity in eucalypt plantations.

Raúl Sardinha has an M.Sc. in Forestry from the University of Lisbon (1962) and a D. Ph. in Wood Science from Oxford University (1974). He was a Full Professor at ISA and retired in 2002. Professor Sardinha directed the Portuguese National Forest Research Station in 1988–1995, was chairman of the installation board of the University of Madeira and Rector of the Piaget University in Guinea-Bissau, and represented Portugal in various international research organizations. He developed research on the properties and suitability of the use of forest products and on natural resources research policy for sustainable management in tropical countries.

Lisbon, Portugal   José M. Cardoso Pereira
May 2020   Professor ISA/UL

Foreword II

The study of environmental physics requires a truly holistic approach forming an important plank in support of the environmental sciences. It provides the basis for the understanding of the complex physical interactions among living beings and their natural environmental components such as the atmosphere, natural or anthropogenic canopies, or water basins. These interactions are based on fluid dynamics and heat transfer processes associated with micrometeorology, airflow turbulence, vertical fluxes of mass and energy, or particle transportation. A major fraction of these phenomena occurs within the atmospheric surface layer of around 1 km depth from that surface, although not independent of the above atmosphere.

This textbook on Environmental Physics follows a first one published in 2014 in the Portuguese Language by Piaget Publisher entitled "Princípios Fundamentais de Ecologia Física" by Abel Rodrigues and Gabriel Pita, after two decades of research on those topics, with a focus on energy budgets in horticultural greenhouses and vertical fluxes of water vapor and carbon dioxide fluxes of cork oak and eucalypt stands, representative of forest stands with lower and higher biomass productivity. The fundamentals of scientific knowledge in the mass and heat transfer were laid in the disciplines of Physical Ecology, lasting between 1993 and 2003, and Heat and Mass Transfer, still existing in the graduate program of Environmental Engineering in the Instituto Superior Técnico, of the University of Lisbon.

The content of the former book was maintained but enriched with information on global carbon budgets and climate change, and principles of mass transfer regarding particles, gases, and vapors. These later environmental topics, despite not restricted to the surface layer, are nowadays of such relevance that its inclusion in a textbook like this is deemed essential because the whole technical and scientific information provides an interdisciplinary view on them.

A comprehensive applied theoretical approach is followed, grounded on fundamentals of fluid turbulent dynamics, heat transfer, or micrometeorology in distinct environments aiming to contribute to addressing applied matters such as vertical flows of sensible heat, evaporation, sediment transport, carbon sequestration, and emissions or climate change. This approach encompasses natural and anthropogenic canopies, such as rough or smooth vegetable or modified canopies, emphasizing mass flows driven by turbulent atmospheric flows and heat transfer processes within the boundary layers involved. The authors aimed to provide a review of the mathematical formulation that supports a range of ecological applications that underlie the evaluation of turbulent fluid flow, heat transfer, transient transfer processes, evaporation, or carbon flows.

The whole content of this book follows a sequential pedagogical strategy, aimed mainly at the graduate and postgraduate students in Environmental, Geographical, Agricultural, and Forestry Engineering or Biological Sciences. The rationale is that the framework of qualitative and quantitative principles of energy and mass transfer presented will complement the biological-focused knowledge of these readers. Thus, they will be better habilitated to integrate and interact with interdisciplinary teams and thereby to assess complex global contemporary environmental issues under a perspective of sustainable development. Such a strategy is multidimensional and requires the involvement of a wide range of sectors. This approach should include a "consolidation and development of the scientific knowledge base and the informed participation of stakeholders". In this sense, the authors believe that this book can contribute to that effort.

The book subjects were organized into eight chapters and two annexes to follow sequentially. Chapter 1 includes a general characterization of the atmospheric mixed boundary layer with an analysis of micro, macro, and synoptic atmospheric scales in the troposphere. A description of its diurnal dynamics with atmospheric stability and vertical adiabatic temperature gradients was carried out. Chapter 2 is dedicated to a characterization of the atmospheric surface boundary layer based on the analogy between eddy diffusivity in turbulent flow and molecular exchange in laminar flow, allowing for flux-gradient assumptions underlying the aerodynamic methodology for quantification of vertical fluxes of mass and energy. Chapter 3 presents a qualitative and quantitative formulation and parameterization of airflow turbulence in the surface layer. A quantitative and qualitative analysis of the continuum spectral turbulent dynamics was made with the discussion of scales of production, inertial and dissipative of turbulent eddies, under distinct atmospheric stability conditions, along with the evaluation of power spectra, autocorrelation, and cross-correlation functions.

Chapter 4 develops the main processes of vertical heat and mass transfer on forest canopies, wherein flux-gradient assumptions may not be valid, with canopy stomatal resistance and drastic intermittent airflow events playing a relevant role in convective dynamics. The proneness of forest canopies to drag processes and their relevance on evapotranspiration and carbon sequestration and thus, the potential mitigation in carbon change is also discussed. Chapter 5 describes the surface boundary layer in urban and modified canopies such as low slope or hill surfaces, with the development of processes such as internal and thermal boundary layers or urban heat islands. Flow phenomena such as extensive wakes, cavities, or horizontal warping horseshoe vortices, speedup profiles in hills or katabatic and descending flows under atmospheric inversion conditions are also assessed. Chapter 6 is dedicated to heat and mass transfer stationary and transient processes of conduction, convection, and radiation for evaluation of local energy budgets and thermal environment control. Relevant topics on particle mass transfer processes which pose serious environmental implications on seed and spore's dispersion, or sediment transport in land erosion and sediment transport in rivers and streams, were also developed. Chapter 7 includes 13 worked-out exercise applications related to microclimate variables, energy budgets, and vertical flows in canopies. Numerical examples to illustrate many of the principles developed to help students to enhance skills in using the equations involved in solving problems of heat flow, dissipation, convection, carbon sequestration as well as measurement issues are presented. It is expected that the resolution of real ecological problems presented in this chapter can be a useful basis for a greater understanding and consolidation of the issues developed in the previous applied theoretical chapters. Following the fundamentals of heat and mass transfer in previous chapters, Chap. 8 addresses the issue of the global carbon budget and its relationships with climate change. These are global phenomena occurring within the convective boundary layer, whose relevance is indisputable. It is a comprehensive chapter that brings together the most current information on the predictions of climate change for various geographies on Earth according to the magnitude of the carbon balance and its distribution in the atmosphere. This chapter ends up with the presentation of the transformation of biomass to biochar, which owed to its stability in the conservation of CO_2, offers a good perspective for carbon sequestration. This is an accessible technology that along with its carbon sequestration potential, and/or reducing biomass fuel enhancer in forest fires, can boost social and economic benefits in rural areas favored by adequate supplies of biomass and derivatives.

Finally, two annexes are added: Annex 1 dedicated to the treatment of instrumentation relevant to the evaluation of micrometeorological parameters and their specificities, precision and accuracy, data acquisition devices, and inherent errors; and Annex 2 reserved for the development and an adequate framework of Newton's laws and the concepts associated with the Moment Conservation law as well as the fundamental topics associated with the conservation of mass and motion, buoyancy, fluid viscosity, evaporation, and psychrometry.

Overall, the main environmental physical and mathematical tools on significant environmental physical subjects are covered, so that the reader can have an intermediate to high-level insight on the issues addressed. Those tools are instrumental to interpret the vast information available and to design strategies on data acquisition and instrumentation to carry out further deepening the knowledge in several areas from more specialized sources. The recommended bibliography is extensively listed, where readers could find more detailed texts if necessary.

Abel Rodrigues
UTI, Instituto Nacional de
Investigação, Agrária e Veterinária
Oeiras, Portugal

Raul Albuquerque Sardinha
Carnaxide, Portugal

Grabiel Pita
IST—Mechanical Engineering
University of Lisbon
Lisbon, Portugal

Acknowledgments

We thank Dr. Maria Ferreira for her English language support. Finally, the authors also express their thanks to the "*Edições Piaget*" and its CEO, Dr. António Oliveira Cruz, for his understanding of fully allowing the use of some of the data contained in the book "*Princípios Fundamentais de Ecologia Física*" published in 2014 by Abel Rodrigues and Gabriel Pita in that publisher.

Contents

1 General Characteristics of the Atmospheric Boundary Layer   1
   References   10

2 Aerodynamic Characterization of the Surface Layer   13
   2.1   General Considerations   13
   2.2   Turbulent Diffusivity Coefficients   20
   2.3   Aerodynamic Method Equations   24
         2.3.1   Direct Form   24
         2.3.2   Iterative Form   29
   References   31

**3 Characterization of Turbulent Flow in the Surface Boundary
   Layer**   33
   3.1   Introduction   33
   3.2   Mean and Flutuation Components for Turbulent Flows   35
   3.3   Taylor's Hypothesis   38
   3.4   Ergodic Conditions   39
   3.5   Introduction to Turbulent Motion Equations   40
         3.5.1   Navier–Stokes Equations for Laminar Flow   40
         3.5.2   Equations for Mean Variables in Turbulent Flow   44
         3.5.3   Equations for Variance   46
         3.5.4   Equations for the Vertical Momentum Flux   50
         3.5.5   Equations for Vertical Flow of Scalar Quantities   51
         3.5.6   Kinetic Energy Budget Equations   54
         3.5.7   The Closure Equation Problem   60
   3.6   Spectral Analysis   63
         3.6.1   Introduction   63
         3.6.2   Fourier Series   63
         3.6.3   Fourier Transforms   66
         3.6.4   Discrete Fourier Transform   69

        3.6.5   Autocorrelation and Cross-Correlation Functions   70

        3.6.6   Spectral Characterization of Turbulence in the Surface
                  Boundary Layer.   72

        3.6.7   Cospectral Analysis   77

  3.7   Eddy Covariance Method   79

  References 101

4 Exchange of Energy and Mass Over Forest Canopies 105

  4.1   Introduction 105

  4.2   Aerodynamic Characterization and Stability in the Rough
        Sublayer. 106

  4.3   Turbulent Transport of Kinetic Energy. 111

  4.4   Evaluation of the Vertical Flows of Heat and Mass Using
        the Bowen Ratio Method 115

  4.5   Evaluation of Evapotranspiration and Energy Coupling
        Using the "Big-Leaf" Approach. 118

  4.6   Carbon Sequestration in Forests. 124

  References 129

5 Flow Over Modified Surfaces 133

  5.1   Introduction 133

  5.2   Internal Boundary Layer 134

  5.3   General Characterization of the Urban Boundary Layer. 137

  5.4   Flow in Urban Areas 138

  5.5   Flow Over Gently Sloping Hills 146

        5.5.1   Flow Under Different Conditions of Stability. 146

        5.5.2   Flow Under Neutral Conditions 150

  5.6   Katabatic Winds Under Stability Conditions. 153

  5.7   Descending Winds Under Inversion 154

  References 156

6 Heat and Mass Transfer Processes 159

  6.1   Conduction. 159

        6.1.1   General Principles 159

        6.1.2   Heat Conduction in the Soil 162

        6.1.3   Thermal Properties of Soils 165

  6.2   Convection. 168

        6.2.1   General Principles 168

        6.2.2   Forced Convection. 172

        6.2.3   Free Convection 177

  6.3   Radiation 180

        6.3.1   General Principles 180

        6.3.2   Spatial Relationships 183

        6.3.3   Radiation Environment. 187

　　　　　6.3.4　Long Wavelength Radiation 195
　　　　　6.3.5　Radiative Properties of Natural Materials 197
　　6.4　Transient Heat Balances 204
　　　　　6.4.1　The Concept of Time Constant 204
　　　　　6.4.2　Transient Responses 206
　　6.5　Mass Transfer 208
　　　　　6.5.1　Gases and Water Vapor Transfer 208
　　　　　6.5.2　Particle Transfer 211
　　　　　6.5.3　Particle Deposition 216
　　　　　6.5.4　Sand and Dust Transfer 218
　　　　　6.5.5　Sediment Transportation in Rivers 224
　　References 234

7　Examples of Applications 237
　　7.1　Concepts of Energy Budget 237
　　7.2　Example 1: Calculation of Energy in a Light Photon 238
　　7.3　Example 2: Estimation of Solar Hourly Radiation 240
　　7.4　Example 3: Estimation of Atmospheric Transmissivity
　　　　　from Solar Radiation Values Measured at the Surface 241
　　7.5　Example 4: Calculation of Short- and Long-Wavelength
　　　　　Incident Radiation on a Given Surface 242
　　7.6　Example 5: Calculation of Radiation Budgets in a Eucalyptus
　　　　　Forest 244
　　7.7　Example 6: Calculation of Sensible Heat Transfer from
　　　　　the Low Canopy to the Adjacent Atmosphere 246
　　7.8　Example 7: Application of the Aerodynamic Method 249
　　7.9　Example 8: Application of the Eddy Covariance Method 251
　　7.10　Example 9: Calculation of Vertical Fluxes Using the Bowen
　　　　　Method 253
　　7.11　Example 10: Calculation of the Solar Radiation Intensity
　　　　　Components and Long Wavelength Incident on a Building
　　　　　with a Known Geometry 256
　　7.12　Example 11: Calculation of Kinetic Energy Budget Components
　　　　　for a Flat Surface 260
　　7.13　Example 12: Calculation of Sedimentation Velocity
　　　　　of a Particle 262
　　7.14　Example 13: Calculation of Variation of Velocity
　　　　　and Pressure in Airflow in a Plain and a Valley 265
　　References 266

8　Fundamentals of Global Carbon Budgets and Climate Change 267
　　8.1　Introduction 267
　　8.2　Topics on GHG Emissions and Global Carbon Budget 272
　　8.3　Alternative Scenarios for Climate Change Predictions 278

8.4   Occurrence of Extreme Events.......................... 282
8.5   Long-Term Climate Change 291
8.6   A Brief Analysis of the Mediterranean Case Study 292
        8.6.1   Introduction................................ 292
        8.6.2   Changes in Temperature........................ 293
        8.6.3   Changes in Precipitation and Moisture Availability 295
        8.6.4   Changes in Extreme Events 297
8.7   The Potential of Biochar in the Modulation of Carbon
        Budget and of Climate Change Mitigation 300
References 304

Annex A1: Instrumentation in Environmental Physics............... 309

Annex A2: Basic Topics on Laws of Motion and Evaporation 333

Bibliography 367

Index 369

Acronyms

a'	Instantaneous fluctuation of arbitrary quantity, A
$a_{w, t, q}$	Empirical constants
$a_{1,2}$	Empirical constants
$b_{1,2}$	Empirical constants
A	Empirical constant; area of a body; area perpendicular to heat and mass transfer
ACF	Autocorrelation function
$\overline{A}$	Mean value of A
A_s	Streamline height
A_f	Acceleration factor
A_h	Shade area on a horizontal surface
A_r	Coordinate rotation matrix
AR4	Fourth Assessment Report of the IPCC
AR5	Fifth Assessment Report of the IPCC
A_s	Amplitude of surface thermal wave
B	Arbitrary quantity
BLC	Loading capacity for biochar
C	Characteristic length
CCF	Cross-correlation function
c_d	Dimensionless drag coefficient
C_D	Global drag coefficient
C_H	Transfer coefficient for sensible heat
CESM	Community Earth System Model
CIMP5	Climate Intercomparison Model Project
C_M	Drag coefficient
c_e^m	Mean value of scalar quantity
cov (A, B)	Covariance between two arbitrary variables
c_p	Specific heat at constant pressure
C_{wk}	Cospectral density or covariance between vertical velocity and a generic variable k
c_s	Sound velocity in air
C_1	Amplitude height
CBC	Closure budget component

COP	Conference of the Parties
D	Damping temperature; characteristic dimension of a body; molecular diffusivity coefficient
D_c	Molecular diffusivity coefficient for water vapor
D_l	Horizontal distance to the contact area of flow point between laminar flow and flat plate
D_v	Molecular diffusivity coefficient for carbon dioxide
D_{tl}	Horizontal distance to the contact area of flow point between the thermal boundary layer and flat plate
DGVM	Dynamic Global Vegetation Model
d	Reference plane for stress concentration; Sun–Earth distance at specified instant
$\bar{d}$	Mean annual Sun–Earth distance
d_i	Differences in carbon fluxes relative to half-hourly periods
d_{ij}	Attenuation/amplification coefficients of means
d_p	Distance between transducers
$d\omega_{ij}$	Solid angle
ΔCat	Average carbon sink relative to atmosphere
e	Partial water vapor pressure
e_*	Friction vapor pressure of friction (equivalent to friction velocity)
e_s	Saturation water vapor pressure
E	Emissive power
E_c	Lower canopy evapotranspiration
E_g	Gray body emissive power
E_f	Forest evapotranspiration
EF	Emission by fossil fuel applications
E_n	Emissive power
ELU	Emission from land use
E_i	Radiative flux per unit area
E_λ	Radiative energy emitted by a black body per unit wavelength
f_c	Nyquist frequency; Coriolis parameter
f_i	Analyzer calibration function
F	Total drag force; resistance function
F_{ij}	Shape factor
F_k	Discrete Fourier transform
FT(f)	Global transfer function
F_r	Froude number
F_x	Cumulative flux
g	Gravity acceleration
G	Stability function; soil heat flux
GCM	Global climate models
GHG	Greenhouse gases
GPP	Gross primary productivity
Gr	Grashof number

G_{sb}	Heat flux conducted in stream bed
H	Sensible heat flux; stability function; building height; hill height
h	Solar hour angle; height of the stable night-time boundary layer; height of superficial elements; Planck constant; stream bed height
h_c	Convective heat transfer coefficient
$\bar{h}_c$	Mean convective heat transfer coefficient
h_d	Mean transfer coefficient
h_x	Dimensionless convective heat transfer coefficient
h/k	Relative submergence
I	Turbulence intensity
IPPC	Intergovernmental Panel on Climate Change
J	Surface irradiance; diffusion flux
K	Turbulent diffusion coefficient
K_M	Turbulent diffusion coefficient for horizontal momentum
K_H	Turbulent diffusion coefficient for sensible heat
K_V	Turbulent diffusion coefficient for water vapor
K_S	Turbulent diffusion coefficient for an arbitrary gas
K'	Radiation intensity emitted from a surface
k	Von Karman constant; roughness height
k	Thermal conductivity coefficient
L	Characteristic length; latent heat of water vapor
L	Monin–Obukhov stability length
LE	Latent heat fluxes
L_e	Lewis number
LES	Large-Scale Simulation
L_u	Ascending long wavelength radiation flux
L_d	Descending long wavelength radiation flux
LE_{eq}	Equilibrium evapotranspiration
LE_{pot}	Evapotranspiration potential
L	Stopping distance
M	Mass dimensional parameter
Md, MAD	Medians for correction of vertical fluxes
m	Air mass
N_{BV}	Brunt–Väisälä frequency
Nu	Nusselt number
n	Dimensionless frequency; empirical constant; number of daily solar hours
n_i	Number of monthly hours under clear sky
P	Air pressure, period (e.g., daily or annual) of thermal surface wave
P_e	Peclet number
P_o	Reference pressure
Pr	Prandtl number
P_w	Wake turbulent energy
Q_c	Carbon dioxide concentration

Q_n	Normalized sand flux
q	Specific humidity; radiative transfer in a given direction; sand mass transport
$\dot{q}$	Internal energy per unit volume
R	Constant for ideal gases; particle radius
R_b	Blackbody emissive power
RCP	Representative Concentration Pathway (of CO_2)
Re	Reynolds number
Re_p	Reynolds number of particles
R_h	Hydraulic radius
Ri_c	Critical Richardson number
Ri_f	Flux Richardson number
Ri_g	Gradient Richardson number
Ri_m	Mass Richardson number
R_n	Radiative balance
R_{ref}	Ecosystem respiration at reference temperature
R(T)	Total respiration
r	Mixing ratio
r_{AB}	Linear correlation between two arbitrary variables
r_{aH}	Aerodynamic resistance to sensible heat transfer
r_M	Aerodynamic resistance to vertical momentum transfer
r_{aS}	Aerodynamic resistance to transfer of a hypothetical gas
r_{aV}	Aerodynamic resistance to water vapor transfer
r_c	Canopy resistance
r_r	Resistance to longwave radiation
r_{hr}	Combined resistance for sensible heat loss and longwave radiation
S	Water channel slope
S_b	Direct solar radiation flux
S_d	Diffuse solar radiation flux
S_e	Rate of energy transfer to stream bed
S_h	Mean monthly total solar radiation flux outside the atmosphere
S_o	Solar constant
S_t	Mean monthly incident total solar radiation
S_v	Rate of energy transfer to atmosphere
S_{bl}	Radiative density flux mean
S_{dt}	Daily irradiance
S_{hd}	Daily irradiance outside the atmosphere
S_{tn}	Total solar radiation on cloudy days
S_p	Direct solar radiation flux
S_t	Total radiative flux
SOM	Soil organic matter
STK	Stokes number
$S_{u_i}, S_{xx}(\omega)$	Spectral power
$S_{xy}(\omega)$	Cospectral power

T	Time-dimensional parameter; absolute temperature
T_*	Friction temperature
T_a	Air temperature
T_c	Canopy temperature
T_E^*	Apparent equivalent temperature
T_f	Effective temperature
T_n	Temperature at cloud bottom
T_o	Regression constant (-46.03 °C)
T_{ref}	Reference temperature
T_S	Sonic temperature (K)
T_v	Virtual temperature
TKE	Turbulent kinetic energy
t	Apparent local solar time; period where Sun is above horizon; Eulerian time scale
t_j	Julian day
U	Mean velocity
UCL	Uptake of CO_2 by land
UCO	Uptake of CO_2 by the oceans
UNEP	United Nations Environment Programme
UNFCCC	United Nations Framework Convention on Climate Change
U_D	Air velocity along linear space between transducers in sonic anemometer
U_{ai}	Unidirectional airflow after contact with transducers
U_i	Unidirectional airflow before contact with transducers
U_N	Air velocity component in normal direction to space over space between transducers in sonic anemometer
UPE	Variable power flow resistance
u_c	Wind velocity constant
u_*	Friction velocity
$\bar{u}_{hc}$	Mean air velocity at canopy height
W	Space between buildings
WMO	World Meteorological Organization
WPL	Webb–Pearman–Leuning coefficient
Z_*	Total height of rough sublayer
z_*	Height of roughness sublayer above level d
z_{0H}	Roughness length for sensible heat
z_{0T}	Roughness length for sensible heat
z_{0V}	Roughness length for latent heat
z_{0M}	Roughness length for linear momentum
z_{01}, z_{02}	Upwind and downwind aerodynamic roughness length from contact point
z_h	Atmospheric boundary layer height
z_i	Day time atmospheric mixed boundary layer height

Greek Symbols

α	Azimuthal angle; thermal diffusivity; hill slope angle; heat capacity per unit area
α_k	Kolmogorov dimensionless constant
α_i	Radiation absorption of ith gas by the analyzer
α_n	Weighting factor
$\alpha(\lambda)$	Absorbance
β	Bowen ratio; thermal expansion coefficient; solar height
γ	Psychrometric constant; hygroscopic growth parameter
γ^*	Modified psychrometric constant
Δz_{max}	Height of maximum acceleration
$\Delta \theta$	Potential temperature variation
$\Delta \omega$	Fundamental infinitesimal frequency
δ	Internal boundary height; solar declination
δ_{ij}	Kronecker's delta
δ_l	Height of laminar boundary layer
δ_{Test}	Height of thermal boundary layer under stability conditions
δ_{tl}	Laminar thermal boundary layer
ε	Viscous dissipation
ε	Emissivity
ε_a	Atmospheric emissivity
η	Kolmogorov microscale; angle concerning the first coordinate rotation
Θ	Angular velocity of Earth's rotation
θ	Potential temperature
θ_c	Critical Shields parameter
θ_v	Virtual potential temperature
κ_H	Laminar diffusion for sensible heat
κ_M	Laminar diffusion for momentum
κ_V	Laminar diffusion for water vapor
Λ	Infinitesimal volume, Eulerian length scale
λ	Wavelength
λ_{BV}	Natural wavelength
λ_{max}	Maximum radiation emission wavelength
μ	Dynamic viscosity
v	Kinematic viscosity; radiation frequency

ζ	Stability factor
ρ	Air density
ρ'	Apparent soil density
$\rho(\lambda)$	Reflectivity
ρ_{hum}	Moist air density
ρ_i	Gas i concentration
ρ_V	Air absolute humidity
σ	Stefan–Boltzman constant
Γ	Dry adiabatic gradient
τ	Momentum vertical flux; instantaneous transmittance to direct radiation; time constant; relaxation time
τ_{ij}	Shear stress matrix
τ_{di}	Diffuse radiation transmissivity
τ_k	Kolmogorov time scale
τ_t	Total transmissivity
ϑ	Angle concerning second rotation of coordinates
φ	Latitude
ϕ_H, ψ_H	Stability function for sensible heat
ϕ_M	Stability function for momentum
ϕ_V	Stability function for water vapour
ϕ_ε	Dimensionless dissipation factor
$\emptyset_{xx}(m)$	Autocorrelation function
$\emptyset_{xy}(m)$	Cross-correlation function
$\Phi(x)$	Radiation flux attenuated by absorption
χ	Absolute air humidity; gas concentration in a surface
ψ	Scalar vectorial quantity, zenith angle
ψ_M	Similarity function for momentum
Ω	Decoupling factor
ω	Solid angle
ω_0	Fundamental angular frequency

General Characteristics of the Atmospheric Boundary Layer

1

Abstract

This chapter includes a general characterization of the atmospheric boundary layer with an analysis of micro, macro, and synoptic atmospheric scales in the troposphere. The dimensional scales and frequencies of average flow, turbulence, and atmospheric waves are discussed in relation to the transport of scalar and vector variables. The conditions of atmospheric stability and instability, associated with the vertical adiabatic gradient of actual and potential temperatures with daily variation of the structure of the atmospheric boundary layer, were evaluated. The dynamics of atmospheric boundary layers, top inversion, mixed layer, and surface layers along with subsidence and divergency play a significant role as discussed.

About 80% of the atmosphere mass is in the troposphere. This layer extends from the Earth's surface up to an average altitude of 11 km and is characterized by a vertical decreasing of temperature with an average gradient of 0.65 °C/100 m.

The atmospheric boundary layer is defined as part of the troposphere that is directly influenced by the Earth's surface and responds to surface perturbations within a timescale of an hour or less (Stull 1994). Some of these perturbations include frictional drag, evapotranspiration, and heat transfer by convection, pollutant emission, and flux changes brought about by surface topographical changes. The thickness of the atmosphere's boundary layer varies over time and space, ranging from hundreds of meters up to one or two kilometers.

Atmospheric motion occurs in scales that vary a great deal (Stull 2000). The size classes usually described are turbulence in the range of 2–200 mm, the constant flux layer that ranges from 200 mm to 200 m, the atmospheric boundary layer up to about 1000–1500 m, the mesoscale dimensions comprising up to 2000 km, the synoptic scale comprising movements of air masses between 2000 and 15,000 km and finally, movements on a planetary scale that extend to heights of about 15,000 km.

Heat and mass exchanges in the biosphere that interface the Earth and the atmosphere enable anthropogenic and natural processes relevant in environmental physics. These processes occur predominantly within the bottom of the atmospheric boundary layer, corresponding to about 10% of its height, known as the surface layer or constant flux layer, or Prandtl layer. When the layering is unstable its height range is between 20 and 200 m, but it may decrease to a few meters when the layering is stable. The designation of constant flux layer arises from the assumption of constant fluxes with height allowing to estimate the fluxes of, e.g., sensible and latent heat within this layer (Foken 2017).

The general structure and dynamics of the atmospheric boundary layer is highly variable and is related to the air temperature and energy balance of the surface adjacent to the atmosphere (Fig. 1.1).

The vertical variation of tangential stresses in the surface layer is constant and the atmospheric flux is determined by surface frictional effects and the air temperature vertical gradient. It is not affected by either the Earth's rotation or large-scale pressure variations. In this layer the wind velocity profile is logarithmic. In the upper layer or Ekman layer of the atmospheric boundary layer, tangential stresses and wind direction are continually changing (Foken 2017) and the flow dynamics are influenced by surface friction, temperature gradient, and the Earth's rotation.

Atmospheric flow can be divided into three broad categories: mean wind, turbulence, and atmospheric waves (Fig. 1.2). Each category may occur separately or in the presence of any of the others (Stull 1994).

The horizontal transport of scalar or vector quantities such as humidity, sensible heat, momentum, and pollutants is mainly due to the mean wind of about 2 to 10 m/s whereas turbulence is largely responsible for the vertical transport of these quantities. The average horizontal wind velocity is higher than the mean velocity in

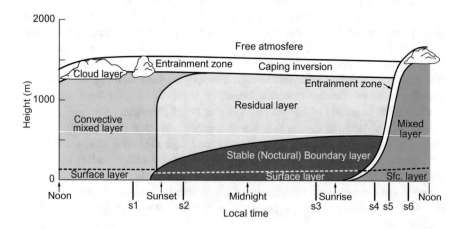

Fig. 1.1 Schematic diagram of the vertical structure of the boundary layer (after Stull 1994)

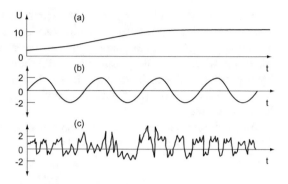

Fig. 1.2 Schematic of **a** mean wind, **b** waves, and **c** turbulence (after Stull 1994)

the vertical direction which is of the order of a few mm or cm/sec. Friction reduces the horizontal flow and decreases the mean wind speed so that it is slowest near the ground.

The atmospheric turbulence of thermal origin coexists with the mechanical turbulence generated by tangential stresses. The atmospheric turbulence is characterized by these thermals or wake-type eddies, which are formed because of forces such as buoyancy or the retarding effect of surface frictional drag. All processes in the atmospheric boundary layer and especially those concerning the micrometeorological range are adjacent to the ground surface and can be simulated and compared easily with laboratory experimental results with wind tunnels (Foken 2017).

Atmospheric waves are often seen in the stable night-time boundary layer and are important for the transport of energy and momentum and have no bearing on a convective transport of scalars such as sensible heat, humidity, and pollutants. Atmospheric waves are caused, for example, by the interaction between the mean flow and surface obstacles or are generated from far away sources such as lightning or explosions (Stull 1994). Atmospheric turbulence can be analyzed by breaking down the main variables into mean components and fluctuations or perturbations. The mean component describes the effects of the mean wind and temperature, while the disturbances or fluctuations relate to the effects of atmospheric waves or to the overlapping turbulent effects.

Atmospheric turbulence is composed of eddies of various sizes, ranging between millimeters and kilometers, and these interact so that the smaller feed on the larger ones, resulting in mass and energy exchange. The larger eddies scale to the depth of the boundary layer varies between 100 and 2000 m in diameter. An approach used to quantify the relative contributions of each eddy in heat and mass exchange processes is the analysis of the turbulence spectrum. Larger eddies with shorter frequencies of 1 Hz or less are highly efficient in heat and mass transport processes in the atmosphere. The smaller eddies of higher frequencies (about the order of 10 Hz) are much weaker because of the dissipating effects of molecular viscosity.

The boundary layer is, in general, thinner in high-pressure regions than in low-pressure regions. The convergence or subsidence and low-level horizontal divergence associated with high-pressure move boundary layer air from high- to low-pressure zones (Fig. 1.3). The shallow depths are often associated with regions that typically lack clouds.

In low-pressure zones, rising motions transport air from the surface of the boundary layer to higher levels of the troposphere. This makes it difficult to clearly define the top of the atmospheric boundary layer. The so-called free atmosphere where the winds blow in geostrophic equilibrium is located above the boundary layer parallel to the isobaric lines. In the free atmosphere, frictional and other surface effects are no longer present, allowing the air to flow without significant levels of turbulence.

An expression that describes the height of the atmospheric boundary layer z_h is given below (Kaimal and Finnigan 1994):

$$Z_h = 0.25 \left(\frac{u_*}{f_c} \right) \tag{1.1}$$

where u_* is the friction velocity and f_c is the Coriolis parameter given by

$$f_c = 2\Theta\pi \sin \phi \tag{1.2}$$

being Θ the Earth's rotation rate, given by the angular velocity, and ϕ the altitude. The units for f_c are 10^{-4} s^{-1}.

Convection is a fundamental physical mechanism in the dynamics of energy exchanges in the lower atmosphere. The type and extent of convective activity are determined by the vertical temperature structure which is expressed in terms of the conditions of neutrality, thermal stability, or instability.

Under conditions of thermal neutrality, the rate of cooling of ascending air parcels, given by the dry adiabatic gradient Γ, (1 °C/100 m drop in temperature

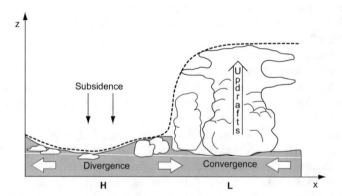

Fig. 1.3 Schematic of variation of boundary layer depth between regions of surface high and low pressure. The shaded area represents the atmospheric boundary layer, and the dotted line shows the height reached by surface-modified air during a one-hour period (after Stull 1994)

with height) is equal to the environmental temperature gradient. As a result, there is no difference in density between these air parcels, the surrounding and ascending air, and circulation until a stationary state is reached which takes place due to internal adiabatic cooling.

The expression for the dry adiabatic gradient is as follows (e.g., Monteith and Unsworth 1991):

$$\Gamma = -\frac{g}{c_p} \tag{1.3}$$

where g is gravity acceleration, and c_p is the specific heat of air at constant pressure.

The Γ parameter describes the vertical temperature gradient induced by gravity that occurs when an air parcel flows in the absence of added external heat. The situation of thermal neutrality is transient. Under thermal instability, commonly found in the convective day time boundary layer, the ambient temperature gradient is greater than the dry adiabatic gradient so that rising air parcels, which cool according to Γ are warmer and less dense than the ambient air and are displaced by buoyancy into regions increasingly further away from the original position.

Under thermal stability, which occurs mainly at night, the environmental temperature gradient is lower than the dry adiabatic gradient, so the rising air parcels, cooler and denser than the surrounding air, move downwards. The quantification of atmospheric stability calls for the concept of potential temperature. The potential temperature is defined as the temperature that an air parcel would show, at absolute temperature (T) and pressure (P), if transported adiabatically to a pressure of 100 kPa.

The potential temperature can be calculated as follows:

$$\theta = T \left(\frac{P_0}{P}\right)^{0.286} \tag{1.4}$$

where P and T are the air pressure and temperature, and the reference pressure P_0 is typically 100 kPa.

The following equation describes the relationship between the actual and potential temperature of the air and their respective gradients:

$$\frac{\partial \theta}{\partial z} = \left(\frac{\partial T}{\partial z} - \frac{g}{c_p}\right) \tag{1.5}$$

which is equivalent to

$$\theta \approx T + \Gamma z \tag{1.6}$$

Since the adiabatic behavior of air parcels is a function of the external pressure, potential temperature Eqs. (1.1–1.5) gives temperatures of air parcels at different pressures.

The virtual potential temperature is defined as the potential temperature dry air must be equal to the density of moist air, which is smaller at a given pressure so that moist air is relative buoyant. For unsaturated air with a mixture ratio r, defined as the ratio of masses of water vapor and dry air, the virtual potential temperature is given by

$$\theta_v = (1 + 0.61r) \tag{1.7}$$

The diurnal boundary layer, also known as the convective or mixed layer, has an unstable surface layer in which $\partial\theta/\partial z < 0$, a convective mixing zone in which vertical gradients are absent, where $\partial\theta/\partial z = 0$ and an upper inversion zone where $\partial\theta/\partial z > 0$. The latter is a buffer that dampens the ascending atmospheric motions (Figs. 1.4 and 1.5).

This dampening effect has a bearing on several phenomena including convective retention involving the descending recirculation of hot air parcels, mechanical mixing and conservation within the boundary layer; sensible and latent heat, and dispersion of pollutants.

The ground surface adjacent to the atmosphere is the zone where greater diurnal absorption and nocturnal dissipation of thermal energy take place and these effects decrease as a function of distance from the surface. The daily variation of the air

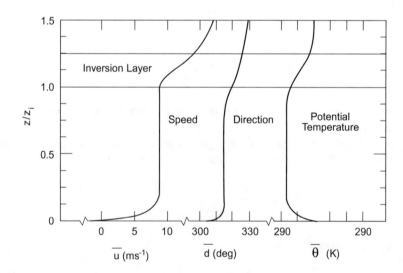

Fig. 1.4 Mean vertical profiles of wind speed, wind direction, and potential temperature in the convective boundary layer (height z is presented as a ratio z/z_i where z_i is the height of the daytime boundary layer) (after Kaimal and Finnigan 1994)

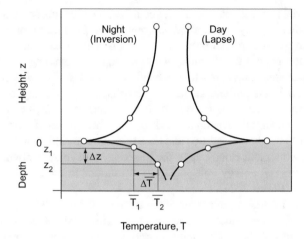

Fig. 1.5 Schematic profiles of average air and soil temperature near the soil-atmosphere interface (Oke 1992)

temperature and of the surface energy balance is essential for soil thermal flow associated with a temperature gradient directed during the day toward the surface and from the surface to atmosphere at night (Fig. 1.5).

The daily heating and cooling of the surface along with fluctuations in solar radiation lead to height variations in the atmospheric boundary layer (Figs. 1.1 and 1.6).

In the day time, the inversion at the top of the atmospheric boundary layer serves as a barrier that prevents vertical motion. The highest average gradients for wind velocity, wind direction, and temperature take place in the surface layer, corresponding to the lower 10% of the boundary layer. In the remaining upper 90% of this layer, vertical gradients of the same magnitude with comparable average values are dampened by turbulence driven motion (Figs. 1.6 and 1.7).

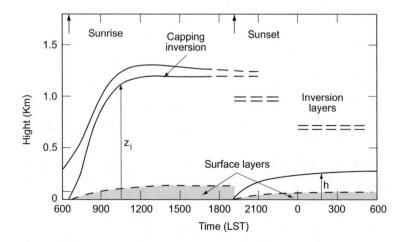

Fig. 1.6 Daily evolution of the convective (day time) and stable (night-time) boundary layers in response to surface heating and cooling (after Kaimal and Finnigan 1994)

For terrestrial and mid-latitude areas, about 30 min after sunrise, a mixed thermal convective boundary layer is formed adjacent to the ground, because of radiative surface heating which in turn promotes heating of the adjacent air. The boundary layer grows throughout the morning, while mixing and retaining air from the less turbulent upper atmosphere, reaching a height of 1 to 2 km by mid-afternoon.

The nocturnal residual and stable boundary layers are quickly destroyed after sunrise with the formation of the mixed layer, especially on days with strong solar radiation. The nocturnal inversion, which prevails before sunrise, evolves into the capping layer to height z_i (Fig. 1.6) and rises with the convective layer as it expands vertically. In winter and cloudy days, the mixed layer development will be damped due to weak thermal energy transport from the surface to the atmosphere and low solar radiation (Foken 2017).

The capping inversion can remain at the same level during the day or may fall in the form of reverse subsidence. The height at which this reversal occurs corresponds to the top of the diurnal atmospheric boundary layer. Figures 1.1 and 1.5 provide schematic representations of the daily evolution of the atmospheric boundary layer.

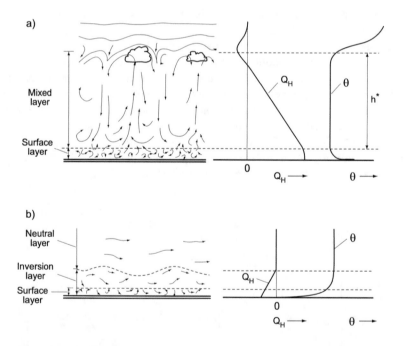

Fig. 1.7 Schematic representation of airflows in the diurnal boundary layer, **a** and night-time boundary layer, **b** (H and θ refer to the sensible heat flux and potential temperature, respectively) (after Oke 1992)

Although the mixed layer may evolve under high wind conditions, more often it results from the process of thermal impulse. Such processes include heat transfer from a heated surface that will generate thermal plumes upward. They also include radiative cooling of the top of the cloudy layer that generates both rising thermal plumes as well as cold air plumes that may occur simultaneously.

These structures have length scalars equivalent to that of the mixed layer and are the main vehicles for transport and mixing of both rising and falling scalar and vector quantities, between the surface and the top layer below the inversion layer. Vigorous mixing of the atmosphere takes place in the mixed layer (Figs. 1.6 and 1.7a)). The potential temperature shows little vertical variation.

Sensible heat flux shows linear vertical variation until it reaches zero near the base of the inversion. This flux reverses the direction in which the nocturnal inversion layer evolves (Figs. 1.6 and 1.7b)). At night, due to thermal stability, the turbulence of mechanical origin is confined to a lower air layer (h in Fig. 1.6) without forming a true mixed layer. This weak turbulence together with radiative cooling gives rise to the night-time potential temperature profile and sensible heat flux. The vertical profiles of the nocturnal potential temperature characterize the surface, reverse and neutral layers (Fig. 1.7b).

Tangential stresses that provoke mechanical turbulence are formed at the interfaces between layers of different densities. These stresses increase until a threshold is reached, and the flow becomes unstable, with the generation of the so-called Kelvin–Helmholtz waves. In these waves, the less dense fluid is entrained in denser fluid, under a mechanism which is statically unstable and discontinuous in space. The different forms of instability generate turbulence that promotes scalars of motion transfer, leading to system homogenization and reduction of shear stresses between the several adjacent fluid layers. All processes in the atmospheric boundary layer, and especially those concerning the micrometeorological range adjacent to the ground surface, can be simulated, and compared easily with laboratory experimental results with wind tunnels (Foken 2017).

The wind velocities are subgeostratospheric throughout the mixed layer with the wind directions crossing the isobars at reduced angles and heading toward low-pressure zones. The middle zone of the mixed layer has a nearly constant average wind speed and direction. The speed decreases with height and approaches zero near the surface (Fig. 1.4). In general, the humidity also decreases with height within the mixed layer, because of surface evapotranspiration and entrainment of dry air from the upper levels. Contaminants tend to concentrate more at lower levels of the mixed layer as thermal plumes cannot penetrate the top of the inversion layer.

During sunset the thermal plumes cease to form, leading to a reduction of turbulence in the mixed layer. The top layer is maintained but weakened forming a discontinuous structure with one or more thinner inversion layers. As the thermal plumes lose their energy near the surface, there is a sudden decrease in turbulent motions, simultaneously with the beginning of radiative cooling. The resultant air layer is called the residual layer (Fig. 1.1), with a height of about 1000 m, which initially has average scalar and vector variables magnitude like that of the mixed layer from which it evolved. The residual layer is neutrally stratified resulting in the turbulence of equal intensity in all directions. In this way, plumes containing chemical agents are evenly dispersed and these may interact and form aerosols or particles that eventually precipitate out (Stull 1994).

Shortly after sunset, along with the formation of the residual layer, the cold night-time air adjacent to the surface is transported upwards by mechanical turbulence, forming a calm and stable boundary layer, which at midnight extends to a height of 100–200 m. A thin and narrow surface layer is found below the nocturnal boundary layer (Fig. 1.1). The flow within the night-time boundary layer is characterized by strong shear stress, small scale turbulent eddies, and activity in the atmospheric waves, mentioned before.

The residual layer, located above the nocturnal boundary layer, is not in a strict sense directly affected by surface effects that lead to turbulence. However, it is usually considered one of the atmospheric boundary layer component, as it lies below the bottom of the inversion layer. The nocturnal residual and stable boundary layers are quickly destroyed after sunrise with the formation of the mixed layer as mentioned above.

Fast super-geostrophic winds or nocturnal jets are formed above the nocturnal boundary layer due to the Earth's rotation. These winds move downward at night. Thus, while the stable air in the nocturnal boundary layer serves to suppress turbulence, nocturnal jets exert an opposite effect in which intense and rapid phenomena promote mixing.

The stable atmospheric boundary layer rarely reaches an equilibrium comparable to that of the convective boundary layer. Wind profiles and average temperatures evolve throughout the night. Also, drainage or katabatic winds from near the surface (adjacent to the ground) caused by the colder air flowing down under the influence of gravity. Thus, at a height of a meter, wind speeds of 1 ms^{-1} occur. Turbulence in the nocturnal boundary layer gradually decreases with height (Kaimal and Finnigan 1994), as it is dampened by thermal stability while decreasing tangential stresses.

The nocturnal boundary layer does not have a sharply defined top, in contrast with the diurnal mixed layer which ends at the top inversion layer. The upper limit of the nocturnal layer is defined as the height at which the turbulence intensity is a small fraction of its surface value. As a rule of thumb, the height of the night-time layer may be defined as that at which turbulence intensity decreases to about 5% of the value at the surface. Alternately, it may be defined as the average height of the inversion layers (Fig. 1.6). At sunrise, these night-time flux perturbations begin to settle, and in this way, the diurnal cycle begins again leading to the formation of the atmospheric boundary layer.

References

Foken, T. (2017). *Micrometeorology*, 2nd ed., Springer, Berlin, 362 pp.
Kaimal, J. C., & Finnigan, J. J. (1994). *Atmospheric boundary layer flows* (p. 289). Their Structure and Measurement: Oxford University Press.

Monteith, J. L., & Unsworth, M. H. (1991). *Principles of environmental physics*. 2nd. Ed., Edward Arnold, 291 pp.

Oke, T. R. (1992). *Boundary layer climates*, 2nd ed., Routledge, 435 pp.

Stull, R. S. (1994). *An introduction to boundary layer meteorology*. Kluwer Academic Publishers, 666 pp.

Stull, R. (2000). *Meteorology for scientist and engineers* (2nd ed., p. 502). Thomson Learning: Brooks/Cole.

Aerodynamic Characterization of the Surface Layer

2

Abstract

In this chapter, an assessment was carried out on aerodynamic boundary surface layer characterization, where vertical vector fluxes of scalar quantities, e.g., momentum or sensible heat, are considered constant. This approach is grounded on the Prandtl mixed layer empirical theory based on the analogy between eddies in turbulent flow and molecules in laminar flow, allowing for flux-gradient assumptions. The similitudes between eddy and turbulent diffusivity coefficients in turbulent and laminar molecular flows were noted here as well as their differences in meaning. It has been shown that within the surface layer, this theory is not strictly applicable in very rough canopies such as forests, since vertical fluxes in these canopies are directed in the opposite direction to the gradients.

An evaluation of the typical vertical logarithmic profile of the average air velocities in the surface layer was performed under different conditions of stability. Included was the empirical treatment of topics such as mass, gradient, Richardson flow numbers, Monin-Obukhov length, dimensionless stability functions or discrete equations for vertical moment fluxes, sensitive and latent heat, or gases in direct or iterative form. This empirical base is instrumental for the assessment of natural and forced convection and heat transfer in environmental systems.

2.1 General Considerations

Most issues relating to environmental physics such as engineering, excepting aeronautical matters, occur in the surface layer. Fortunately, the characterization of turbulence and vertical profiling of mean variables in this layer is straightforward. The surface or constant flow layer is the lower layer of the atmospheric boundary layer, corresponding to 10% of the total height, where the vertical heat and mass fluxes vary up to 10%. In this layer, convective fluxes of momentum, evapotran-spiration, water vapor, and convective heat flux, also known as sensible heat, can be considered constant.

Typically, heat and mass fluxes are higher in the surface layer, decreasing to zero at the top of the planetary boundary layer (PBL). The height of the surface layer varies over daytime and nighttime periods. On clear nights, in the absence of wind, the height of the surface layer can be down to only 10 m, which limits its practical value. In contrast, during the daytime under normal wind conditions, its height can be about 100 m or more.

In surfaces with vegetation or urban canopies, rates of mass and energy transfer between these areas and the atmosphere are determined by measuring vertical fluxes in the atmospheric boundary layer. The surface layer consists of two sublayers (Fig. 2.1) which are the roughness and inertial sublayers. The roughness sublayer includes a zone with individual elements of the rough surface plus the adjacent air zone that is influenced by wakes caused by the individual elements. The structure of that sublayer is influenced by the distribution of the rough elements. In the roughness sublayer, for example, in forested areas, flux-gradient principles typically do not apply, and these aerodynamic anomalies occur because fluxes occur in a direction opposite to the gradient.

The inertial sublayer is located above the roughness sublayer. In this sublayer, the atmosphere is more stable depending on parameters such as friction velocity, u_*, defined later in this chapter (Eq. 2.10), that quantify turbulent velocity fluctuations in the atmosphere and height of surface elements. To calculate mass and energy

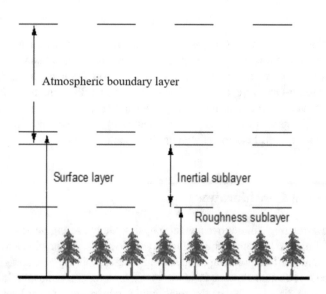

Fig. 2.1 Structure of the surface layer (after Valente 1999)

fluxes from gradients, measurements of vertical gradients of scalar and vector quantities need to be carried out in the inertial sublayer.

In simple terms, incompressible turbulent flow in the surface layer follows Prandtl's theory for the mixed layer. This theory described in detail elsewhere (e.g., Tennekes and Lumley 1980), presumes that certain assumptions hold true, which in real situations is not the case. Correct use of the mixed layer theory, adapted from the kinetic theory for gases, should be checked to see if the momentum transport model holds true using a velocity gradient. However, this is only possible if the length scales for turbulent transport flows are lower than the length scales in which mean gradients vary significantly. In biophysical systems such as forestry ecosystems, the opposite occurs and transport phenomena are mainly due to intermittent downwind structures or eddies with length scales several times the tree height (Blanken et al. 1998).

It is also assumed that momentum is conserved when particles move between two distinct points, although, in practice, this does not hold true. Also, a relationship between stress and strain in fluid flow, parameterized by a proportionality constant, the eddy diffusion constant K (e.g. Tennekes and Lumley 1980), is not always applicable as the velocity standard deviation and length scales vary a great deal, and so that the proportionality constant varies across the velocity fields. For the strict application of these principles, a fundamental assumption of equality of the length scales is made, despite that eddies of distinct class sizes participate in energy and mass transportation processes. Notwithstanding these limitations, in most situations, simplified analysis of the surface layer can be used to analyze fluxes and mass-energy balances of microsystems in the realm of environmental physics.

As mentioned before, mass and energy vertical fluxes produced by turbulence near the surface are retarded because of the friction effect of the terrain on the horizontal wind. This friction delay is a continuous process of absorption of linear wind momentum that is responsible for the downward vertical flow of motion quantities. In this way, tangential stresses are produced resulting from the strong interaction between the mean flow and fluctuations in turbulent velocity. Thus, the flux of momentum in combination with velocity profiles and roughness parameters serve to evaluate turbulence efficiency for the vertical transport of heat and mass.

The laminar boundary layer contacts with surfaces such as terrain, or obstacles such as plants, animals, and houses. In this thin layer, with a thickness of the order of several millimeters, the air adjacent to the surfaces moves in laminar flow, and the respective streamlines are parallel to the surface. Among practical examples of laminar flow, are a layer of smoke from a smoldering cigarette sliding along its surface and a thin layer of water flowing slowly from a faucet.

The thickness of the boundary layer gradually increases until a critical combination of factors (flow velocity, distance, and viscosity) induce instability in the flow, leading to a radical change characterized by eddies and wave motion typical of turbulent flow. The Reynolds number (Eq. 6.16), which is the ratio between inertial and viscous forces, can be used as a criterion to mark the transition between the laminar and turbulent flow.

The laminar sublayer remains on the surface and its thickness is dependent on-air velocity and surface roughness. In this sublayer, there is no convection, and therefore, any exchange of non-radiative energy occurs through molecular diffusion. The following equations can be used to quantify the vertical flux of sensible heat, F_H, water vapor, F_{LE}, and linear momentum, τ:

$$F_H = -\rho c_p k_H \frac{\partial T}{\partial z} \tag{2.1}$$

$$F_{LE} = -\rho c_p k_V \frac{\partial \rho_v}{\partial z} \tag{2.2}$$

$$\tau = -\rho k_M \frac{\partial u}{\partial z} \tag{2.3}$$

where ρ, c_p, T, ρ_v, u, k_H, k_V, and k_M, respectively, are the air density (kgm^{-3}), specific heat at constant pressure (Jkg^{-1}K^{-1}), air temperature, air absolute humidity (kgm^{-3}), air velocity (ms^{-1}) and the coefficients of molecular diffusion for sensible heat, water vapor and linear motion expressed in m^{2}s^{-1}. These diffusion coefficients are small and rather constants of around 10^{-5} m^{2}s^{-1} and varying slightly with temperature. The laminar layer serves as a barrier between the contact surface and air in the adjacent surface layer being of practical interest only for studies on heat and mass transfer from small obstacles. Further details about the relationship between tangential stresses and molecular viscosity are referred to in Annex 2.

The surface layer is characterized by turbulent flow wherein heat and mass transfer processes are far more efficient than in laminar flow. The mixed layer theory allows for a simplified analysis based on an analysis between atmospheric eddies, that induce convective turbulent diffusion, and fluid molecules responsible for molecular diffusion in the laminar boundary layer. Those molecules form continuous sliding layers one on top of the other. Thus, flux-gradient equations related to mixed layer theory, like Eqs. (2.1), (2.2), and (2.3) for the laminar boundary layer, can be used. However, in this case, the turbulent diffusivity coefficients, K, are not constant but vary over time and space. These coefficients vary with eddies height, and thus with the height above the surface. Molecular and turbulent diffusion coefficients within the boundary layer range between 10^{-5} and 10^2 m^{2}s^{-1} (Oke 1992).

This methodology for aerodynamic measurements of vertical fluxes in the surface layer, based on the analogy between eddies under turbulent flow and molecules in laminar flow allows for flux-gradient assumptions. These are useful for quantifying or modeling scalar fluxes that cannot be measured directly or complement of results of the turbulence covariance methodology based on measurements of instantaneous fluctuations (Mölder et al. 1999).

Quantification of the vertical flux of scalar quantities such as water vapor and sensible heat is simpler using the aerodynamic model insofar that, unlike linear momentum and in the absence of buoyancy, these do not interact with mean velocity fields, being, therefore, passive elements subjected to fluid flow.

Under conditions of thermal neutrality, the vertical profile of mean air velocity in the surface layer is logarithmic, with the slope (du/dz) higher near the ground, as shown in Fig. 2.2.

A graphic representation of u vs. logarithm of z is a straight line through the data points according to the equation

$$u(z) = A \ln z + B \tag{2.4}$$

where A and B are constant independent of height. B can be replaced by $-A\ln z_{oM}$, where z_{oM} is the value for z in Eq. (2.4) when the velocity reaches zero. This can be expressed as

$$u(z) = A \ln(z/z_{oM}) \tag{2.5}$$

The z_{oM} parameter, denoted the roughness length, corresponds to the value in Eq. (2.5) where the wind velocity becomes zero and is represented by the point at which the experimental data points are extrapolated to intersect with the logarithmic ordinate axis, $\ln z$.

Roughness length can be described as the length scales representative of the efficiency of a surface in removing momentum from the mean flow. Equation (2.4) then gives

$$\partial u / \partial z = A/z \tag{2.6}$$

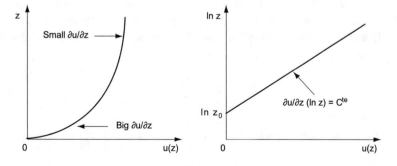

Fig. 2.2 Vertical velocity gradient and logarithmic wind profile in the constant flux layer (after Thom 1975)

It follows that the vertical gradient of the air velocity $(d(u)/d(z))$ is inversely proportional to the distance above the surface. The drag force per unit area of soil caused by horizontal air motion is the shear or tangential stress τ, also called skin friction. The physical dimensions of τ are mass times acceleration/area, that is

$$\tau \approx MLT^{-2}/L^2 \tag{2.7}$$

From Eq. (2.7), the dimensions of tangential stress are equivalent to motion flux, that is, (mass $\times$ velocity) per unit area and per unit time (Thom 1975)

$$\tau \approx MLT^{-1}/L^2 T \tag{2.8}$$

The wind drag force on a given surface is thus an outcome of the continuous downward flux of horizontal momentum between the air in motion and the surface. This flux is related to mechanically driven eddies in the turbulent boundary layer.

Equation (2.8) can be written as follows:

$$\tau \approx \left(M/L^3\right)\left(LT^{-1}\right)^2 \tag{2.9}$$

showing that shear stress is equivalent to the product of specific mass M/L^3, times the velocity LT^{-1}, squared. This specific mass refers to air moving over a surface, and the velocity is associated with the speed that the horizontal momentum from the mean flux reaches at the surface, depending on the effectiveness of vertical turbulent transport processes.

Equation (2.9) can be written as

$$\tau = \rho u_*^2 \tag{2.10}$$

where ρ is the air density and u_* is the friction velocity associated with the momentum flux, τ. The friction velocity is proportional to the tangential stress of eddy rotation resulting from frictional drag.

In a situation with a logarithmic wind profile such as that expressed in Eqs. (2.5) and (2.6), the A parameter with velocity dimensions, is proportional to friction velocity and independent of height. Then, by defining the proportionality constant as $1/k$, Eq. (2.6) becomes

$$\frac{\partial u}{\partial t} = \frac{u_*}{kz} \tag{2.11}$$

being k the von Karman constant of 0.41, regardless of the type of surface.

Replacing A with u_*/k in Eq. (2.5) gives

$$u(z) = \frac{u_*}{k} \ln \frac{z}{z_{oM}} \qquad (2.12)$$

The kz product can be identified as the size of the eddy or mixing length l, at height z

$$l = k\,z \qquad (2.13)$$

It can be seen from Eq. (2.12) that for a given value of $u(z)$, u_* will be higher on a rough surface exhibiting a roughness length greater than a smooth surface. As a result, the effectiveness of turbulent transfer for a given surface will vary directly with the degree of aerodynamic roughness length specified by z_{oM}.

For vegetated canopies with uniform height h, a good approximation for z_{oM} is given by

$$z_{oM} = 0.1h \qquad (2.14)$$

Within a plant community, analysis of turbulent flow is done by assuming that vertical elements are concentrated at a distance d, from the ground. The d parameter is known as the stress concentration plane or zero-plane displacement. Thus, a reference plane is defined at a distance d, such that the distribution of shear stresses on the individual elements is aerodynamically equivalent to the total stress at height d. The mean size of eddies over a layer of vegetation is then proportional to a distance above d, and not to the total height, h. The parameter d appears in the wind velocity profiling to consider the vegetation height. It follows that Eqs. (2.11), (2.12), and (2.13) can be substituted for

$$\frac{\partial u}{\partial z} = \frac{u_*}{k(z-d)} \qquad (2.15)$$

$$u(z) = \frac{u_*}{k} \ln \frac{z-d}{z_{oM}} \qquad (2.16)$$

$$l = k(z-d) \qquad (2.17)$$

each one valid only for conditions where $z \geq h$. In practice, the parameter d can be estimated by graphical analysis, plotting $u(z)$ versus $\ln(z\text{-}d)$, for d values ranging between 0.6 and 0.8h. Provided that wind speed measurements refer to neutral thermal stability, the value of d will give a straight line when $u(z)$ versus $\ln(z-d)$ is plotted. The corresponding $\ln z_{oM}$ value where the velocity u is zero is obtained from the intercept with the ordinate axis (Thom 1975). Figure 2.3 shows a schematic representation of the boundary layer adjacent to the forest canopy, indicating the heights for d and z_{oM}.

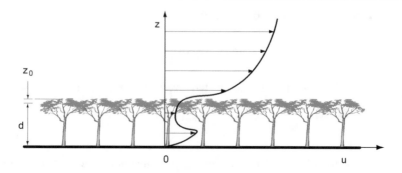

Fig. 2.3 Flow over forest canopy showing mean speed, u, as a function of height, z. The height of d and z_{OM} are shown (after Stull 1994)

2.2   Turbulent Diffusivity Coefficients

The eddy diffusivity coefficient K, of a given quantity at any point in a fluid medium, can be defined as the ratio between the flux property through the medium and the mean concentration gradient, in the same direction through a given point (Thom 1975; Monteith and Unsworth 1991). Therefore, for any property with physical dimensions, Q

$$K = \left\{ (Q/L^2 T)/(Q/L^3)/L \right\} = L^2 T^{-1} \tag{2.18}$$

The physical dimensions of K are area per unit time. This means that for any diffusive process, the magnitude value for K is related to the area affected per second, by the diffusion of a given scalar or vectorial property in a small fluid sample. During laminar airflow, when diffusion is only molecular in nature, κ, the equivalent molecular diffusion coefficient in laminar flow, the turbulent diffusivity coefficient, is only about 10–20 mm^{2}s^{-1}. In the case of turbulent flux over plant communities, the diffusivity coefficient K can be of the order of 1 m^{2}s^{-1} (Oke 1992; Thom 1975). Molecular diffusivity is a physical property of the fluid, independent of its location. Turbulent diffusivity is a property of the flow directly proportional to eddy size and to the distance from the ground, minus height d, as defined above.

In the case of the horizontal momentum, its turbulent diffusivity coefficient K_M, or turbulent viscosity is defined as the ratio of the momentum flux τ and its concentration gradient $(\partial(\rho u/\partial(z))$. As the flow is incompressible, the following is obtained:

$$\tau = K_M(\partial u/\partial z) \tag{2.19}$$

This equation is like Eq. (2.3), which is related to momentum flux and vertical gradient velocity in the laminar boundary layer. Taking Eq. (2.19) and combining it with Eqs. (2.10), (2.15), and (2.17) gives

$$K_M = ku_*(z - d) = lu_* \qquad (2.20)$$

if z exceeds d at a given height, the turbulent diffusivity is equal to the product of friction velocity and size of eddies. Thus, in the surface layer directly above vegetation, the magnitude of a specified gradient will decrease in height to compensate for the linear increase in height of turbulent diffusivity.

Definitions of turbulent diffusivity coefficients for sensible heat, water vapor or any other gas, K_H, K_V and K_s, are analogous

$$K_{H=-H/[\partial(\rho c_p T)/\partial z]} \qquad (2.21)$$

$$K_v = -LE/[\partial(\rho q)/\partial z] \qquad (2.22)$$

$$K_s = -F/[\partial(\rho S)/\partial z] \qquad (2.23)$$

In Eq. (2.22), variable ρ represents the density of moist air and variable q is the specific humidity defined as the mass of water vapour per unit mass of humid air (see Eqs. (4.13) and (4.14), pg. 116). Specfic humidity is a variable useful for issues of water vapour transport at lower atmosphere, because it is independent of temperature (see Eq. 3.23 in pg. 38). Variables $\partial \rho q/\partial z$, $\partial \rho S/\partial z$, H, LE, and F are, respectively, the gradients for absolute humidity, concentration of a hypothetical gas (e.g. CO_2), sensible heat flux, latent heat flux, and evapotranspiration and flux of a hypothetical gas.

Analysis of aerodynamic drag is based on concepts that are analogous to Ohm's Law in electrical circuits

$$Electrical\ resistance = \frac{potencial\ difference}{current\ intensity} \qquad (2.24)$$

If the potential difference is substituted by the concentration (quantity per unit volume) and the intensity of the current substituted by the flux (quantity per unit area per unit time), for the linear momentum, then Eq. (2.24) becomes

$$Aerodynamic\ resistance = \frac{Concentration\ difference}{flux} \qquad (2.25)$$

For a scalar or vector quantity of dimension Q, the concentration dimensions are Q/L and those for the flux are Q/TL^2, so that the aerodynamic resistance dimensions become T/L or $(velocity)^{-1}$.

To analyze aerodynamic resistance, it is necessary to define overall drag F, on an isolated obstacle (e.g., area of a flat leaf) with an area A, as follows (Thom 1975):

$$F = \rho u^2 A C_M \tag{2.26}$$

where ρ is the air density and C_M is a dimensionless proportionality constant denoted the drag coefficient, a function of the orientation of the obstacle and air-speed, u. The overall drag is equal to the rate of change of linear momentum in the airflow, and therefore, the C_M value is a measure of the efficiency of the obstacle in absorbing momentum from the atmospheric flow. The C_M parameter may thus be defined as the momentum transfer coefficient. This drag, also termed as form drag, is a force applied to bodies immersed in moving fluids, additional to skin friction, in the flow direction because of the fluid acceleration and is dependent on the shape and orientation of bodies (Annex 2).

The momentum concentration Q_M, in an incident air flow, is given by:

$$Q_M = \rho u \tag{2.27}$$

with units as (ML^{-3}) (LT^{-1}) = MLT^{-1}/L^3, (linear momentum/volume). Using Eq. (2.25) to calculate aerodynamic drag r_M, for linear momentum transfer between the incident air and the surface of an individual element where the momentum concentration is zero, gives

$$r_M = \frac{\rho u A}{F} \tag{2.28}$$

where F/A, corresponding to the overall drag applied per unit area of the obstacle, is representative of the momentum flux over the obstacle. From Eq. (2.26) comes

$$r_M = \frac{1}{u C_M} \tag{2.29}$$

Drag force per unit of the horizontal area of the vegetative canopy can be expressed in a way equivalent to Eq. (2.26) giving

$$\tau = \rho u^2 C_{aM} \tag{2.30}$$

where C_{aM} is the drag coefficient per unit horizontal area for the entire canopy. Equations (2.10), (2.16), and (2.30) can be written as

$$C_{aM} = \left\{ \frac{u_*}{u(z)} \right\}^2 = \frac{K^2}{\left(\ln(z - d)/z_{OM} \right)^2} \tag{2.31}$$

It follows that C_{aM} increases with z_{0M}, that is, the drag per unit area is larger for rougher canopies.

By repeating the same approach, for calculating r_M, based on Eqs. (2.25) and (2.27), it is possible to determine the aerodynamic resistance of the canopy z_{aM}, to the momentum transfer between a height z, in the boundary layer of a uniform canopy, where the concentration is $\rho u(z)$, and the height of the elements at the surface where the concentration is zero, is given by

$$r_M = \frac{\rho u(z)}{\tau} \tag{2.32}$$

where τ is the downward momentum flux. The Eq. (2.32) describing the resistance of momentum transfer to a canopy, is equivalent to Eq. (2.29) for the resistance of momentum transfer towards a single element.

By combining Eqs. (2.30), (2.31), and (2.32), the following equation can be derived:

$$r_M = \frac{1}{u(z)C_{aM}} = \frac{u(z)}{u_*^2} = \frac{\{\ln(z-d)/z_{oM}\}^2}{K^2 u(z)} \tag{2.33}$$

In general, according to Eq. (2.33), the aerodynamic resistance, r_M, decreases with increased wind speed and surface roughness. Oke (1992) provided values for aerodynamic resistance for typical terrain types. Some of these in sm^{-1} units were: 200 for aqueous surfaces; 70 for turf; 20–50 for agricultural crops; and 5–10 for forests. Measurements for cork oak forests in Portugal gave values of about 15.2 sm^{-1} (Rodrigues 2002).

Equation (2.18) shows that the units for turbulent diffusivity are $L^2 T^{-1}$. The units for $1/K$ are either TL^{-2} or TL^{-1}/L, equivalent to the aerodynamic resistance per unit length, or aerodynamic resistivity. For the momentum transfer, the integral of $1/K_M$ between heights z_1 and z_2, where the airspeed u has values u_1 and u_2, equals the aerodynamic resistance to the momentum transfer between these two heights, given by $r_M (z_1, z_2)$. Indeed, by combining Eqs. (2.10), (2.19), and (2.33), the following is obtained (Thom 1975):

$$\int_{z_1}^{z_2} \frac{dz}{K_M} = \frac{1}{u_*^2} \int_{u_1}^{u_2} du = \frac{u_2 - u_1}{u_*^2} = r_M(z_1 - z_2) \tag{2.34}$$

so that the analogy between $1/K_M$ and r_M is confirmed.

2.3   Aerodynamic Method Equations

2.3.1   Direct Form

The aerodynamic method for calculating vertical atmospheric fluxes of mass and energy can be used under a variety of conditions. These include temperature stability, steady-state, and homogeneous surfaces without changes in radiative or wind fields during the observation period, and fluxes constant in height. The disturbances in the diurnal boundary layer make the application of aerodynamic method mainly possible in homogeneous terrain with uniform fetch, which is defined as a distance from the measurement point to a change in surface properties or obstacle (Foken 2017).

The similarity of transfer coefficients is also assumed, for example, for water vapor, gases, thermal convection, and momentum K_v, K_S, K_H, and K_M, respectively

$$K_v = K_S = K_H = K_M \tag{2.35}$$

or, from the relationship with aerodynamic resistances

$$r_{M(z_1, z_2)} = r_{aH}(z_1 - z_2) = r_{aV}(z_1, z_2) = r_{aS}(z_1, z_2) \tag{2.36}$$

To calculate profiles and fluxes using the aerodynamic method, under variable conditions of stability, or in the absence of thermal neutrality, dimensionless corrective functions of stability need to be used, which adjust the flux estimates by profiling the effects of thermal stability. Thus, the effects of variation in thermal stability are made through a generalized Eq. (2.15) as follows:

$$\frac{\partial u}{\partial z} = \frac{u_*}{k(z - d)} \phi_M \tag{2.37}$$

where in the dimensionless stability function for momentum transfer ϕ_M, is greater or less than one, under conditions of stability and instability, respectively. In both cases, the absolute value is dependent on the degree of stability or instability of the flow.

The equations for temperature and humidity profiles, quantified by partial vapor pressure e, are obtained in a similar way

$$\frac{\partial T}{\partial z} = -\frac{T_*}{k(z - d)} \phi_H \tag{2.38}$$

$$\frac{de}{dz} = -\frac{e_*}{k(z - d)} \phi_V \tag{2.39}$$

where T_* and e_* parameters are given by

$$T_* = \frac{H}{\rho c_p u_*} \tag{2.40}$$

and:

$$e_* = \frac{\gamma LE}{\rho c_p u_*}$$ (2.41)

where ϕ_H, ϕ_V, ρ and c_p are stability functions for the temperature profile, and water vapor partial pressure, air density, and the specific heat of air at constant pressure, respectively.

After conversion, the following coefficients of momentum diffusivity, heat, and water vapor, respectively, are obtained

$$K_M = k_{u_*}(z - d)\phi_M^{-1}$$ (2.42)

$$K_H = k_{u_*}(z - d)\phi_H^{-1}$$ (2.43)

$$K_V = k_{u_*}(z - d)\phi_V^{-1}$$ (2.44)

The general equation for the diffusivity coefficient for any hypothetical gas is derived in the same way. These last three equations make it possible to infer that the ratios among stability functions are equal to ratios between the respective thermal diffusivity coefficients. Therefore, turbulent changes for sensible heat and water vapor, for example, are similar and this holds true for the exchange of any atmospheric scalar quantity.

The dimensionless Richardson number Ri is a measure of thermal stability of atmospheric flow along the surface. A given volume of flowing air has a kinetic energy component of mechanical origin, because of the interaction between eddies with characteristic dimension l, with mean flow $(ke)_i$, and an additional force due to buoyancy, $(ke)_b$. The gradient Richardson number expresses the ratio $((ke)_b/(ke)_i)$. Under thermal instability, Ri is negative, whereas under thermal stability Ri is positive. A zero Ri value corresponds to thermal neutrality conditions.

If the temperature and airspeed are known at two heights, z_1 and z_2, the gradient Richardson number Ri_g, for the layer between these heights, given in terms of finite-difference is (Thom 1975)

$$Ri_g = \frac{g}{T}\frac{(T_2 - T_1)(z_2 - z_1)}{(u_2 - u_1)^2}$$ (2.45)

The advantage of using the gradient Richardson number is that it depends on mean gradient variables, which are easy to obtain.

Above a critical Richardson value Ri_c (0.25 in inviscid flow), the flow changes from laminar to turbulent. For values between 0 and Ri_c, turbulence is mechanically generated by tangential interactions between turbulent fluctuations and the mean flow fields; for $Ri_g < 0$, turbulence is of mixed mechanical and convective origin.

Additional uses for the Richardson number are the flux Richardson number Ri_f, and the mass Richardson number Ri_m, (Kaimal and Finnigan 1994)

$$Ri_f = \frac{g}{T} \frac{\overline{W'T'}}{\overline{u'w'}(\partial\bar{u}/\partial z)} \tag{2.46}$$

$$Ri_m = \frac{g}{T} \frac{(T(z) - T(0))/2}{(u(z)/z)^2} \tag{2.47}$$

In Eq. (2.46), the numerator and denominator represent, respectively, (i) the production or destruction of turbulence by thermal stratification and (ii) turbulence production, resulting from the interaction between shear stress and the mean flow field. In this way, the Ri_f parameter combines concepts on turbulence correlation with mean gradient parameters to characterize the effect of flow thermal stratification on the occurrence of turbulent events.

The unit value of Ri_f is critical as values below, especially negative, indicates that the flow is turbulent. When Ri_f values are between 0 and 1, thermal stability is insufficient to prevent mechanical turbulence. For Ri_f values greater than unity the flow is laminar.

Under calm air conditions, Ri_m parameter is a good indicator of thermal stability near the ground. In Eq. (2.47), $T(z)$, and $T(0)$ refer to the mean temperatures at height z, which can be located above the canopy and ground surface, respectively. The term $u(z)$ represents the mean wind at height z.

Differential form of Eq. (2.45) defines the Richardson number gradient

$$Ri_g = \frac{g}{T} \frac{\left(\frac{\partial T}{\partial z}\right)}{\left(\frac{\partial u}{\partial z}\right)^2} \tag{2.48}$$

and by combining Eqs. (2.37), (2.38), and (2.40) with it, yields

$$Ri_g = -k\frac{g}{T}\frac{H}{\rho c_p}\frac{(z-d)}{u_*^3}\frac{\phi_H}{\phi_M^2} \tag{2.49}$$

As Ri_g, ϕ_M and ϕ_H are dimensionless, a new parameter L, known as the Monin-Obhukov stability length, can be defined from Eq. (2.49) as follows:

$$L = -\frac{u_*^3}{k(g/T)(H/\rho c_p)} = -\frac{u_*^2 T}{kgT_*} \tag{2.50}$$

Equation (2.49) can then be expressed as

$$R_i = \frac{(z-d)}{L}\left(\frac{\phi_H}{\phi_M^2}\right) \tag{2.51}$$

Thus, a relationship is established between $(z-d)/L$ and Ri, which are evaluation parameters for thermal stability. The L parameter takes on length dimensions including variables related to free and forced convection (Ch. 6), without taking height into account.

Under conditions of atmospheric instability, L value can be considered the height above $z = d$ in which free convection becomes the main heat transfer mechanism and Ri_g reaches a corresponding value of -1 (Thom 1975). The variable $\xi = (z\text{-}d)/L$ is zero or negative under conditions of thermal neutrality and instability, whereas it is positive under thermal stability conditions.

Under conditions of stability or slight thermal instability, the semi-empirical functions ϕ, can be written as follows (Webb 1970):

$$\phi_M = \phi_H = \phi_V = (1 - 5Ri)^{-1} \quad Ri \geq -0.1 \tag{2.52}$$

where ϕ_M, ϕ_V, and ϕ_H are stability functions for fluxes of linear momentum, water vapor, and sensible heat.

On the other hand, under conditions of thermal instability, the most frequently used functions were described by Dyer and Hicks (1970). These functions can be expressed as follows:

$$\phi_M^2 = \phi_H = \phi_V = (1 - 16R_i)^{-(1/2)} Ri < -0.1 \tag{2.53}$$

Illustrates changes in the geometry of turbulent eddies under different conditions of thermal stability Fig. 2.4

Stability functions can be obtained directly from Eqs. (2.37), (2.38), and (2.39) based on direct measurements of gradients, fluxes, and friction parameters at various levels of ξ. This provides an alternative to the empirical approach using Eqs. (2.52) and (2.53).

To calculate mass and energy using the aerodynamic method directly to obtain flow-gradient type equations, an assumption is made about the possible similarity between the turbulent diffusion coefficients under conditions of thermal neutrality (Eq. 2.35). From the definitions of these coefficients (Eqs. 2.19, 2.21, 2.22, and 2.23) and generalizations for distinct thermal stability conditions, the following equations for fluxes of sensible heat, latent heat, and hypothetical gases (e.g. CO_2), in discrete form, are obtained:

$$H = -\rho c_p k^2 \left(\frac{\Delta u \Delta T}{(\ln((z_2 - d)/(z_1 - d)))^2}\right)(\phi_M \phi_H)^{-1} \tag{2.54}$$

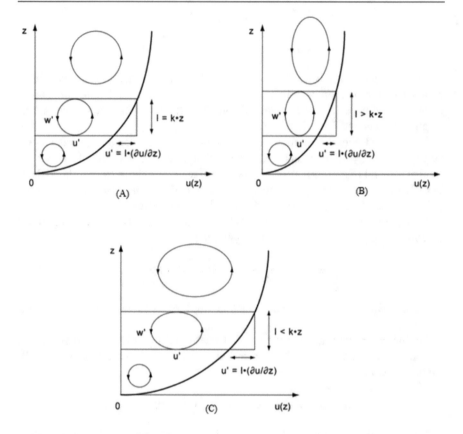

Fig. 2.4 Vertical wind profiles in the form of eddies under different conditions of thermal stability: **a** thermal neutrality, **b** thermal instability, and **c** thermal stability

$$LE = -Lk^2 \left(\frac{\Delta u \Delta \rho_v}{\left(\ln((z_2 - d)/(z_1 - d)) \right)^2} \right) (\phi_M \phi_v)^{-1} \qquad (2.55)$$

$$F_C = -k^2 \left(\frac{\Delta u \Delta Q_c}{\left(\ln((z_2 - d)/(z_1 - d)) \right)^2} \right) (\phi_M \phi_c)^{-1} \qquad (2.56)$$

where k is the von Karman constant (of the order of 0.41), L is the latent heat of vaporization of water (of the order of 2.44 MJkg^{-1}), ρ_V and Q_c are water vapor and CO_2 concentrations (vapour or gas mass per air volume – units kgm^{-3}) measured at heights z_1 and z_2, u is air velocity and $(\phi_M \phi_H)^{-1} = (\phi_M \phi_V)^{-1} = (\phi_M \phi_C)^{-1}$ are

$$F_{est} = \left(1 - 5Ri_g\right)^2 \quad Ri_g \geq -0.1 \tag{2.57}$$

$$F_{est} = \left(1 - 16Ri_g\right)^{0.75} \quad Ri_g < -0.1 \tag{2.58}$$

In practice, Eqs. (2.54), (2.55), and (2.56) can be solved by considering half hours averages of the differences on the numerators of equations.

2.3.2   Iterative Form

Integrating Eqs. (2.37) and (2.38) relative to vertical profiles, in order of z, according to Arya (1988) and Garrat (1994), yields

$$u(z) = (u_*/k)[ln((z-d)/z_{om}) - \psi_M((z-d)/L)] \tag{2.59}$$

$$T(z) - T(0) = (T_*/k)[ln((z-d)/z_{oT}) - \psi_H((z-d)/L)] \tag{2.60}$$

where T (0) is the surface temperature corresponding to the temperature at height $d + z_{0T}$, the equations ψ_M and ψ_H, are the similarity functions. Assuming for the unstable conditions $\xi < 0$, those functions will be expressed by the following forms:

$$\psi_M = 2ln((1+x)/2) + ln\left((1+x^2)/2\right) - 2arctg(x) + \frac{\pi}{2} \tag{2.61}$$

$$\psi_H = 2\ln\left((1+x^2)/2\right) \tag{2.62}$$

where

$$x = (1 - 16\xi)^{1/2} \tag{2.63}$$

Under conditions of thermal stability, $\xi > 0$, the ψ_M and ψ_H functions are as follows:

$$\psi_M = \psi_H = -5\xi \tag{2.64}$$

The calculation of Monin-Obhukov length L is carried out through e.g. Equation (2.50). However, this equation requires the values of u_* and T_* and (Eqs. (2.59), and (2.60) requires the knowledge of the stability factor $\xi = (z-d/L)$ needed for calculating the ψ functions (Eqs. (2.61) and (2.62)). Therefore, the calculation of fluxes must be iterative.

Calculation of u_* and T_*, using Eqs. (2.59) and (2.60), can be performed from measurements made at more than one level through the application of linear least-squares regression principles. Two lines with known slopes are obtained wherein the ordinates are $u(z)$ or $T(z)-T(0)$ and the abscissas are given by

$$[(ln(z - d)/z_{oM}) - \psi_M((z - d)/L)]$$

or

$$[(ln(z - d)/z_{oT}) - \psi_H((z - d)/L)], \text{ respectively.}$$

By using an arbitrary initial value for L, u_* and T_* can be calculated using Eqs. (2.59) and (2.60), with the ψ functions obtained through Eqs. (2.61) and (2.62), and thereafter L is calculated using Eq. (2.50). Using the calculated L value, the procedure is repeated until a value equal or close to the pre-defined L value is obtained. The final values for u_* and T_* are lastly calculated using Eqs. (2.59) and (2.60). According to Eq. (2.40), the sensible heat flux, H, is given by

$$H = \rho c_p u_* T_* \tag{2.65}$$

Figure 2.5. illustrates the iterative procedure for calculating sensible heat flux.

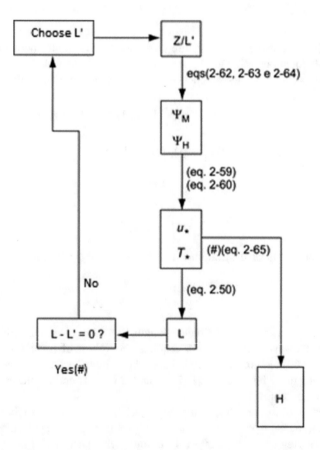

Fig. 2.5 Schematic representation of the iterative aerodynamic method for the calculation of sensible heat

The procedure for calculating the latent heat flux is done in a similar way. Eqs. (2.59) and (2.60) can be written to give a straight line in the form y = mx + b

$$u(z) = (u_*/k)[(ln(z - d)) - \psi_M((z - d)/L)] - (u_*/k)lnz_{oM} \qquad (2.66)$$

and

$$T(z) = (T_*/k)[ln(z - d) - \varphi_H((z - d)/L)] + T(0) - (T_*/k)lnz_{0T} \qquad (2.67)$$

making it possible to see that for calculating u_* and T_*, the values for z_{OM} and z_{oT} are not required. In addition, the roughness length for momentum z_{OM} and the term $((TO)(T_*/k)lnz_T)$ can be obtained from the ordinate intercept of Eqs. (2.66) and (2.67).

The friction velocity u_* can be obtained through the application of eddy covariance method, discussed in Chap. 3, improving thereby the iterative approach calculation of fluxes and aerodynamic variables (Foken 2017).

For the calculation of the roughness length of z_{0T} and z_{0V}, empirical expressions such as the one described by Campbell and Norman (1998) can be used

$$z_{0M} = 0.13h \qquad (2.68)$$

$$z_{0T} = z_{0V} = 0.2z_{0M} \qquad (2.69)$$

where z_{0M}, z_{0T}, and z_{0V} are the roughness lengths for momentum, sensible, and latent heat.

References

Arya, S. P. (1988). *Introduction to micrometeorology*. International Geophysics Series 42, Academic Press, 307 pp.

Blanken, P. D., Black, T. A., Neumann, H. H., Hartog, G. D., Yang, P. C., Nesic, Z., et al. (1998). Turbulent flux measurements above and below the overstorey of a Boreal Aspen forest. *Boundary Layer Meteorology, 89*, 109–140.

Campbell, G. S., & Norman, J. M. (1998). *An introduction to environmental biophysics*. Springer, 293 pp.

Dyer, A. J., & Hicks, B. B. (1970). Flux-gradient relationships in the constant flux layer. *Quarterly Journal of the Royal Meteorological Society* 96: 715–721.

Foken, T. (2017). *Micrometeorology*, 2nd ed., Springer, Berlin, 362 pp.

Garrat, J. R. (1994). *The atmospheric boundary layer*. Cambridge University Press, 316 pp.

Kaimal, J. C., & Finnigan, J. J. (1994). *Atmospheric boundary layer flows* (p. 289). Their Structure and Measurement: Oxford University Press.

Mölder, M., Grelle, A., Lindroth, A., & Halldrin, S. (1999). Flux profile relationships over a Boreal Forest-Roughness sublayer corrections. *Agricult Forest Meteorol, 98–99*, 645–658.

Monteith, J. L., & Unsworth, M. H. (1991). *Principles of environmental physics*, 2nd. Ed., Edward Arnold, 291 pp.

Oke, T. R. (1992). *Boundary layer climates*, 2nd ed., Routledge, 435 pp.

Rodrigues, A. M. (2002). *Fluxos de Momento, Massa e Energia na Camada Limite Atmosférica em Montado de Sobro*. Ph.D. Thesis (Environment, Energy profile) Instituto Superior Técnico, U.T.L., Lisbon, 235 pp. [in portuguese].

Stull, R. S. (1994). *An introduction to boundary layer meteorology*. Kluwer Academic Publishers, *666*, 1.

Tennekes, H., & Lumley, J. L. (1980). *A first course in turbulence*. MIT Press, 300 pp.

Thom, A. S. (1975). *Momentum, mass and heat exchange of plant communities*, pp. 57–109. In: J. L. Monteith (ed.), *Vegetation and atmosphere*, Vol I. Academic Press, 277 pp.

Valente, F. M. R. T. (1999). *Intercepção da Precipitação em Povoamentos Esparsos. Modelação do Processo e Características Aerodinâmicas dos Cobertos Molhados*. Ph.D. Thesis, Universidade Técnica de Lisboa, Instituto Superior de Agronomia, Lisbon, (in portuguese).

Webb, E. K. (1970). Profile relationships: The log-linear range, and extension to strong stability. *Quarterly Journal of the Royal Meteorological Society, 96*, 67–90.

Characterization of Turbulent Flow in the Surface Boundary Layer

3

Abstract

This chapter aimed to make a characterization of turbulent airflow in the surface boundary layer, following qualitative and quantitative approaches. The former is based on a generalization of Navier–Stokes equations, applied to flow mean and fluctuation components, for obtaining budgets of vectorial and scalar quantities. The latter is based on similarity relationships dependent on atmospheric stability for evaluation of the components of kinetic energy budget equations. A spectral and cospectral frequency characterization of the turbulent flow, aiming to analyze the spectral structure of production, transport, inertial and dissipative scales was performed, grounded on a brief introduction on fundamentals of Fourier analysis. Comparison of measured and calculated spectra following empirical similarity principles, particularly in slopes of curves in the inertial region, is fundamental for quality control assessment of atmospheric measurements and for evaluation of turbulent dynamics under distinct atmospheric stability conditions. The assessment of the power spectrum, autocorrelation, and cross-correlation functions enhances the potential of frequency analysis of predominant turbulent eddies. Finally, a discussion is presented about eddy covariance methodology to obtain vertical fluxes, with measurements of fluctuations of scalar and vectorial quantities. The methodology is applied under turbulent transport frequencies, considering quality control proceedings and applications on local carbon budgets.

3.1 Introduction

As mentioned in previous chapters, atmospheric flows are turbulent in nature. Flowing fluid moves in a highly disordered and chaotic way, making the velocity fields difficult to be accurately reproduced under experimental conditions.

© The Author(s), under exclusive license to Springer Nature Switzerland AG 2021   33
A. Rodrigues et al., *Fundamental Principles of Environmental Physics*,
https://doi.org/10.1007/978-3-030-69025-0_3

Turbulent flows occur at high Reynolds numbers, wherein inertia forces associated with convective effects predominate over viscous forces associated with diffusivity. These convective effects are primarily responsible for large diffusivity of turbulence that causes a huge increase in transfer processes for heat, mass, and linear momentum. Turbulent transport is, therefore, far more effective than transport by molecular diffusion.

Turbulent flow is based on random velocity fluctuations and thus the respective instantaneous values u_I must be a sum of the average of the mean velocity $\bar{u}$ over a period with fluctuations around this mean value. Turbulence is a characteristic of the flow and not of the fluid itself. Even small length scalars that are dynamically significant are larger than intermolecular distances or the molecules themselves. The main features of turbulence prevalent at high Reynolds number conditions are not conditioned by the molecular properties of the fluids.

Turbulent flow is rotational, three dimensional, and continuous, and can be characterized using equations of fluid mechanics. Its random and non-linear nature makes for a complex mathematical treatment of turbulence because of the closure problem, in which there are more unknowns than equations.

Viscous shear stresses dissipate kinetic energy from turbulent flow, increasing the internal energy of the fluid. The viscous dissipation prevents the infinitely small eddies from forming by converting the energy in these low-dimensional scales into heat (Tennekes and Lumley 1980; Shaw 1995a). Maintaining the turbulent flow characteristics requires a mechanism that will continuously supply energy, to compensate for losses through viscosity. As noted above, this energy can be mechanically derived from tangential stresses of the mean flow or from buoyancy.

Turbulent eddies have characteristic dimensional and time scales, ranging from molecular dimensions lasting fractions of a second, to millimeters or to kilometers, lasting hours. Turbulent eddies can be considered as air parcels with uniform thermodynamic properties, with small-scale eddies coalescing to form bigger ones, due to surface roughness and flow velocity. The largest eddies are atmospheric pressure systems (Foken 2017). Since the equations for motion are non-linear, each individual flow pattern will depend significantly on the initial and boundary conditions. Thus, despite common properties, each turbulent flow is different, depending on the specificities of the surrounding environment. The interaction between this environment and turbulence results in a situation of permanent adjustment or dynamic equilibrium that is never truly reached.

In practice, for simplifying the analysis of turbulence empirical concepts such as mixing length can be used, by analogy with the kinetic theory of gases. Such concepts can be valid for hypothetical conditions where the length scales and velocities are constant, easily characterized, and of reduced dimension, as compared with scalars of typical dimensions for the mean flow. Under laboratory conditions, for example, using flat surfaces, one can identify distinct stages in the transition from laminar to turbulent flow. The initial stage creates primary instability with the formation of small eddies. This instability leads to pronounced highly unstable tri-dimensional secondary movements that amplify locally into three-dimensional waves and strong tangential stresses develop. These scattered structures coalesce forming a continuous turbulent field.

Turbulence can be described as an integral part of the atmosphere and is mainly random, requiring statistical methods. It follows that there should be a separation between the mean and the turbulent flow components to determine the means and standard deviations of scalar and vector quantities in calculating statistical parameters. Among these parameters are variance representative of the intensity of turbulence or kinetic energy, and covariance for vertical fluxes or shear stress.

These parameters can be represented using algebraic summation equations of budget, e.g., mass, linear momentum, variances, or kinetic energy. The terms of the equations make it possible to introduce various parameters that collectively make up the budget.

Spectral analysis is another mathematical tool that makes it possible to analyze the sizes and frequencies of eddies that make up the turbulent velocity field.

3.2 Mean and Flutuation Components for Turbulent Flows

Atmospheric turbulent flow is characterized by a continuous variation of the vector or scalar parameters, inherent to the dynamics of movement around a mean value calculated from data obtained over periods of about half hour. Any time-dependent quantity A can then be written as follows:

$$A = \overline{A} + a' \tag{3.1}$$

where $\overline{A}$ represents the mean value and a' the instantaneous fluctuation.

Some of the rules for defining means for time-dependent quantities of products or derivatives, considering that fluxes and variances are products of fluctuations are as follows:

$$\overline{c} = c \tag{3.2}$$

where c is a constant.

$$\overline{cA} = c\overline{A} \tag{3.3}$$

$$\overline{\left(\overline{A}\right)} = \overline{A} \tag{3.4}$$

$$\overline{\left(\overline{A}B\right)} = \overline{A}\,\overline{B} \tag{3.5}$$

where B is an arbitrary quantity.

$$\overline{(A + B)} = \overline{A} + \overline{B} \tag{3.6}$$

$$\overline{\left(\frac{dA}{dt}\right)} = \frac{\overline{dA}}{dt} \tag{3.7}$$

Applying these rules to the variables composed of mean values and fluctuations (Reynolds means) as in Eq. (3.1):

$$\overline{(\overline{A})} = \overline{(\overline{A} + A')} = \overline{(\overline{A})} + \overline{a'} = \overline{(\overline{A})} \tag{3.8}$$

as the mean fluctuations:

$$a' = A - \overline{A} \tag{3.9}$$

is zero.

It follows then that

$$\overline{(A\,B)} = \overline{(\overline{A} + a')\,(\overline{B} + b')} = \overline{(\overline{A}\overline{B} + a'\overline{B} + \overline{A}b' + a'b')} = \overline{(\overline{A}\overline{B})} + \overline{(a'\overline{B})} + \overline{(\overline{B}\,b')} + \overline{(a'\,b')} = \overline{(\overline{A}\overline{B})} + \overline{(a'\,b')} = \overline{A}\overline{B} + \overline{(a'\,b')}$$

$$\tag{3.10}$$

The mean of the fluctuation's product is a non-linear variable, in principle non-zero.

Other non-linear variables are, for example, $\overline{(a'^2)}$, $\overline{(a'b'^2)}$, $\overline{(a'^2b')}$, or $\overline{(a'^2b'^2)}$. These variables are important for the characterization of atmospheric turbulence. It is necessary to consider the definitions of variance for any scalar or vector quantity where for large data sets $1/N \approx 1/(N-1)$:

$$\sigma_A^2 = \frac{1}{N} \sum_{i=0}^{N-1} (A_i - \overline{A})^2 = \frac{1}{N} \sum_{i=0}^{N-1} (a_i')^2 = \overline{(a'^2)} \tag{3.11}$$

The standard deviation is the square root of the variance:

$$\sigma_A = \left(\overline{a'^2}\right)^{1/2} \tag{3.12}$$

The standard deviation of velocity is a measure of the magnitude of the deviation or dispersion of the measured values around the mean. Thus, the intensity of turbulence I can be defined as (Stull 1994)

$$I = \sigma_V / \overline{V} \tag{3.13}$$

The covariance of two random variables A and B is defined as

$$\text{cov}\,(A,\ B) = \frac{1}{N}\sum_{i=0}^{N-1}(A_i - \overline{A})(B - \overline{B}) = \frac{1}{N}\sum_{i=0}^{N-1}a_i'b_i' = \overline{a'\,b'} \tag{3.14}$$

The concept of linear correlation coefficient r_{AB} between two random variables is closely associated with covariance:

$$r_{AB} = \frac{\overline{a'\,b'}}{\sigma_A \sigma_B} \tag{3.15}$$

This statistic varies between -1, when the variables vary oppositely, and 1, when the variables vary in the same way. Variables without any common pattern of variation have a zero r_{AB} value.

The turbulent kinetic energy (TKE) is another fundamental variable for the characterization of turbulence in terms of energy transfer among eddies of different dimensional scales. The TKE budget allows comparison between the terms for production either through thermal buoyancy or shear stresses, with the terms for heat dissipation due to viscosity. TKE per unit air mass can be defined in two ways depending on whether the TKE average flow component is related with:

$$TKE/m = \frac{1}{2}\left(\overline{U}^2 + \overline{V}^2 + \overline{W}^2\right) \tag{3.16}$$

or fluctuations:

$$e = \frac{1}{2}\left(\overline{u'^2} + \overline{v'^2} + \overline{w'^2}\right) \tag{3.17}$$

Turbulent eddies carry scalar and vector quantities, associated with different flows, such as density ρ, vertical velocity W, and volumetric content of the scalar quantity k to which the flow refers. The average value of W is zero, because in a flat location that is sufficiently large and geometrically uniform, there is no preferential vertical flow, and because the rising air mass equals the downward air mass during a reasonable period (at least 10 min). The mean wind velocity in the constant flux layer can therefore be considered horizontal. The fluctuation value of W and w' will be positive in the case of upward movement, and negative if in the opposite direction.

The mean flow of the flux quantity F then becomes:

$$F = \overline{(\overline{\rho} + \rho')(\overline{w} + w')(\overline{k} + k')} \tag{3.18}$$

$$F = \left(\overline{\overline{\rho}\overline{w}\overline{k}} + \overline{\overline{\rho}\overline{w}k'} + \overline{\overline{\rho}w'\overline{k}} + \overline{\overline{\rho}w'k'} + \overline{\rho'\overline{W}\overline{k}} + \overline{\rho'\overline{W}k'} + \overline{\rho'w'\overline{k}} + \overline{\rho'w'k'}\right) \quad (3.19)$$

In the last equation, the mean fluctuations, by definition, equal to zero. Fluctuations of ρ are also zero because the fluid does not vary in density and is incompressible. The average value of w is likewise zero.

Thus, Eq. (3.19) can be simplified:

$$F = \rho \,\overline{w'k'} \quad (3.20)$$

Equation (3.20) applied to various momentum fluxes, sensible heat, latent heat, and carbon dioxide, is given by

$$\tau = -\rho\overline{u'w'} \quad (3.21)$$

$$H = \rho \, c_p \, \overline{w'T'} \quad (3.22)$$

$$LE = \rho \, L\overline{w'q'} \quad (3.23)$$

$$C = \rho \, \overline{w'c'} \quad (3.24)$$

Combining Eq. (2.10) $\tau = \rho u_*^2$ from Chap. 2, with Eq. (3.21) we have

$$\tau = -\rho\overline{u_1'u_3'}$$

which gives the defining equality of friction velocity u_*^2

$$u_*^2 = -\overline{u_1'u_3'} \quad (3.25)$$

To measure fluctuations described in these equations, sensitive instrumentation coupled to a data acquisition system is needed. Sensors need to be synchronized and should be of sufficiently high frequency to record fluctuations of the properties of smaller eddies capable of turbulent transport. For example, in forested areas, aerodynamic studies require sensors capable of operating between 0.1 and 10 Hz.

3.3 Taylor's Hypothesis

This hypothesis can be used mainly to replace analysis of micrometeorological parameters in a large region of space region at an instant in time, with analysis of a single point in space, over an extended period. This is the basis for installing observation towers, at a given point on terrain, equipped with appropriate instrumentation to characterize microclimates and for the measurement of vertical fluxes of energy and mass. Taylor's hypothesis assumes that for a given property Φ, a turbulent eddy does not change in space, or "freezes", as it advects along a path. This "freezing" means that the variations observed in continuous wind speed measurements are considered a function of the advection time and are not a consequence of spatial variation along the characteristic dimension of the eddy.

The Taylor simplification is useful in cases where eddy properties evolve over a longer time by comparison to the displacement of its entire characteristic dimension. Under these circumstances, the internal changes of the eddies are minimal throughout the dislocation period, and all the characteristic dimensions are retained at the measurement point. Taylor's hypothesis can be expressed as follows:

$$\frac{\partial \Phi}{\partial t} = -u\frac{\partial \Phi}{\partial x} - v\frac{\partial \Phi}{\partial y} - w\frac{\partial \Phi}{\partial z} \tag{3.26}$$

where Φ is a scalar or vectorial variable and u, v, and w, are the wind speed components in x, y, and z directions.

Turbulence is frozen when:

$$\sigma_v \leq 0.5U \tag{3.27}$$

where U is the mean wind speed and σ its standard deviation (Willis and Deardoff 1976). The Taylor hypothesis can be used when the turbulence intensity is low relative to the mean flow.

3.4 Ergodic Conditions

It is convenient to simplify the measurement of time-dependent variables using rules and averages (Eqs. 3.1, 3.10, and 3.18) for obtaining continuous microclimatic data, using sensors installed at a given point. Taylor's premise for calculating the means should, in principle, deal with time and space separately. That is, for a given variable $B(t, s)$ which is a function of time and space, the average time over a total period of measurement t will be

$$^t\overline{B(s)} = \frac{1}{N}\sum_{i=0}^{N-1}A(i, s) \tag{3.28}$$

where the i index is the number of time increments Δt and $t = i\Delta t$. The corresponding spatial mean becomes

$$^s\overline{B(t)} = \frac{1}{N} \sum_{j=0}^{N-1} A(t,j) \tag{3.29}$$

where the index j is the number of spatial increments Δs and $s = j \, \Delta s$.

The joint space-time averaging corresponding to the sum of N identical experiments, completely random in space, is

$$^c\overline{B(t,s)} = \frac{1}{N} \sum_{i=0}^{N-1} A_i(t,s) \tag{3.30}$$

Although this combined means would be ideal, in practice it is difficult to obtain. The spatial mean would be fine if statistically identical turbulent flows at each point in space were obtained, but this is not possible in the real atmosphere (Stull 1994).

Mobile systems with sensors mounted on a moving platform could, in principle, be used to study the linear variation of the velocity fields but in practice these are not used. When dealing with non-linear variability such systems are a simplification, and indeed sensor displacement would have to be particularly fast to be able to discriminate temporal variations along the velocity field. The more feasible option involves calculating time averages, and sensors are typically installed in an observation station at a fixed point. In this situation, the ergodic assumption can be made, that is, flow is stationary in time and homogeneous in space (Stull 1994). The ergodic condition presupposes equality of time, space, and combined averages, and is useful in the experimental study of turbulence in the surface boundary layer.

3.5 Introduction to Turbulent Motion Equations

3.5.1 Navier–Stokes Equations for Laminar Flow

The Navier–Stokes equations are partial differential equations that can be solved by numerical methods and are used for mass conservation and linear momentum analysis in laminar flow.

The general principle of the conservation of mass is that, for a given control volume, the sum of the budget of the mass flowing through its surface with the time variation of the mass flow within such a volume is zero. For an infinitesimal control volume element of a Newtonian fluid with mass dm and volume $d\Lambda = dxdydz$, the following equation for the mass budget is obtained:

$$\left[\frac{\partial \rho u}{\partial x} + \frac{\partial \rho v}{\partial y} + \frac{\partial \rho w}{\partial z} + \frac{\partial \rho}{\partial t} \right] = 0 \tag{3.31}$$

Considering the vector operator ∇:

$$\nabla = \vec{i}\,\frac{\partial}{\partial x} + \vec{j}\,\frac{\partial}{\partial y} + \vec{k}\,\frac{\partial}{\partial z} \tag{3.32}$$

Equation (3.31) can then be written as

$$\nabla \rho \vec{V} + \frac{\partial \rho}{\partial t} = 0 \tag{3.33}$$

Equation (3.33) can be simplified in cases where the flow is stationary and incompressible.

Under these conditions:

$$\frac{\partial \rho}{\partial t} = 0 \text{ and } \frac{\partial(\rho u_j)}{\partial x_j} = \rho\,\frac{\partial(u_j)}{\partial x_j} \tag{3.34}$$

so that Eq. (3.31) becomes:

$$\left[\rho\,\frac{\partial u}{\partial x} + \rho\,\frac{\partial v}{\partial y} + \rho\,\frac{\partial w}{\partial z} \right] = 0 \tag{3.35}$$

Applying Einstein's summation notation gives:

$$\frac{\partial(u_j)}{\partial x_j} = 0 \tag{3.36}$$

To obtain the linear momentum conservation equations, Newton's second law is applied to a set of fluid particles:

$$\vec{F} = \left.\frac{d\vec{P}}{dt}\right)_{set} \tag{3.37}$$

where the linear momentum $\vec{P}$ of the system is given by:

$$\vec{P}_{system} = \int_{mass} \vec{V} dm \tag{3.38}$$

with $\vec{V}$ being fluid velocity vector.

Considering a particle of mass dm, Newton's second law can be written as:

$$\vec{F} = dm \frac{d\vec{V}}{dt}\Bigg)_{set} \tag{3.39}$$

or

$$\vec{F} = dm\frac{d\vec{V}}{dt} = dm\left[u\frac{\partial\vec{V}}{\partial x} + v\frac{\partial\vec{V}}{\partial y} + \frac{\partial\vec{V}}{\partial z} + \frac{\partial\vec{V}}{\partial t}\right] \tag{3.40}$$

where $u = \frac{dx}{dt}$, $v = \frac{dy}{dt}$, and $w = \frac{dz}{dt}$, and the term in the square brackets in Eq. (3.40), correspond to the total acceleration of the fluid particle. In this term, the sum of the first three terms on the left side is known as the convective acceleration component and the fourth term is the local acceleration component.

Forces acting on a fluid particle can be classified as volume forces (e.g., gravity) and tangential forces (e.g., shear stresses). Atmospheric air can be considered a Newtonian fluid in which the stresses are directly proportional to strains. Therefore, developing the calculations for Newton's second law for an infinitesimal cubic element of mass dm and volume $dV = dx\,dy\,dz$ gives the following equations (Fox and McDonald 1985):

$$\rho\frac{dv}{dt} = \rho f V - \frac{dp}{dx} +$$
$$\frac{\partial}{\partial x}\left[\mu\left(2\frac{\partial u}{\partial x} - \frac{2}{3}\nabla.\vec{V}\right)\right] + \frac{\partial}{\partial y}\left[\mu\left(\frac{\partial u}{\partial y} + \frac{\partial v}{\partial x}\right)\right] + \frac{\partial}{\partial z}\left[\mu\left(\frac{\partial w}{\partial x} + \frac{\partial u}{\partial z}\right)\right] \tag{3.41}$$

$$\rho\frac{dv}{dt} = \rho f U - \frac{dp}{dy} +$$
$$\frac{\partial}{\partial x}\left[\mu\left(\frac{\partial u}{\partial y} + \frac{\partial v}{\partial x}\right)\right] + \frac{\partial}{\partial y}\left[\mu\left(2\frac{\partial u}{\partial y} - \frac{2}{3}\nabla.\vec{V}\right)\right] + \frac{\partial}{\partial z}\left[\mu\left(\frac{\partial v}{\partial z} + \frac{\partial w}{\partial y}\right)\right] \tag{3.42}$$

$$\rho\frac{dw}{dt} = \rho g - \frac{dp}{dz} +$$
$$\frac{\partial}{\partial x}\left[\mu\left(\frac{\partial w}{\partial x} + \frac{\partial u}{\partial z}\right)\right] + \frac{\partial}{\partial y}\left[\mu\left(\frac{\partial v}{\partial z} + \frac{\partial w}{\partial y}\right)\right] + \frac{\partial}{\partial z}\left[\mu\left(2\frac{\partial w}{\partial z} - \frac{2}{3}\nabla.\vec{V}\right)\right] \tag{3.43}$$

These three equations can be written in the following summarized form:

$$\rho\frac{du_i}{dt} = -u_j\frac{\partial u_i}{\partial x_i} - \delta_{i3}\,g + f_c\,\varepsilon_{ij3}\,u_j - \frac{1}{\rho}\left(\frac{\partial p}{\partial x_i}\right) + \frac{1}{\rho}\left(\frac{\partial \tau_{ij}}{\partial x_j}\right) \tag{3.44}$$
$$\quad\text{I}\qquad\qquad\text{II}\qquad\ \ \text{III}\qquad\text{IV}\qquad\ \ \text{VI}\qquad\quad\ \text{VI}$$

where f_c is the Coriolis parameter, δ_{ij} is the Kronecker delta unit value if $i = j$ and of zero value if $i \neq j$; ε_{ij3} is the symbol for the permutation value and is $+1$ if $i = 1$ and $j = 2$, -1 if $i = 2$ and $j = 1$, zero if $j = 3$ or $i = j$. The τ_{ij} symbol represents the matrix for viscous forces.

In this equation, term I is the inertial term representing momentum storage; term II is the advective term for motion quantity; term III represents the term for vertical gravity force; term IV represents the Coriolis effect due to the Earth's rotation; term V is the gradient for static pressure forces; and term VI is the influence of viscous stress of a molecular nature. Term IV's variable f_c of the order of 10^{-4}s^{-1} relates to the Coriolis force and is defined as

$$f_c = \left(1.45 \times 10^{-4}\text{s}^{-1}\right)\sin\phi \tag{3.45}$$

where ϕ is site latitude.

The matrix for the viscous forces τ_{ij}, is given by:

$$\tau_{ij} = \mu\left(\frac{\partial u_i}{\partial x_j} + \frac{\partial u_j}{\partial x_i}\right) - \frac{2}{3}\mu\frac{\partial u_k}{\partial x_k}\delta_{ij} \tag{3.46}$$

where the dynamic viscosity of air μ at atmospheric pressure and at ambient temperatures is very low (1.983×10^{-5} kgm^{-1} s). After some manipulation and if viscosity does not vary with position, term VI becomes

$$VI = \left(\frac{\mu}{\rho}\right)\left\{\frac{\partial^2 u_i}{\partial x_j^2} + \frac{\partial}{\partial x_i}\left[\frac{\partial u_j}{\partial x_j}\right] - \left(\frac{2}{3}\right)\frac{\partial}{\partial x_i}\left[\frac{\partial u_k}{\partial x_k}\right]\right\} \tag{3.47}$$

Under conditions of incompressibility Eq. (3.36) is valid, and variations in viscosity are negligible, Eq. (3.44) becomes (Stull 1994):

$$\frac{du_i}{dt} + u_j\frac{\partial u_i}{\partial x_i} = -\delta_{i3}g + f_c\varepsilon_{ij3}u_j - \frac{1}{\rho}\left(\frac{\partial p}{\partial x_i}\right) + v\left(\frac{\partial^2 u_i}{\partial x_j}\right) \tag{3.48}$$

where v is the kinematic viscosity given by μ/ρ. Equation (3.48) is the equation for linear momentum conservation in laminar flow. This is used as a starting point for obtaining the main turbulence equations.

These equations are non-linear, as advective terms are included which are the product of the velocity components and their respective derivatives. These equations are simplified in homogeneous flows with zero spatial derivatives. However, highly complex atmospheric flows, with three-dimensional eddies, waves, and coherent structures vary in a highly random and chaotic way. Under such conditions velocity gradients need to be analyzed using three coordinates as this input is needed to solve the system of equations.

An additional difficulty in solving the equations is the presence of viscous forces encompassed in some of the terms. In simplified analysis, the presence of viscosity is not considered (inviscid flow). Viscous forces are essential in flows along the contact surfaces as well as to estimate the energy dissipated via the turbulence cascade mechanism.

3.5.2   Equations for Mean Variables in Turbulent Flow

Equations that characterize turbulent flow result from a generalization of the Navier–Stokes equations for laminar flow of Newtonian fluids, with addition of terms for random velocity fluctuations. This generalization involves subjecting the velocity field u_j to the Reynolds decomposition into mean components $\bar{u}_j$, and fluctuations u_j' with a zero average.

The Boussinesq approximation makes it possible to neglect density fluctuations in the various terms in the linear momentum budget equation generic for turbulent flow, except for the term for gravity because of gravitational acceleration.

As a rule of thumb (Stull 1994), the practical application of the Boussinesq approximation in linear momentum equation for turbulent flow involves simply replacing every occurrence of ρ with $\bar{\rho}$ and, g with $\left(g - \left(\theta_v'/\bar{\theta}_v\right)\right)$, where θ_v is the virtual potential temperature.

3.5.2.1   Continuity Equation

The continuity equation expresses the mass conservation principle in differential form, under conditions of incompressibility, by the following expression:

$$\frac{\partial(u_j)}{\partial x_j} = \frac{\partial\left(\bar{u} + u_j'\right)}{\partial x_j} = 0 \tag{3.49}$$

or

$$\frac{\partial \bar{u}}{\partial x_j} + \frac{\partial u_j'}{\partial x_j} = 0 \tag{3.50}$$

so, calculating the means we have:

$$\overline{\frac{\partial \bar{u}}{\partial x_j} + \frac{\partial u'j}{\partial x_j}} = \frac{\partial \bar{u}}{\partial x_j} + \frac{\partial \overline{u'_j}}{\partial x_j} = \frac{\partial \bar{u}}{\partial x_j} = 0 \tag{3.51}$$

The continuity equation for the mean flow becomes:

$$\frac{\partial \bar{u}}{\partial x_j} = 0 \tag{3.52}$$

Thus from Eqs. (3.51) and (3.49), the continuity equation for the instantaneous velocity fluctuations becomes:

$$\frac{\partial u_j'}{\partial x_j} = 0 \tag{3.53}$$

3.5.2.2 Equation for Linear Momentum Conservation

Equation (3.53) results from the generalization of Eq. (3.48) for linear momentum conservation for turbulent flow. It considers the decay of velocity components, the average values, and fluctuations to quantify the increase in motion quantities formed in the control volume because of mechanical or thermal turbulence.

Then, expanding the velocity and pressure variables into the mean and turbulent components gives:

$$
\frac{\partial(\overline{u_i}+u_i')}{\partial t} + \left(\overline{u_j}+u_j'\right)\frac{\partial(\overline{u_i}+u_i')}{dx_j} = -\delta_{i3}\left(g-\frac{\theta_v'}{\theta_v}\right) + f_c\varepsilon_{ij3}\left(u_j+u_j'\right) -
$$
$$
-\frac{1}{\rho}\frac{\partial(\overline{p}+p')}{\partial x_i} + v\frac{\partial^2(\overline{u_i}+u_i')}{\partial x_j^2}
\tag{3.54}
$$

or

$$
\frac{\partial\overline{u_i}}{\partial t} + \frac{\partial u_i'}{\partial t} + \overline{u_j}\frac{\partial\overline{u_i}}{\partial x_j} + \overline{u_j}\frac{\partial u_i'}{\partial x_j} + u_j'\frac{\partial\overline{u_i}}{\partial x_j} + u_j'\frac{\partial u_i'}{\partial x_j} = -\delta_{i3}g + \delta_{i3}g\left(\frac{\theta_v'}{\theta_v}\right) +
$$
$$
+ f_c\varepsilon_{ij3}\overline{u_j} + f_c\varepsilon_{ij3}u_j' - \frac{1}{\rho}\frac{\partial\overline{p}}{\partial x_i} - \frac{1}{\rho}\frac{\partial p'}{\partial x_i} + v\frac{\partial^2 u_i}{\partial x_j^2} + v\frac{\partial^2 u_i'}{\partial x_j^2}
\tag{3.55}
$$

Averaging Eq. (3.55) gives u_i

$$
\overline{\frac{\partial\overline{u_i}}{\partial t}} + \overline{\frac{\partial u_i'}{\partial t}} + \overline{\overline{u_j}\frac{\partial\overline{u_i}}{\partial x_j}} + \overline{\overline{u_j}\frac{\partial u_i'}{\partial x_j}} + \overline{u_j'\frac{\partial\overline{u_i}}{\partial x_j}} + \overline{u_j'\frac{\partial u_j'}{\partial x_j}} = -\overline{\delta_{i3}g} + \delta_{i3}g\,\overline{\frac{\theta_v'}{\theta_v}} +
$$
$$
+ \overline{f_c\varepsilon_{ij3}\overline{u_j}} + \overline{f_c\varepsilon_{ij3}u_j'} - \overline{\frac{1}{\rho}\frac{\partial\overline{p}}{\partial x_i}} + \overline{\frac{1}{\rho}\frac{\partial p'}{\partial x_i}} + \overline{v\frac{\partial^2 u_j}{\partial x_j^2}} + v\overline{\frac{\partial^2 u_j'}{\partial x_j^2}}
\tag{3.56}
$$

The means for the terms with fluctuations are zero, so that Eq. (3.55) can be written as

$$
\frac{\partial\overline{u_i}}{\partial t} + \overline{u_j}\frac{\partial\overline{u_i}}{\partial x_j} + \overline{u_j'\frac{\partial u_i'}{\partial x_j}} = -\delta_{i3}g + \overline{f_c\varepsilon_{i3}\overline{u_j}} - \frac{1}{\rho}\frac{\partial\overline{p}}{\partial x_i} + v\frac{\partial^2\overline{u_i}}{\partial x_j}
\tag{3.57}
$$

Then, by multiplying the continuity equation with the turbulent motions Eq. (3.53) for u_i', and using the means gives

$$
\overline{u_j'\frac{\partial u_j'}{\partial x_j}} = 0
\tag{3.58}
$$

and by adding Eq. (3.58) to the third term on the left of Eq. (3.57) gives (Stull 1994):

$$
\frac{\partial u_i}{\partial t} + \overline{u_j}\frac{\partial\overline{u_i}}{\partial x_j} + \frac{\partial\overline{\left(u_i'u_j'\right)}}{\partial x_j} = \delta_{i3}g + f_c\varepsilon_{ij3}\overline{u_j} - \frac{1}{\rho}\frac{\partial\overline{p}}{\partial x_i} + v\frac{\partial^2 u_i}{\partial x_j^2}
\tag{3.59}
$$

or:

$$\frac{\partial \bar{u}_i}{\partial t} + \bar{u}_j \frac{\partial \bar{u}_i}{\partial x_j} = -\delta_{i3}g + \overline{f_c \varepsilon_{lj3} \bar{u}_j} - \frac{1}{\rho}\frac{\partial \overline{p}}{\partial x_i} + v\frac{\partial^2 u_i}{\partial x_j^2} - \frac{\partial \left(\overline{u_i' u_j'}\right)}{\partial x_j}$$

$$\text{I} \qquad\quad \text{II} \qquad\quad\quad \text{III} \quad\;\; \text{IV} \qquad\; \text{V} \qquad\; \text{VI} \qquad \text{VII}$$

In Eq. (3.59) term I represents the time variation of mean momentum; term II describes advection of mean momentum by the mean wind; term III quantifies the effects of gravity; term IV is the Coriolis effects relating to the Earth's rotation; term V is the mean pressure gradient forces; term VI represents the influence of viscous stresses; and term VII represents the influence of shear stress (Reynolds stress) in the mean flow.

Term VII or friction due to tangential stresses is often larger than many of the other terms in the equation characterizing mean flow. This indicates that the turbulent motions need to be considered when studying turbulent flow parameters, regardless of whether these are three dimensional or unidimensional. The generic product of term VII, $\overline{u_i' u_j'}$, with stress units is representative of exchanges of linear momentum, $\rho u_i'$ per unit time through the surface of an element that is perpendicular to u_i', in the u_j' direction. This product is a shear stress and a non-diagonal matrix element of the Reynolds stresses.

3.5.3   Equations for Variance

Equations for variance and covariance budgets for instant fluctuations of scalar and vector quantities are now analyzed. Variance gives an indication of the energy and intensity of turbulence, while covariance describes the turbulent fluxes of energy and mass. In theory, the equations used to estimate the balance of the instantaneous fluctuations can be useful in predicting turbulent phenomena such as gusts. However, the average time of validity of such phenomena is very short, proportional to their lifetime, ranging from seconds to a few minutes, for practical application (Stull 1994). Thus, equations to estimate fluctuations are mainly used for developing equations relating to the prognosis of variance and covariance of the variables.

To formulate the equation of fluctuations of instantaneous velocity, first subtract Eq. (3.59) representing the budget of the respective average components from Eq. (3.55) on the budget of average flow components. This gives a representative equation on the budget for turbulent fluctuations:

$$\frac{\partial u_i'}{\partial t} + \bar{u}_j \frac{\partial u_i'}{\partial x_j} + u_j' \frac{\partial \bar{u}_i}{\partial x_j} + u_j' \frac{\partial u_i'}{\partial x_j} = \delta_{i3}\left(\frac{\theta_v'}{\bar{\theta}_v}\right)g + f_c \varepsilon_{ij3} u_j' -$$
$$-\frac{1}{\rho} + \bar{u}_j \frac{\partial p'}{\partial x_i} + v\frac{\partial^2 u_i'}{\partial x_j^2} + \frac{\partial \left(\overline{u_i' u_j'}\right)}{\partial x_j}$$

$$(3.60)$$

Equation (3.55) can now be used for calculating variance for the flow components. Multiplying the terms of this equation with $2u_i'$ gives

$$
\begin{aligned}
&2u_i'\frac{\partial u_i'}{\partial t} + 2u_i'\bar{u}_j\frac{\partial u_i'}{\partial x_j} + 2u_i'u_j'\frac{\partial u_i}{\partial x_j} + 2u_i'u_j'\frac{\partial u_i'}{\partial x_j} = 2u_i'\delta_{i3}\left(\frac{\theta_v'}{\bar{\theta}_v}\right)g + \\
&+ 2f_c\varepsilon_{ij3}u_i'u_j' - 2\left(\frac{u_i'}{\rho}\right)\frac{\partial p'}{\partial x_i} + 2u_i'\nu\frac{\partial^2 u_i'}{\partial x_j^2} + 2u_i'\frac{\partial\left(\overline{u_i'u_j'}\right)}{\partial x_j}
\end{aligned}
\tag{3.61}
$$

Now using the rule for the derivative product to convert $2u_i'\;\partial u_i'/\partial t$ into $\partial(u_i')^2/\partial t$, gives

$$
\begin{aligned}
&\frac{\partial\left(u_i'\right)^2}{\partial t} + \bar{u}_j\frac{\partial\left(u_i'\right)^2}{\partial x_j} + 2u_i'u_j'\frac{\partial\bar{u}_i}{\partial x_j} + u_j'\frac{\partial\left(u_i'\right)^2}{\partial x_j} = 2u_i'\delta_{i3}\left(\frac{\theta_v'}{\bar{\theta}_v}\right)g + \\
&+ 2f_c\varepsilon_{ij3}u_i'u_j' - 2\left(\frac{u_i'}{\rho}\right)\frac{\partial p'}{\partial x_i} + 2u_i'\nu\frac{\partial^2 u_i'}{\partial x_j^2} + 2u_i'\frac{\partial\left(\overline{u_i'u_j'}\right)}{\partial x_j^2}
\end{aligned}
\tag{3.62}
$$

Applying averages for the terms in Eq. (3.62) we have

$$
\begin{aligned}
&\frac{\overline{\partial\left(u_i'\right)^2}}{\partial t} + \bar{u}_j\frac{\overline{\partial\left(u_i'\right)^2}}{\partial x_j} + 2\overline{u_i'u_j'\frac{\partial\bar{u}_i}{\partial x_j}} + \overline{u_j'\frac{\partial\left(u_i'\right)^2}{\partial x_j}} = \overline{2u_i'\delta_{i3}\left(\frac{\theta_v'}{\bar{\theta}_v}\right)}g + \\
&+ \overline{2f_c\varepsilon_{ij3}u_i'u_j'} - \overline{2\left(\frac{u_i'}{\rho}\right)\frac{\partial p'}{\partial x_i}} + \overline{2u_i'\nu\frac{\partial^2 u_i'}{\partial x_j^2}} + \overline{2u_i'\frac{\partial\left(\overline{u_i'u_j'}\right)}{\partial x_j^2}}
\end{aligned}
\tag{3.63}
$$

The last term of Eq. (3.63) is zero because the mean for u_i' is zero. Adding the term $\overline{u_i'^2\partial u_j'/\partial x_j} = 0$ to the left side of the equation, the last term of this side can be written as $\partial\left(\overline{u_j'u_i'^2}\right)/\partial x_j$

$$
\begin{aligned}
&\frac{\partial\left(u_i'\right)^2}{\partial t} + \bar{u}_j\frac{\partial\overline{\left(u_i'\right)^2}}{\partial x_j} + 2\overline{u_i'\;u_j'}\frac{\partial\bar{u}_i}{\partial x_j} + \frac{\partial\overline{\left(u_j'\;u_i'^2\right)}}{\partial x_j} = \\
&= 2\delta_{i3}\;\overline{u_i'\left(\frac{\theta_v'}{\bar{\theta}_v}\right)}g + 2f_c\varepsilon_{ij3}\;\overline{u_i'u_j'} - 2\overline{\prime\left(\frac{u_i'}{\rho}\right)\frac{\partial p'}{\partial x_i}} + 2\nu\;\overline{u_i'\frac{\partial^2 u_i'}{\partial x_j^2}}
\end{aligned}
\tag{3.64}
$$

Equation (3.64) is the general equation for the variance of the wind speed components, $\overline{\left(u_i'\right)^2}$. For an expedite characterization of turbulent flow in the boundary layer this equation may be simplified in several ways. For the last term on the right-hand side of Eq. (3.64), the following equality can be obtained (Stull 1994):

$$
\overline{2u_i'\nu\frac{\partial^2 u_i'}{\partial x_j^2}} = \nu\frac{\overline{\partial^2\left(u_i'\right)^2}}{\partial x_j^2} - 2\nu\overline{\left(\frac{\partial u_i'}{\partial x_j}\right)^2}
\tag{3.65}
$$

From the development of the equation:

$$\frac{\partial^2 \left(\overline{u_i^2}\right)}{\partial x_j^2} = \frac{\partial}{\partial x_j}\left[\frac{\partial \left(u_i'\right)^2}{\partial x_j}\right] = \frac{\partial}{\partial x_j}\left[2\,u_i'\frac{\partial u_i'}{\partial x_j}\right] = \left[2\,\frac{\partial u_i'}{\partial x_j}\frac{\partial u_i'}{\partial x_j}\right] +$$
$$+ \left[2\,u_i'\frac{\partial^2 u_i'}{\partial x_j^2}\right] = 2\left[\left(\frac{\partial u_i'}{\partial x_j}\right)^2\right] + \left[2\,u_i'\frac{\partial^2 u_i'}{\partial x_j^2}\right]$$

(3.66)

and multiplying both terms on the left and right side of Eq. (3.66) by v gives Eq. (3.65). In the surface boundary layer, the first term on the right side of Eq. (3.65), of the order of 10^{-7} m^2s^{-3}, is indicative of molecular diffusion of velocity variance and its variability throughout the atmospheric boundary layer. The second term on the right side of Eq. (3.65), representative of the tangential shear stresses, is of a higher magnitude of about 10^{-2} m^2s^{-3} in the surface boundary layer.

The following can then be written

$$\overline{2u_i'v\frac{\partial^2 u_i'}{\partial x_j^2}} \cong -2v\,\overline{\left(\frac{\partial u_i'}{\partial x_j}\right)^2}$$

(3.67)

The viscous dissipation ε is a positive term defined by

$$\varepsilon = v\,\overline{\left(\frac{\partial u_i'}{\partial x_j}\right)^2}$$

(3.68)

so, its use in Eq. (3.64) in the form of Eq. (3.67) represents the energy loss. The smaller the size of the eddies that participate in the dissipative process, the greater the loss. For smaller eddies, the turbulent movements are eliminated by viscosity and irreversibly converted into heat. However, the rate of heating from the dissipation of kinetic energy of eddies with smaller dimensions is low and can be neglected in the equations for the conservation of sensible heat (Stull 1994). The units for ε are $[L^2T^{-3}]$ (Tennekes and Lumley 1980).

The third term of the right side of Eq. (3.64) is the pressure that can be developed as follows:

$$-2\,\overline{\frac{u_i'}{\rho}\frac{\partial p\prime}{\partial x_i}} = -\left(\frac{2}{\rho}\right)\frac{\partial \overline{(u_i'p')}}{\partial x_i} - 2\left(\frac{p'}{\rho}\right)\left[\frac{\partial u_i'}{\partial x_i}\right]$$

(3.69)

The expression in the straight brackets in the second term on the right side represents Eq. (3.53), for continuity of turbulent fluctuations. This zero-value expression is the sum of three terms $\partial u_1'/\partial x_1$, $\partial u_2'/\partial x_2$, and $\partial u_3'/\partial x_3$, which individually promote redistribution of kinetic energy from components with more energy to those with less energy. Thus, the second term on the right side is the pressure redistribution term. This term does not change the total variance but does tend to redistribute the kinetic energy in the turbulent field, which becomes more isotropic, and for this reason it is referred to as the return-to-isotropy term.

The values of terms $\partial u_i / \partial x_i$ are higher for smaller eddies, so that the isotropy increases in smaller dimensional scales corresponding to higher spectral frequency range.

The second term on the right-hand side of Eq. (3.64) is the Coriolis number that can be described in the following way:

$$2f_c\, \varepsilon_{ij3}\, \overline{u_i'\, u_j'} = 2f_c\, \varepsilon_{213}\, \overline{u_2'\, u_1'} + 2f_c\, \varepsilon_{123}\, \overline{u_1' u_2'} =$$
$$= -2f_c\, \overline{u_2' u_1'} + 2f_c\, \overline{u_1' u_2'} = 0 \tag{3.70}$$

The equality Eq. (3.70) means that the Coriolis force does not generate variance or turbulent kinetic energy (TKE). The Coriolis term only promotes internal redistribution of kinetic energy at a rate which is about of three orders of magnitude lower than the other terms of Eq. (3.64). Thus, this term can be neglected.

After these simplifications, Eq. (3.64) becomes:

$$\underbrace{\frac{\partial \overline{(u_i')^2}}{\partial t}}_{\text{I}} + \underbrace{\overline{u}_j \frac{\partial \overline{(u_i')^2}}{\partial x_j}}_{\text{II}} =$$

$$= \underbrace{2\delta_{i3} \overline{\left(\frac{u_i' \theta_{v'}}{\theta_v}\right)} g}_{\text{III}} - \underbrace{\frac{\partial \overline{\left(u_j' u_i'^2\right)}}{\partial x_j}}_{\text{IV}} - \underbrace{\left(\frac{2}{\overline{\rho}}\right) \frac{\partial \overline{(u_i' p')}}{\partial x_i}}_{\text{V}} - \underbrace{2\overline{u_i' u_j'} \frac{\partial \overline{u}_i}{\partial x_j}}_{\text{VI}} - \underbrace{2\varepsilon}_{\text{VII}} \tag{3.71}$$

Term I in Eq. (3.71) represents the local storage of the velocity fluctuations variation; term II is the advection of variance by the average wind; term III refers to the effects of buoyancy and thermal instability; term IV refers to the transport of variance $\overline{u_i'^2}$ by turbulent eddies $\overline{u_i'^2}$; term V refers to the redistribution or transport of variance by pressure fluctuations associated with phenomena such as variability of thermal stability or turbulent structures; term VI refers to positive variance production resulting from the product between a negative sign and the momentum flux, usually negative downward flow; and term VII represents the viscous dissipation of velocity variance.

For a given specific velocity component, e.g. u_1, Eq. (3.71) becomes

$$\underbrace{\frac{\partial \overline{(u_1')}^2}{\partial t}}_{\text{I}} + \underbrace{\overline{u}_j \frac{\partial \overline{(u_1')}^2}{\partial x_j}}_{\text{II}} =$$

$$= \underbrace{-\frac{\partial \overline{\left(u_j' u_1'\right)}^2}{\partial x_j}}_{\text{III}} - \underbrace{\left(\frac{2}{\overline{\rho}}\right) \frac{\partial \overline{(u_1' p')}}{\partial x_1}}_{\text{IV}} - \underbrace{2\overline{u_1' u_j'}\,' \frac{\partial \overline{u}_1}{\partial x_j}}_{\text{V}} + \underbrace{2\left(\frac{p'}{\overline{p}}\right)\overline{\left[\frac{\partial u_1'}{\partial x_1}\right]}}_{\text{VI}} - \underbrace{2\nu \overline{\left(\frac{\partial u_1'}{\partial x_j}\right)^2}}_{\text{VII}}$$

$$\tag{3.72}$$

In this equation, the various terms are equivalent to those corresponding to Eq. (3.71). Term VI is the redistributive term of return to isotropy. This term must be inserted in the equations equivalent to Eq. (3.72), relative to the variance of each fluctuation of the wind speed components. It should be noted that if Eq. (3.72) is applied to u_3, then term III of Eq. (3.71) for quantification of vertical buoyancy effects should also be considered.

3.5.4 Equations for the Vertical Momentum Flux

The equations for turbulent flow of momentum, sensible heat, water vapor, and other gases, such as carbon dioxide, $(\overline{u_i'u_k'},\ \overline{u_i'T'},\ \overline{u_i'E'},\ \text{or}\ \overline{u_i'c'})$ are obtained by adding equations for instantaneous fluctuations. By taking Eq. (3.60), multiplying each term by u_k' and calculating the means gives

$$
\overline{u_k'\frac{\partial u_i'}{\partial t}} + \overline{u_k'\bar{u}_j\frac{\partial u_i'}{\partial x_j}} + \overline{u_k'u_j'\frac{\partial \bar{u}_i}{\partial x_j}} + \overline{u_k'u_j'\frac{\partial u_i'}{\partial x_j}} = \overline{\delta_{i3}u_k'\left(\frac{\theta_v'}{\theta_v}\right)}g +
$$
$$
+ f_c\varepsilon_{ij3}\overline{u_k'u_j'} - \frac{\overline{u_k'\rho\frac{\partial p}{\partial x_i}}}{+} + \overline{vu_k'\frac{\partial^2 u_i'}{\partial x_j^2}}
$$
(3.73)

then, by exchanging the indices i and k in Eq. (3.73) gives:

$$
\overline{u_i'\frac{\partial u_k'}{\partial t}} + \overline{u_i'\bar{u}_j\frac{\partial u_k'}{\partial x_j}} + \overline{u_i'u_j'\frac{\partial \bar{u}_k}{\partial x_j}} + \overline{u_i'u_j'\frac{\partial u_k'}{\partial x_j}} = \overline{\delta_{k3}u_i'\left(\frac{\theta_v'}{\theta_v}\right)}g +
$$
$$
+ f_c\varepsilon_{kj3}\overline{u_i'u_j'} - \frac{\overline{u_i'\frac{\partial p'}{\partial x_k}}}{\rho} + \overline{vu_i'\frac{\partial^2 u_k'}{\partial x_j^2}}
$$
(3.74)

From the addition of Eqs. (3.73) and (3.74) and applying the product derivative rule to expressions of the type $u_i'\partial u_k'/\partial t + u_k'\partial u_i'/\partial t = \partial(u_i'u_k')/\partial t$, we have

$$
\frac{\overline{\partial u_i'u_k'}}{\partial t} + \bar{u}_j\frac{\overline{\partial u_i'u_k'}}{\partial x_j} + \overline{u_i'u_j'\frac{\partial \bar{u}_k}{\partial x_j}} + \overline{u_k'u_j'\frac{\partial \bar{u}_i}{\partial x_j}} + \frac{\overline{u_j'\partial u_i'u_k'}}{\partial x_j} =
$$
$$
= \overline{\delta_{i3}u_k'\left(\frac{\theta_v'}{\theta_v}\right)}g + \overline{\delta_{k3}u_i'\left(\frac{\theta_v'}{\theta_v}\right)}g + f_c\varepsilon_{ij3}\overline{u_k'u_j'} + f_c\varepsilon_{kj3}\overline{u_i'u_j'} -
$$
$$
- \overline{\left(\frac{u_i'}{\rho}\right)\frac{\partial p'}{\partial x_k}} - \overline{\left(\frac{u_k'}{\rho}\right)\frac{\partial p'}{\partial x_i}} + \overline{vu_k'\frac{\partial^2 u_i'}{\partial x_j^2}} + \overline{vu_i'\frac{\partial^2 u_k'}{\partial x_j^2}}
$$
(3.75)

The most common simplification of Eq. (3.75) involves vertical transport of momentum ($i = 1$, $k = 3$). For this application, it is assumed that the mean vertical component is zero, the horizontal plane is homogeneous, and that the coordinate system is aligned with the average wind. This gives (Stull 1994; Tennekes and Lumley 1980)

$$
\frac{\partial \overline{(u_1'u_3')}}{\partial t} = -\overline{u_3'^2}\frac{\partial \bar{u}_1}{\partial x_3} - \frac{\partial \overline{(u_1'u_3'^2)}}{\partial x_3} + \frac{g}{\theta_v}\overline{u_1'\theta_v'} + \left[\overline{p'\left(\frac{\partial u_1'}{\partial x_3} + \frac{\partial u_3'}{\partial x_1}\right)}\right]\frac{1}{\rho}
$$

$$
\quad\text{I}\qquad\qquad\text{II}\qquad\quad\text{III}\qquad\qquad\text{IV}\qquad\qquad\qquad\text{V}
$$

$$- 2v \overline{\left(\frac{\partial u_3' \partial u_1'}{\partial x_3^2} \right)} \tag{3.76}$$
$$\text{VI}$$

In Eq. (3.76), term I corresponds to the storage of momentum flux; term II represents the mechanical production of momentum via tangential stresses; term III is the turbulent transport of linear momentum; term IV represents momentum production or consumption through buoyancy; term V is the redistribution of momentum via return to isotropy; and term VI is the viscous dissipation.

Kaimal and Finnigan (1994) describe an equation for vertical transport to the surface boundary layer as follows:

$$\underset{\text{I}}{\frac{\partial \overline{(u_1' u_3')}}{\partial t}} = \underset{\text{II}}{-\overline{u_3'^2} \frac{\partial \overline{u}_1}{\partial x_3}} + \underset{\text{III}}{\frac{g}{\theta_v} \overline{u_1' \theta_v'}} - \underset{\text{IV}}{\frac{\partial \overline{(u_1' u_3'^2)}}{\partial x_3}} -$$

$$\underset{\text{V}}{\frac{1}{\rho} \left[\overline{u_3' \left(\frac{\partial p'}{\partial x_1} \right)} + \overline{u_1' \left(\frac{\partial p'}{\partial x_3} \right)} \right]} \tag{3.77}$$

These authors consider that the dissipation of turbulence occurs mostly by pressure forces, and so exclude viscous dissipation forces from the budget in Eq. (3.77). Wyngaard et al. (1971) consider likewise that term III for turbulent transport is low in the surface boundary layer. The final budget is then established between the term V, for pressure destruction and terms II for production of mechanical turbulence, and III for buoyancy.

3.5.5   Equations for Vertical Flow of Scalar Quantities

For the calculation of the budget for the covariances $\overline{u_i' \theta'}$ of sensible heat, the starting point is the budget for instantaneous temperature fluctuations obtained from energy conservation in the surface boundary layer. This considers the convective transport of sensible heat and the change in the water vapor phase which either releases or absorbs latent heat (Stull 1994):

$$\underset{\text{I}}{\frac{\partial \theta}{\partial t}} + \underset{\text{II}}{u_j \frac{\partial \theta}{\partial x_j}} = \underset{\text{III}}{k_\theta \frac{\partial^2 \theta'}{\partial x_j^2}} - \underset{\text{IV}}{\frac{1}{\rho c_p} \left(\frac{\partial R_{nj}}{\partial x_j} \right)} - \underset{\text{V}}{\left(\frac{LE}{\rho c_p} \right)} \tag{3.78}$$

where κ_θ is the molecular thermal diffusivity, LE is the latent heat associated with the change in phase flux of water vapor; and R_{nj} is the component of the radiative budget in j direction.

In Eq. (3.78) I, II, and III represent the terms for storage, advection, and molecular diffusion and terms IV and V represent the sources (or sinks) of heat from radiation and change in the water vapor phase.

Averaging for converting variables into the mean and turbulent components, gives

$$\frac{\partial\bar\theta}{\partial t} + \frac{\partial\theta\prime}{\partial t} + \bar u_j\frac{\partial\bar\theta}{\partial x_j} + \bar u_j\frac{\partial\theta\prime}{\partial x_j} + u_j'\frac{\partial\bar\theta}{\partial x_j} + u_j'\frac{\partial\theta\prime}{\partial x_j} = \kappa_\theta\frac{\partial^2\bar\theta}{\partial x_j^2} + \kappa_\theta\frac{\partial^2\theta\prime}{\partial x_j^2} - $$
$$- \frac{1}{\bar\rho c_p}\left(\frac{\partial\bar R_{nj}}{\partial x_j}\right) - \frac{1}{\bar\rho c_p}\left(\frac{\partial R'_{nj}}{\partial x_j}\right) - \left(\frac{LE}{\bar\rho c_p}\right) \tag{3.79}$$

Calculating again the means for Eq. (3.79), we have

$$\underset{\text{I}}{\frac{\partial\bar\theta}{\partial t}} + \underset{\text{II}}{\bar u_j\frac{\partial\bar\theta}{\partial x_j}} = \underset{\text{III}}{\kappa_\theta\frac{\partial^2\bar\theta}{\partial x_j^2}} - \underset{\text{IV}}{\frac{\partial\left(\overline{u_j'\theta'}\right)}{\partial x_j}} - \underset{\text{V}}{\frac{1}{\bar\rho c_p}\left(\frac{\partial\bar R_{nj}}{\partial x_j}\right)} - \underset{\text{VI}}{\left(\frac{\overline{LE}}{\bar\rho c_p}\right)} \tag{3.80}$$

The new term IV (nonexistent in Eq. 3.79) is the vertical flow of sensible heat. Subtracting now Eq. (3.80) from Eq. (3.79) yields a budget equation for the temperature fluctuations:

$$\frac{\partial\theta}{\partial t} + \bar u_j\frac{\partial\theta\prime}{\partial x_j} + u_j'\frac{\partial\bar\theta}{\partial x_j} + u_j'\frac{\partial\theta\prime}{\partial x_j} = \kappa_\theta\frac{\partial^2\theta'}{\partial x_j^2} + \frac{\partial\left(\overline{u_j'\theta'}\right)}{\partial x_j} - \frac{1}{\bar\rho c_p}\left(\frac{\partial R'_{yj}}{\partial x_j}\right) \tag{3.81}$$

and multiplying the terms of Eq. (3.81) by fluctuations u_i' and applying the means

$$\overline{u_i'\frac{\partial\theta\prime}{\partial t}} + \overline{\bar u_j u_i'\frac{\partial\theta\prime}{\partial x_j}} + \overline{u_j'u_i'\frac{\partial\bar\theta}{\partial x_j}} + \overline{u_j'u_i'\frac{\partial\theta\prime}{\partial x_j}} = \overline{\kappa_\theta u_i'\frac{\partial^2\theta\prime}{\partial x_j^2}} + $$
$$+ \overline{u_i'\frac{\partial\left(\overline{\theta'u_j'}\right)}{\partial x_j}} - \left(\frac{1}{\bar\rho c_p}\right)\overline{u_i'\left(\frac{\partial R'_{nj}}{\partial x_j}\right)} \tag{3.82}$$

Taking now Eq. (3.60) relative to the balance of fluctuations u_i', multiplying by fluctuation θ_i and applying averages, it will come

$$\overline{\theta'\frac{\partial u_i'}{\partial t}} + \overline{\theta'\bar u_j\frac{\partial u_i'}{\partial x_j}} + \overline{\theta'u_j'\frac{\partial\bar U_i}{\partial x_j}} + \overline{\theta'u_j'\frac{\partial u_i'}{\partial x_j}} = \delta_{i3}\overline{\theta'}\left(\frac{\theta'_v}{\overline\theta_v}\right)g + f_c\varepsilon_{ij3}\overline{u_j'\theta'} - $$
$$- \overline{\frac{\theta'}{\bar\rho}\frac{\partial p'}{\partial x_i}} + v\overline{\theta'\frac{\partial^2 u_i'}{\partial x_i^2}} \tag{3.83}$$

Assuming $\kappa_\theta = v$ and combining Eqs. (3.82) and (3.83), yields

$$\underbrace{\frac{\partial\left(\overline{u_i'\theta'}\right)}{\partial t}}_{\text{I}} + \underbrace{\bar{u}_j\frac{\partial\left(\overline{u_i'\theta'}\right)}{\partial x_j}}_{\text{II}} = \underbrace{-\overline{\theta'u_j'}\frac{\partial\overline{U_i}}{\partial x_j}}_{\text{III}} - \underbrace{\overline{u_i'u_j'}\frac{\partial\overline{\theta}}{\partial x_j}}_{\text{IV}} + \underbrace{\delta_{i3}\left(\frac{\overline{\theta'\theta_v}}{\overline{\theta}_v}\right)g-}_{\text{V}}$$

$$\underbrace{-\frac{\partial\left(\overline{\theta'u_i'u_j'}\right)}{\partial x_j}}_{\text{VI}} + \underbrace{f_c\varepsilon_{ij3}\left(\overline{u_j\theta}\right)}_{\text{VII}} - \underbrace{\frac{1}{\rho}\left[\frac{\partial\left(\overline{p'\theta'}\right)}{\partial x_i}\right.}_{\text{VIII}} - \underbrace{\left.\overline{p'\frac{\partial\theta'}{\partial x_i}}\right]}_{\text{IX}} + \underbrace{v\frac{\partial^2\left(\overline{u_i'\theta'}\right)}{\partial x_j^2}}_{\text{X}}$$

$$\underbrace{-2v'\overline{\left(\frac{\partial u_i'}{\partial x_j}\right)\left(\frac{\partial\theta'}{\partial x_j}\right)}}_{\text{XI}} - \underbrace{\overline{\left(\frac{u_i'}{\overline{\rho}c_p}\right)\left(\frac{\partial R_{nj}'}{\partial x_j}\right)}}_{\text{XII}} \tag{3.84}$$

Terms I and II relate to storage and to advection correlation between the fluctuations of velocity and temperature; terms III, IV, and V reflect production or consumption by average flow, tangential tensions due to turbulence and buoyancy; term VI deals with transport by turbulent motions; term VII represents the Coriolis forces; terms VIII and IX represent a redistribution due to the return to isotropy and the transport by the pressure correlation term; terms X and XI represent the molecular diffusion and viscous dissipation, respectively; term XI can be represented by $2\varepsilon_{u_i\theta}$, following the definition of ϵ at Eq. (3.68). Lastly, the term XII is related to the correlations between velocity and the radiative budget fluctuations.

For the surface boundary layer, the equation can be simplified by eliminating the Coriolis forces as well as term XII (including radiative balance fluctuations) and molecular diffusion:

$$\frac{\partial\left(\overline{u_i'\theta'}\right)}{\partial t} + \bar{u}_j\frac{\partial\left(\overline{u_i'\theta'}\right)}{\partial x_j} = -\overline{\theta'u_j'}\frac{\partial\bar{u}_i}{\partial x_j} - \overline{u_i'u_j'}\frac{\partial\bar{\theta}}{\partial x_j} + \delta_{i3}\left(\frac{\overline{\theta'\theta_v}}{\overline{\theta}_v}\right)g - \frac{\partial\left(\overline{\theta'u_i'u_j'}\right)}{\partial x_j} - \frac{1}{\rho}$$
$$\left[\frac{\partial\left(\overline{p'\theta'}\right)}{\partial x_i} - \overline{p'\frac{\partial\theta'}{\partial x_i}}\right] - 2\varepsilon_{u_i\theta} \tag{3.85}$$

In the case of vertical flow of sensible heat, $(i = 3)$, given the conditions of homogeneity in the horizontal plane and absence of advective effects (Stull 1994), we have

$$\frac{\partial\left(\overline{u_3'\theta'}\right)}{\partial t} = -\overline{u_3'}^2\frac{\partial\bar{\theta}}{\partial x_3} - \frac{\partial\left(\overline{\theta'u_3^2}\right)}{\partial x_3} + $$
$$+ \left(\frac{\overline{g\theta'\theta_v'}}{\overline{\theta}_v}\right) - \frac{1}{\rho}\left[\frac{\partial\left(\overline{p'\theta'}\right)}{\partial x_3} - \overline{p'\frac{\partial\theta'}{\partial x_3}}\right] - 2\varepsilon_{u_3\theta} \tag{3.86}$$

or for the surface boundary layer (Kaimal and Finnigan 1994):

$$\frac{\partial\left(\overline{u_3'\,\theta'}\right)}{\partial t} \approx -\overline{u_3'^2}\frac{\partial\overline{\theta}}{\partial x_3} + \overline{\left(\frac{g\theta'^2}{\overline{\theta_v}}\right)} - \frac{\partial\left(\overline{\theta'u_3'^2}\right)}{\partial x_3} - \frac{1}{\rho}\overline{\left[\theta'\frac{\partial p'}{\partial x_3}\right]} \qquad (3.87)$$

$$\text{I} \qquad\qquad\qquad \text{II} \qquad\quad \text{III} \qquad\quad \text{IV} \qquad\quad \text{V}$$

Equation (3.87) is like Eq. (3.77) where term II represents the production of shear stresses; term III is the production by buoyancy; term IV is the turbulent transport by tangential tensions and term V refers to pressure destruction.

As with Eq. (3.77), the elimination of the correlations is mainly due to pressure forces, so the viscous dissipation component can be neglected in this equation. Likewise, term II for turbulent transport in the surface boundary layer is low. Thus, the final budget relates to term V for pressure destruction and terms of production of turbulence II, III mechanical and buoyancy, respectively. For the vertical flow of another scalar, such as carbon dioxide or moisture, the algebraic developments are analogous, and equations like Eqs. (3.86) and (3.87) can be obtained.

3.5.6   Kinetic Energy Budget Equations

The TKE budget, expressed per unit mass, makes it possible to characterize the context of the production processes, transport, and loss of turbulent fluctuations. As previously mentioned, TKE is a measure of turbulence intensity. In the inertial sublayer, TKE is mechanically derived from shear stresses in the mean flow and thermally produced by buoyancy forces. This kinetic energy is transferred via the inertial cascade from larger to smaller eddies and converted into heat by viscous dissipation.

From Eq. (3.71) on the budget for the variance for velocity fluctuations, the kinetic energy equations can be deduced (Stull 1994; Tennekes and Lumley 1980):

$$\frac{\partial\overline{e}}{\partial t} + \overline{u}_j\frac{\partial\overline{e}}{\partial t} =$$

$$\text{I} \qquad\quad \text{II}$$

$$= \delta_{i3}\frac{g}{\overline{\theta_v}}\left(\overline{u_i'\theta_v'}\right) - \frac{\partial\left(\overline{u_j'e}\right)}{\partial x_j} - \left(\frac{1}{\rho}\right)\frac{\partial\overline{(u_i'p')}}{\partial x_i} - 2\overline{u_i'u_j'}\frac{\partial\overline{u}_i}{\partial x_j} - \varepsilon \qquad (3.88)$$

$$\text{III} \qquad\qquad \text{IV} \qquad\quad \text{V} \quad\;\; \text{VI} \qquad\quad \text{VII}$$

where $\overline{e}$ is the turbulent kinetic energy per unit mass given by $0.5\left(u_1^2 + u_2^2 + u_3^2\right)$. In this equation, term I is the rate of storage of kinetic energy; term II is the advection by the average kinetic energy field speed; term III is the term production or consumption by buoyancy. This term is either production or loss, depending on whether the heat flux is positive (daytime) or negative (night time); term IV is the turbulent kinetic energy transport caused by u_j'; and term V is the transport or pressure correlation which describes how the turbulent kinetic energy is redistributed by pressure fluctuations. This term is associated with the flow of large eddies. Term VI refers to production/consumption by tangential stresses in the surface boundary layer. Its sign is generally opposite that of the mean velocity vector because of downward dissipation of linear momentum; and term VII corresponds to viscous dissipation and thermal conversion of kinetic energy. On average, the transformation of kinetic energy into heat is about $2\ \mathrm{Wm}^{-2}$, a very low power in the overall energy equation (e.g., Foken 2017).

At a given location, TKE storage varies with the time of day. It increases from morning to the afternoon and thereafter decreases from afternoon to night time, when the terms for losses, e.g., by dissipation, exceed terms for production. The storage term in the surface boundary layer can vary throughout the daily cycle by two orders of magnitude between values of the order of 5×10^{-5} and $5\times10^{-3}\ \mathrm{m^2s^{-3}}$. The advective term may in practice, be considered zero in a homogeneous area, assuming steady-state conditions. Term III in Eq. (3.88) representing production/consumption of buoyancy is positive in the surface boundary layer, with maximum values of about $10^{-2}\ \mathrm{m^2s^{-3}}$ (Stull 1994), corresponding to thermal generation of turbulence when the surface is warmer than the surrounding air.

On cloudy days, surface heating is lower negatively influencing the formation of turbulence. The thermal output term only affects the vertical component of the turbulent kinetic energy budget and is anisotropic. The terms for return to isotropy in Eq. (3.72) cause part of the kinetic energy to move in different horizontal directions. Loss of kinetic energy due to thermal effects clearly occurs under conditions of thermal stability, for example, during situations of nighttime inversion to the surface when colder than the surrounding air.

Term IV for vertical turbulent transport can be either loss or gain of kinetic energy, depending on whether the flux is divergent or convergent. On the surface, vertical transport of turbulent kinetic energy predominates relative to fluctuations of the two horizontal components. In contrast, in the intermediate region of the mixed layer, transport of the vertical component of wind velocity fluctuations predominates.

Static pressure fluctuations in the surface boundary layer (term V) are negligible of the order of 0.005 kPa. The pressure terms are, as mentioned below, usually obtained by the difference from the remaining terms. Term VII, molecular dissipation, is more significant for smaller eddies corresponding to higher scalars of atmospheric turbulence (Shaw 1995b). The rates for turbulent kinetic energy dissipation during daytime are higher in the superficial boundary layer, where the turbulence generation rates are also higher. During night time, or in the presence of thermal inversions, turbulence is caused by tangential stresses as there is less dissipation and the turbulent kinetic energy is lower. After sunrise, production of turbulence through buoyancy increases, in addition to mechanical production, with a concomitant increase in the dissipation rate. The variation in the dissipation rate ranges from $10^{-5}\ \mathrm{m^2s^{-3}}$ at night to $10^{-1}\ \mathrm{m^2s^{-3}}$ during the day (Stull 1994).

Whenever turbulent tangential stresses combine and interact with the mean flow, term VI of Eq. (3.88) turbulence is always produced (e.g., Kaimal and Finnigan 1994) as the negative sign which precedes this term is multiplied by another negative sign. The latter product results from the negative flux of downward momentum by the mean velocity gradient of the horizontal wind which is in general positive. The maximum production of mechanical turbulence always occurs in the atmospheric layer adjacent to the surface.

The contributions of terms for turbulent production through buoyancy and tangential stresses indicate the predominant type of convection, which is free when buoyancy predominates and forced under mechanical turbulent transport (Chap. 6). The order of magnitude of the tangential stresses in the surface boundary layer will be greater on a windy day and lower on a calm day, and turbulence in this layer may be of both thermal and mechanical origin. Both the mechanically and thermally driven turbulences are anisotropic with mechanical turbulence being predominantly horizontal, whereas thermal turbulence is predominantly vertical.

Dimensional variability of the terms for the kinetic energy budget (Eq. 3.88) divided by w_*^3/z_i, where z_i is the height of the convective boundary layer, is illustrated in Fig. 3.1.

In a coordinate system aligned with the direction of the mean wind under conditions of horizontal homogeneity, zero mean vertical velocity, and no advection, Eq. (3.88) applied to the vertical component of wind velocity becomes

$$\frac{\partial \overline{e}}{\partial t} = \frac{g}{\theta_v}\left(\overline{u_3'\theta_v'}\right) - \frac{\partial\left(\overline{u_3'e}\right)}{\partial x_3} - \left(\frac{1}{\overline{\rho}}\right)\frac{\partial\left(\overline{u_3'p'}\right)}{\partial x_3} - \overline{u_1'u_3'}\frac{\partial \overline{u_1}}{\partial x_3} - \epsilon \tag{3.89}$$

$$\text{I} \qquad \text{III} \qquad \text{IV} \qquad \text{V} \qquad \text{VI} \qquad \text{VII}$$

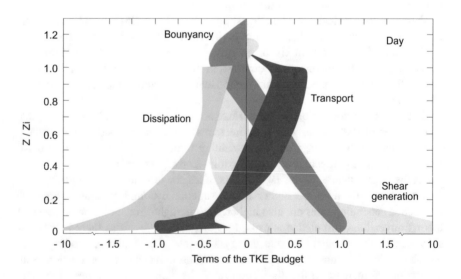

Fig. 3.1 Variability of dimensionless terms for kinetic energy budget (TKE) (Stull 1994)

In this equation, the terms on the right are negative when there is kinetic energy loss and positive when there are gains. Term VII is the loss by viscous dissipation of turbulence whenever the TKE is non-zero. Eq. (3.89) can be used to analyze the various components terms of the TKE energy balance in the surface boundary layer, where these terms are of greater magnitude, and therefore easier to measure, as can be seen in Fig. 3.1. Under conditions of thermal neutrality, without energy production by buoyancy effects and when the pressure redistribution components add up to zero, turbulent mechanical production in Eq. (3.89) is balanced by dissipation viscous, ε (Kaimal and Finnigan 1994):

$$\varepsilon = -\frac{1}{\overline{u_1' u_3'}} \frac{\partial \overline{u_1}}{\partial x_3} \tag{3.90}$$

Under steady-state conditions $\left(\partial \overline{e}/\partial t\right) = 0$, Eq. (3.89) can be written as the sum of dimensionless parameters. Multiplying the terms of this equation by $\left(k(z-d)/u_*^3\right)$ gives (Kaimal and Finnigan 1994):

$$0 = -\underbrace{\frac{z-d}{L}}_{\text{II}} - \underbrace{\frac{k(z-d)}{u_*^3}\frac{d\left(\overline{u_3'e}\right)}{\partial x_3}}_{\text{III}} - \underbrace{\frac{k(z-d)}{u_*^3}\left(\frac{1}{\overline{\rho}}\right)\frac{\partial\left(\overline{u_3'p'}\right)}{\partial x_3}}_{\text{IV}} + \underbrace{\frac{k(z-d)}{u_*}\frac{\partial\overline{u}_1}{\partial x_3}}_{\text{V}} -$$

$$-\underbrace{\frac{k(z-d)\varepsilon}{u_*^3}}_{\text{VI}} \tag{3.91}$$

$$0 = -\underbrace{\frac{z-d}{L}}_{\text{II}} - \underbrace{\phi_t}_{\text{III}} + \underbrace{\phi_p}_{\text{IV}} + \underbrace{\phi_M}_{\text{V}} - \underbrace{\phi_\varepsilon}_{\text{VI}} \tag{3.92}$$

wherein L is the Monin–Obukhov length, and the remaining terms are given by

(II)
$$\frac{z-d}{L} = \xi = -\left(\frac{g}{\overline{\theta}}\right)\frac{\left(\overline{u_3'\theta'}\right)}{u_*^3/k(z-d)} \tag{3.93}$$

(III)
$$\phi_t = \begin{cases} -\xi & \xi \le 0 \\ 0 & \xi > 0 \end{cases} \tag{3.94}$$

(IV) ϕ_p defined under conditions of neutrality or thermal instability, as the difference between the remaining terms.

(V) $$\phi_M = (k(z-d)/u_*)(\partial \overline{u}/\partial z) \tag{3.95}$$

(VI) $$\varphi_\varepsilon = \left(k\,(z-d)\,\varepsilon/u_*^3 \right) \tag{3.96}$$

Figure 3.2 illustrates the variation of dimensionless factors with the ratio z/L, when d is zero.

Each of the dimensionless terms of Eq. (3.91) corresponding to known dimensionless parameters, follow the Monin–Obukhov similarity theory, discussed later. The corresponding empirical functions for dependence on thermal stability were obtained from field data (experiments in Kansas, e.g., Businger et al. 1971).

These functions are as follows (Kaimal and Finnigan 1994):

$$\phi_M = \left| \begin{array}{ll} (1+16|z-d|/L)^{(-1/4)} & \xi \le 0 \\ (1+5(z-d)/L) & \xi > 0 \end{array} \right. \tag{3.97}$$

$$\phi_t = \left\{ \begin{array}{ll} -\xi & \xi \le 0 \\ 0 & \xi > 0 \end{array} \right. \tag{3.98}$$

$$\phi_\varepsilon = \left\{ \begin{array}{ll} \left(1+0.5|\xi|^{(2/3)}\right)^{(3/2)} & \xi \le 0 \\ 1+0.5|\xi| & \xi > 0 \end{array} \right. \tag{3.99}$$

In the surface boundary layer under conditions of instability the ϕ_t term is positive, reflecting the upward transport of the turbulent kinetic energy production

Fig. 3.2 Similarity functions in the surface layer (φ_H, dimensionless factor for the sensible heat flux) (after Kaimal and Finnigan 1994)

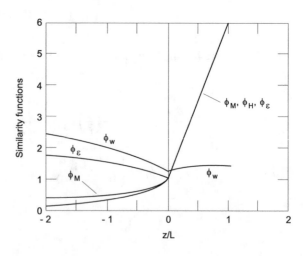

due to buoyancy. Under conditions of atmospheric instability, the ϕ_p term obtained from field measurements is equal to the difference between the production of kinetic energy by tangential stresses and dissipation (Kaimal and Finnigan 1994).

Under conditions of thermal stability, measurements carried out in the field (Kaimal and Finnigan 1994) have shown that $\phi_t = 0$ and $\phi_M \approx \phi_\varepsilon$. Thus, ϕ_p will be of the same order of magnitude as production by buoyancy ξ, and negative under these conditions.

Production of TKE involves mechanical production of turbulence by tangential interaction among turbulent eddies and the mean flow (VI term of Eq. 3.89). Thus, there will be a transfer of kinetic energy from the mean flow to the instantaneous fluctuations (Stull 1994).

The equation for kinetic energy conservation from the mean flow can be obtained from the product of $\bar{u}_i$ using the terms of Eq. (3.57):

$$\frac{\partial \bar{u}_i}{\partial t} + u_j \frac{\partial \bar{u}_i}{\partial x_j} = -\delta_{i3} g + f_c \epsilon_{ij3} \bar{u}_j - \frac{1}{\bar{\rho}} \frac{\partial \bar{p}}{\partial x_i} + v \frac{\partial^2 u_i}{\partial x_j^2} - \frac{\partial \left(\overline{u_i' u_j'} \right)}{\partial x_j} \tag{3.100}$$

giving

$$\underbrace{\frac{\partial (0.5\bar{u}_i)^2}{\partial t}}_{\text{I}} + \underbrace{\bar{u}_j \frac{\partial (0.5\bar{u}_i)^2}{\partial x_j}}_{\text{II}} = \underbrace{-\delta_{i3} g\bar{u}_i}_{\text{III}} + \underbrace{f_c \varepsilon_{ij3} \bar{u}_i \bar{u}_j}_{\text{IV}} - \underbrace{\frac{\bar{u}_i}{\bar{\rho}} \frac{\partial \bar{P}}{\partial x_i}}_{\text{V}} + \underbrace{v \bar{u}_i \frac{\partial^2 \bar{u}_i}{\partial x_i^2}}_{\text{VI}} -$$

$$\underbrace{\bar{u}_i \frac{\partial \left(\overline{u_i' u_j'} \right)}{\partial x_j}}_{\text{VII}} \tag{3.101}$$

In Eq. (3.98), term I represents the storage of kinetic energy from the mean flow; term II is the transport of kinetic energy from the mean flow by advection by the mean velocity fields; term III is the effect of gravity acceleration; term IV is the effects of the Coriolis force; term V is representative of the kinetic energy production due to the effect of acceleration of the mean flow through the mean pressure gradients; term VI represents the effect of molecular dissipation; and term VII is that the interaction between the mean flow and turbulent fluctuations. This term can be written as follows:

$$-\bar{u} \frac{\partial \overline{u_i' u_j'}}{\partial x_j} = \overline{u_i' u_j'} \frac{\partial (\bar{u}_i)}{\partial x_j} - \frac{\partial (\overline{u_i' u_i'} \bar{u}_i)}{\partial x_j} \tag{3.102}$$

Substituting Eq. (3.102) in Eq. (3.101) gives

$$\frac{\partial (0.5\overline{u}_i)^2}{\partial t} + \overline{u}_j \frac{\partial (0.5\overline{u}_i)^2}{\partial x_j} = -\delta_{i3}\, g\overline{u}_i + f_c \varepsilon_{ij3}\overline{u}_i\overline{u}_j - \frac{\overline{u}_i}{\rho}\frac{\partial \overline{P}}{\partial x_i} + \nu\, \overline{u}_i \frac{\partial^2 \overline{u}_i}{\partial x_i^2} +$$
$$+ \overline{u_i'u_j'}\frac{\partial (\overline{u}_i)}{\partial x_j} - \frac{\partial (\overline{u_i'u_j'}\,\overline{u}_i)}{\partial x_j} \tag{3.103}$$

The $\overline{u_i'u_j'}\frac{\partial (\overline{u}_i)}{\partial x_j}$ term describing mechanical production of turbulent kinetic energy by tangential interaction between the mean flow and the turbulent fluctuations then have opposite signs in Eqs. (3.88) and (3.103). This confirms that the kinetic energy losses from the mean flow are gains by the turbulent field and vice versa.

3.5.7 The Closure Equation Problem

To separate the effects of mean flow from the mean effects of flow on the surface boundary layer, equations were developed based on general partial derivatives that are representative of the various components of turbulent transport of several variables under study such as variances, fluxes, or turbulent kinetic energy. However, because of its non-linearity, conversion of the equations for mean flow of motion into turbulent flow gives rise to a system of equations with more unknowns than equations. A variable is considered unknown if there is no diagnostic or prognostic equation defining it. If new equations are introduced into the system, then there are more unknowns than equations. The closure problem arises because a statistical description of atmospheric turbulence requires an infinite number of equations (Tennekes and Lumley 1980; Stull 1994).

The closure problem appears because some of the information on the correlation among variables is lost when means are determined and to close the system of equations, this information needs to be restored back into the system. For example, Eq. (3.59) for the mean velocity $\partial \overline{u}_i/\partial t$ is associated with an unknown term such as $\partial \left(\overline{u_i'u_j'}\right)/\partial t$ (second moment) or Eq. (3.76) for $\partial \left(\overline{u_1'u_3'}\right)/\partial t$ gives rise to a term such as $\partial \left(\overline{u_1'u_3'u_3'}\right)/\partial x_3$ (third moment) and so on. In an equation for determining means, the emergence of new terms for covariance will inevitably lead to loss of information.

To use a selected finite number of equations, assumptions are needed to be able to calculate unknowns. The closure methods are usually considered local if an unknown quantity at a given point in space is parameterized by values and/or gradients of known quantities at this point. Local closure thus assumes that the turbulence is analogous to molecular diffusion, using the diffusion-gradient approach, discussed in Chap. 2, where the turbulent diffusion is proportional to the concentration gradient.

One possible closure technique involves Eq. (3.91) for TKE in the dimensionless form using Eqs. (3.93) to (3.96) as empirical functions. Another approach sometimes referred to as half order is the use of global transfer coefficients. For example, in the case of sensible heat flux:

$$\frac{H}{\rho c_p} = \overline{u_3' T'} = C_H \overline{u}(T_0 - T) \tag{3.104}$$

where C_H is the overall coefficient of transfer of sensible heat between the surface at temperature T_0 and the atmosphere at temperature T. For the linear momentum, the overall drag coefficient C_D, is given by an expression of type (Shaw 1995c):

$$C_D = \left(\frac{u_*}{\overline{u}}\right) = \frac{k^2}{\left[\ln\left(\frac{z}{z_0}\right) + \varnothing_m\left(\frac{z}{L}\right)\right]^2} \tag{3.104a}$$

obtained from principles previously discussed in Chap. 2. The overall transfer coefficient for the heat transfer C_H can be expressed in terms of roughness length z_o and an equivalent length for heat transfer efficiency z_{oH} in a similar way:

$$C_{HI} = k^2 / [\ln(z/z_0) + \phi_m(z/L)]$$

$$C_H = C_{H1} / [\ln(z/z_{0H}) + \phi_h(z/L)] \tag{3.105}$$

where ϕ_h is a function of thermal stability.

The method for first-order closure in the surface boundary layer is based on the gradient-diffusion principle. The first-order closure is then:

$$\overline{u_i' \psi'} = -K \frac{\partial \overline{\psi}}{\partial z} \tag{3.106}$$

where in ψ' is a general designation of a scalar vectorial quantity. Equation (3.106) can be applied to fluxes of momentum, carbon, sensible heat, among others.

The flow-gradient principle may also apply to the second moment as follows (Shaw 1995c):

$$\overline{u_i' u_j'} = -K_m \left(\frac{\partial \overline{u}_i}{\partial x_j} + \frac{\partial \overline{u}_j}{\partial x_i}\right) \tag{3.107}$$

which is a representative expression showing that vertical flux of linear momentum is directly proportional to the mean velocity gradient and where the negative sign indicates that this flux is directed from higher to lower velocities. The K_m coefficient is the momentum diffusivity coefficient. Eq. (3.107) can be used for closure of equations when higher order terms are equivalent to the left side of the equation. In this case, an assumption about the equality of the turbulent diffusion coefficients can be made:

$$K_m = K_H = K_w = K_s \tag{3.108}$$

The analogy between turbulent and molecular diffusion coefficients is reflected by the fact (see Chap. 2) that both have $L^2 T^{-1}$ dimensions, indicative of the product of a velocity and a length. The mixing length relates to the full-length scales of turbulence, being a function of the system geometry (e.g. characteristic dimension) as well as of the type of thermal stratification of the atmosphere. In this context, the friction velocity $\overline{u}_*$ is a scalar of turbulent velocity obtained from Eq. (3.25):

$$u_*^2 = -\overline{u_1' u_3'}$$
(3.25)

or by combining Eq. (2.15) with Eq. (2.17):

$$u_*^2 = l \frac{\partial u}{\partial z}$$
(3.108a)

Limitations of some of the assumptions of the mixed layer theory should be pointed out (see for e.g. Chap. 2 and later in this chapter on the application of the eddy covariance WPL correction method). For example, scalar quantities transported can affect the transport process and eddy length scales can be about the same order of magnitude as those associated with the distribution of the quantity to be transported. Accordingly, it may not be accurate to express the vertical flux in terms of the local gradient of the property. The use of turbulent diffusion coefficients that serve to simplify the complex flow, while simultaneously retaining the equations of second moments, discussed above, makes it possible the analysis of terms related to the dissipation and transport of kinetic energy. In this way, some information inherent to the covariance terms is restored (Shaw 1995d).

As an example of the modeling of the components for equations of higher order, the return-to-isotropy term (term V in Eq. 3.76) can be simplified by the equation:

$$\frac{1}{\rho} \overline{\left[p' \left(\frac{\partial u_i'}{\partial x_k} + \frac{\partial u_k'}{\partial x_i} \right) \right]} = -\frac{u_*}{\lambda} \left(\overline{u_i' u_j'} - \frac{\delta_{ij}}{3} u_*^2 \right)$$
(3.109)

where λ is a length scalar and u_* the friction velocity. Equation (3.109) indicates that for normal forces $i = j$, the momentum budget is proportional to the difference between the variance of velocity and 1/3 the friction velocity.

Similarly, for the term for viscous dissipation of linear momentum, analogous to Eq. (3.68), the equation can be written (Shaw 1995d):

$$2v \overline{\left(\frac{\partial u_i'}{\partial x_k} \frac{\partial u_j'}{\partial x_k} \right)} = -\frac{u_*^3}{l} \delta_{ij}$$
(3.110)

The Eq. (3.110) is based on the principle that the dissipation rate is of the order of u_*^3/l, where l is the characteristic length scale of the energetic eddies.

When large eddies predominate in the transport processes, further developments of numerical techniques for the simulation of large-scale turbulent flows (LES), numerical methods with higher order closure equations (2nd or 3rd) or non-local closure, are given in references such as Stull (1994).

3.6   Spectral Analysis

3.6.1   Introduction

Spectral analysis or Fourier analysis is a mathematical tool that describes data series in terms of contributions arising from different time scales. For example, a time series for the air temperature at a mid-latitude location will be highly variable over a day-long period, not only because of variation in the radiative cycle but also due to seasonal changes over a year-long period. This variation in the time domain translates into a concomitant variation in the frequency domain, meaning that a significant variation in time series over periods of 24 h and 8760 h (24 x 365 days) is analogous to a significant variation in the frequencies of $1/24$ h $= 0.0417$ h^{-1} and $1/8760 = 0.000114$ h^{-1}. The absolute frequency is usually expressed in s^{-1} or Hz.

The frequency analysis (harmonic analysis) involves representing fluctuations or variations of time series, as summations or integrals of trigonometric functions (sines and cosines). These harmonics are trigonometric functions in the sense that they comprise integer multiples of the fundamental frequency determined by the sample size of the data series. Similarly, spectral analysis can be carried out to represent time series of an atmospheric parameter with turbulent fluctuations.

3.6.2   Fourier Series

In the nineteenth century Fourier established a paradigmatic principle "there is no function f(x) or part of a function that cannot be expressed in trigonometric series" (Morrison 1994). A time function can then be represented by a Fourier series:

$$f(t) = a_0 + \sum_{n=1}^{\infty} a_n \cos(n\omega_o t) + \sum_{n=1}^{\infty} b_n \sin(n\omega_o t) \qquad (3.111)$$

where $\omega_0 = 2\pi/T$, the fundamental angular frequency of the function, expressed in rad/s and a_n and b_n are the amplitude coefficients (height of the oscillations) that are given by:

$$a_n = \frac{2}{T} \int_{-T/2}^{T/2} f(t) \cos(n\omega_o t) dt \quad (n = 0, 1, 2 \ldots) \qquad (3.112)$$

$$b_n = \frac{2}{T} \int_{-T/2}^{T/2} f(t) \sin(n\omega_o t) dt \quad (n = 0, 1, 2 \ldots) \qquad (3.113)$$

The associated frequencies $2\omega_0$, $3\omega_0$, $4\omega_0$, ... are the harmonics.

The term a_o is given by the mean value of $f(t)$ in a period T:

$$a_o = \frac{1}{T} \int_{-T/2}^{T/2} f(t)dt \qquad (3.114)$$

The Fourier series can also be expressed in terms of exponential functions, using Euler's formula:

$$e^{iy} = cos(y) + isin(Y) \qquad (3.115)$$

$$e^{-iy} = cos(y) - isin(Y) \qquad (3.116)$$

or:

$$cos(y) = \left(e^{iy} + e^{-iy}\right)/2$$

or:

$$sin(y) = \left(e^{iy} - e^{-iy}\right)/2i$$

Substituting in the above Fourier series:

$$f(t) = a_0 + \sum_{n=1}^{\infty} a_n \cos(n\omega_o t) + \sum_{n=1}^{\infty} b_n \sin(n\omega_o t) \qquad (3.117)$$

gives:

$$f(t) = a_0 + \sum_{n=1}^{\infty} a_n \frac{e^{in\omega_0 t} + e^{-in\omega_0 t}}{2} + \sum_{n=1}^{\infty} b_n \frac{e^{in\omega_0 t} - e^{-in\omega_0 t}}{2i} \qquad (3.118)$$

given that $1/i = -i$

$$f(t) = a_0 + \sum_{n=1}^{\infty} a_n \frac{e^{in\omega_0 t} + e^{-in\omega_0 t}}{2} - ib_n \frac{e^{inw_0 t} - e^{-inw_0 t}}{2} \qquad (3.119)$$

equivalent to:

$$f(t) = a_0 + \sum_{n=1}^{\infty} e^{in\omega t} \frac{a_n - ib_n}{2} + e^{-in\omega t} \frac{a_n + ib_n}{2} \qquad (3.120)$$

After some manipulation, the expression for the Fourier series as the summation of exponential functions is obtained:

$$f(t) = \sum_{n=-\infty}^{+\infty} c_n e^{in\omega_0 T} \tag{3.121}$$

with:

$$c_n = \frac{1}{T} \int_{-T/2}^{T/2} f(t) e^{-in\omega_0 t} dt \tag{3.122}$$

Equations (3.121) and (3.122) are often denoted synthesis and analysis equations, respectively (Morrison 1994).

The development described above for the Fourier series is a time analysis that can be represented in a graph where the ordinate represents continuous variation of the value of oscillating functions obtained by Fourier decomposition, and the abscissa represents time. Another alternative is the analysis of line spectra, in which the abscissa is the frequency of the sinusoids and the ordinate their amplitudes. A three-dimensional graph can be drawn with the values of the function $f(t)$ as the ordinate and with two perpendicular axes in the normal plane relative to the ordinate axis, representing time and frequency, respectively (Fig. 3.3a). This pattern of sinusoidal variation reflecting a harmonic can be applied for characterizing transient heat balances in environmental systems.

According to frequency analysis, a given sinusoid can be considered a line at distance $f = 1/T$ from the origin and the height with the amplitude C_1 (Fig. 3.3b) varying parallel to the time axis. Thus, the time variation can be regarded as a projection of the three-dimensional curve, on the time plane and the sinusoidal variation can be regarded as a projection of the same curve on the frequency plane. In frequency analysis, the height line C_1 will not be negative due to the symmetry of the curve. Fig. 3.3c shows both the frequency and the amplitude of the sinusoid.

A phase diagram (Fig. 3.3d) should indicate the phase shift of the curve for example, at instant t = 0. The phase angle is determined by the ordinate distance (in radians) between zero and the point where positive peak occurs. If the maximum occurs after the origin of the function, it is referred to as forwarded, and conversely if the peak occurs before, it is referred to as delayed. In the case of Fig. 3.2, the phase angle is $\pi/2$. Clearly, the phase shift angles can be different (Fig. 3.4).

Spectral analysis of amplitude lines and phase (Fig. 3.3c, d) is useful when functions have complex configurations. Figure 3.5 represents the spectral lines in phase and frequency, corresponding to a continuous quadrature wave function with several harmonic components.

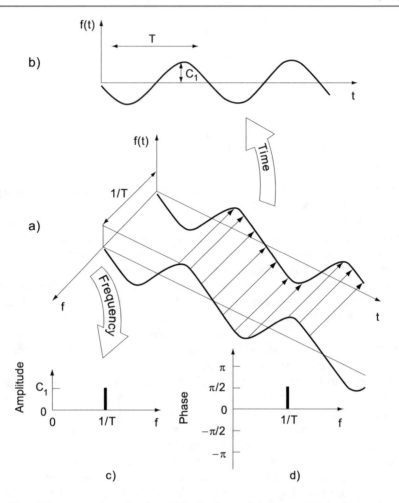

Fig. 3.3 **a** 3D representation of a sinusoid in the time and frequency domain, **b** time domain, **c** frequency amplitude, and **d** phase frequency (adapt. Chapra and Canale 1989)

3.6.3   Fourier Transforms

Fourier series are useful for spectral analysis of a periodic function, with different configurations of other functions that are not repeated regularly. For example, thunder can occur once or repeatedly at irregular intervals, causing interference with electronic or electrical equipment operating over a wide frequency range. This suggests intuitively that a discrete signal produced by thunder has a spectrum of diverse frequencies. When such discrete phenomena are analyzed, the Fourier integral can be used instead of the Fourier series. The $F(i\omega)$ function denoted the f (t) transform, the Fourier integral $f(t)$, or spectral density function $f(t)$ is defined by

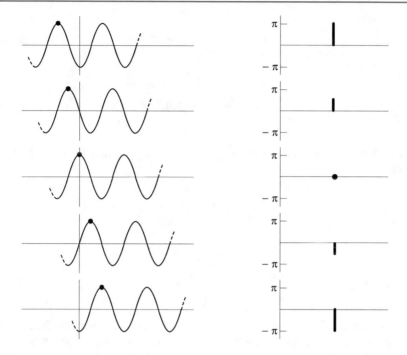

Fig. 3.4 Representation of phase shifts (after Chapra and Canale 1989)

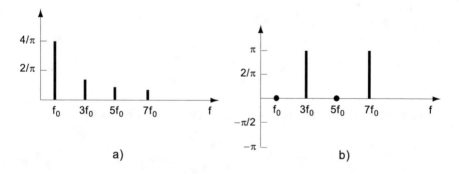

Fig. 3.5 Spectral lines for **a** amplitude and **b** phase (after Chapra and Canale 1989)

Eq. (3.123). The $f(t)$ function defined by Eq. (3.124) is denoted as the inverse Fourier transform or Fourier integral approximation of $f(t)$. Equations (3.123) and (3.124) are known as the Fourier transform pairs.

$$F(i\omega) = \int\limits_{-\infty}^{+\infty} f(t)e^{-i\omega t}dt \tag{3.123}$$

$$f(t) = \frac{1}{2\pi} \int\limits_{-\infty}^{+\infty} F(i\omega)e^{i\omega t}d\omega \tag{3.124}$$

$F(i\omega)$ is obtained from the Fourier series in complex form (Stearns and Hush 1990):

$$f(t) = \sum\limits_{n=-\infty}^{+\infty} c_n e^{in\omega_0 T} \tag{3.125}$$

$$c_n = \frac{w_0}{2\pi} \int\limits_{-\pi/w_0}^{-\pi/w_0} f(t)e^{-in\omega_0 t}dt \tag{3.126}$$

To obtain the Fourier transform, the fundamental frequency of the function w_0 defining the interval between the frequencies of the harmonics, tends to zero, and can then be represented as Δw. Thus, the integral lines of a discrete spectrum converge tend toward a continuous spectrum. In other words, the period $T = 2\pi/w_0$ tends to infinity, which means that the time function is no longer repeated, and becomes aperiodic.

Equation (3.125) can be written as

$$f_p(t) = \sum\limits_{n=-\infty}^{+\infty} c_n e^{inw_0 T} \tag{3.127}$$

where p index indicates the hypothetical periodicity. From Eq. (3.126)

$$f_p(t) = \sum\limits_{n=-\infty}^{+\infty} \left(\frac{\Delta\omega}{2\pi}\right) \int\limits_{-\pi/\Delta\omega}^{-\pi/\Delta\omega} f(\tau)e^{-in\Delta\omega\tau}d\tau\, e^{in\Delta\omega t} \tag{3.128}$$

$$= \frac{1}{2\pi} \sum\limits_{n=-\infty}^{+\infty} \left[\int\limits_{-\pi/\Delta\omega}^{-\pi/\Delta\omega} f(\tau)e^{-in\Delta\omega\tau}d\tau \right] e^{in\Delta\omega t}\, \Delta\omega \tag{3.129}$$

where τ is a transitory dummy variable.

At the limit, with the time T tending to infinity and $\Delta\omega$ tending to zero, the function $f_p(t)$ is no longer periodic and is then simply designated $f(t)$:

$$f(t) = \frac{1}{2\pi} \lim_{\Delta\omega \to 0} \sum_{n=-\infty}^{+\infty} \left(\left[\int_{-\pi/\Delta\omega}^{-\pi/\Delta\omega} f(\tau)e^{-in\Delta\omega\tau}d\tau \right] e^{in\Delta\omega t} \Delta\omega \right) \quad (3.130)$$

and converting the sum into integral:

$$f(t) = \frac{1}{2\pi} \int_{-\infty}^{\infty} \left[\int_{-\infty}^{\infty} f(\tau)e^{-in\Delta\omega\tau}d\tau \right] e^{i\omega t} \Delta\omega \quad (3.131)$$

The Fourier transform $F(i\omega)$ defined by Eq. (3.123) becomes the term in brackets in Eq. (3.131), substituting the terms c_n for the coefficients of the Fourier series. Note that the product $n\Delta w$ is ω, since it is the abscissa corresponding to the frequency of the harmonics.

In this context, the Fourier transform, $F(i\omega)$ in Eq. (3.123), allows to analyse time equation function $f(t)$ in infinite frequencies (spectral density of frequencies) containing all its information. Conversely, the inverse Fourier transform Eq. (3.124), makes it possible to obtain the time function $f(t)$ from the spectral components.

The transform pair allows conversion between time and frequency regimes of non-periodic functions and possibly limited domain in the same way that the Fourier series does for periodic functions defined for an unlimited time interval.

In the above context, the Fourier series converts a continuous periodic time function into discrete spectral values in the form of lines, in the frequency domain. Conversely, Fourier transforms can be applied to continuous functions over a specific time interval, eventually infinitesimal pulses, to obtain continuous spectra in the frequency domain.

3.6.4 Discrete Fourier Transform

In practice, environmental physics functions and databases are given as finite data sets or converted into the discrete form when recorded continuously. When considering a continuous function between instants 0 and T, this interval can be divided into N intervals with widths $\Delta t = T/N$ so that f_n becomes the value of the continuous function at instant t_n.

A discrete Fourier transform then takes the form:

$$F_k = \sum_{n=0}^{N-1} f_n e^{-ik\omega_0 n} \quad (3.132)$$

and the inverse transform:

$$f_n = \frac{1}{N} \sum_{n=0}^{N-1} F_k e^{-ik\omega_0 n} \tag{3.133}$$

where $\omega_0 = 2\pi/N$ is the angular frequency.

Equations (3.132) and (3.133) are the discrete analogues of Eqs. (3.123) and (3.124), respectively.

This complex processing of these equations can be carried out with appropriate software packages. The starting point for such programming is Euler's identity:

$$e^{\pm ib} = cosb \pm isinb \tag{3.134}$$

making it possible to write Eqs. (3.132) and (3.133) as

$$F_k = \frac{1}{N} \sum_{n=0}^{N-1} [f_n \cos(k\omega_0 n) - i f_n \sin(k\omega_0 n)] \tag{3.135}$$

and

$$f_n = \sum_{n=0}^{N-1} [F_k \cos(k\omega_0 n) - i F_k \sin(k\omega_0 n)] \tag{3.136}$$

A simple way to perform this calculation is to use the Fast Fourier Transform (FFT) algorithm. This has been developed in references such as Chapra and Canale (1989) and Lynn and Fuerst (1998).

3.6.5 Autocorrelation and Cross-Correlation Functions

The autocorrelation function (ACF) given in the time domain is important for studying the properties of a discrete random function, as it allows statistical relationships to be analyzed between successive values of the function.

The autocorrelation function $\phi_{xx}(m)$, is defined as

$$\phi_{xx}(m) = \lim_{N \to \infty} \frac{1}{2N+1} \sum_{n=-N}^{N} x(n)x(n+m) \tag{3.137}$$

showing that the ACF is the mean product of a sequence of values of a function x (n) with a time-lagged version at m instants.

Figure 3.6 shows the autocorrelation of the instantaneous fluctuations of the components u and w of wind velocity for 13 min data series and 21 Hz sampling rate of the surface boundary layer of cork oak stands in Portugal (Rodrigues 2002).

The cross-correlation function (CCF) $\varphi_{xy}(m)$, is defined as

$$\phi_{xy}(m) = \lim_{N \to \infty} \frac{1}{2N+1} \sum_{n=-N}^{N} x(n)y(n+m) \tag{3.138}$$

showing that, the CCF is the mean of the product of the sequence of values of a function $x(n)$, with a version $y(n)$ lagged in time by m instants. The ACF and the CCF can be compared because the lag times for one or two functions allows for the assessment of the temporal structure of these functions. These functions are calculated for a set of lag instants.

The ACF establishes the lagged phase products of instantaneous fluctuations, making it possible, for example, to evaluate the formation and duration of a turbulent phenomenon. In this case, the fact that ACF tends to zero, indicates that the eddy (an example of a random process) forms without being either permanent or recurrent. In the case where $m = 0$, the autocorrelation is equal to the square of the function $x(n)$:

$$\phi_{xx}(0) \approx x(n)^2/N \tag{3.139}$$

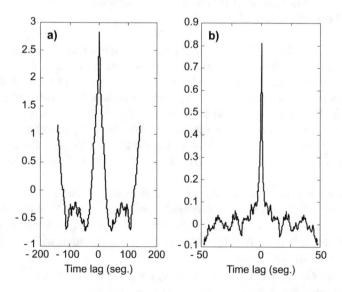

Fig. 3.6 Autocorrelation of data for instantaneous fluctuations for wind velocity (after Rodrigues 2002)

The power spectral function $S_{xx}(\omega)$, is defined as the Fourier transform of the ACF of random processes (discrete functions):

$$S_{xx}(\omega) = \int\limits_{-\infty}^{\infty} \phi_{xx}(\tau)\, e^{-j\omega\tau} d\tau \qquad (3.140)$$

where the ACF $\phi_{xx}(\tau)$, the inverse transform:

$$\phi_{xx}(\tau) = \frac{1}{2\pi} \int_{-\infty}^{\infty} S_{xx}(\omega) e^{j\omega t} d\omega \qquad (3.141)$$

Similarly, cospectral power $S_{xy}(\omega)$ is defined as the Fourier transform of the CCF of random processes:

$$S_{xy}(\omega) = \int\limits_{-\infty}^{\infty} \phi_{xy}(\tau)\, e^{-j\omega t} d\tau \qquad (3.142)$$

$$\phi_{xy}(\tau) = \frac{1}{2\pi} \int_{-\infty}^{\infty} S_{xy}(\omega) e^{j\omega t} d\omega \qquad (3.143)$$

3.6.6 Spectral Characterization of Turbulence in the Surface Boundary Layer

The turbulence at the surface boundary layer is, as above mentioned, made up of a set of eddies of various sizes, ranging from several millimeters to hundreds of meters, making spectral analysis an essential tool for the assessment of the dominant frequency of the turbulent flow. Spectral analysis of fluctuations of any scalar or vectorial flow quantity aims to study the variation in the spectral power function with the frequency. Turbulence spectra depends on site parameters, fluxes and micrometeorological conditions, and its knowledge is relevant form choosing of sensors and definition of optimal sensing strategy for specific atmospheric conditions (Foken 2017).

Spectral analysis is useful to assess the time and length scales of the flow, distribution of TKE on the set of frequencies and as a criterion to assess the quality of collected data. The total power in a time function $x(j)$, corresponding to a field of turbulent fluctuations is given by the respective autocorrelation with zero lag, ACF (0) or $(x(j)^2/n)$ by Eq. (3.139).

However, as $x'(j)$ is the fluctuations field, it follows that the mean $\overline{x'(j)}$ is zero so that the respective variance $S_{xy}(\omega)$, can be expressed as:

$$\sigma^2 x'(j) = \frac{1}{n}\sum_n^{j=1}\left(x'(j) - \overline{x'(j)}\right)^2 = \frac{1}{n}\sum_n^{j=1}(x'(j))^2 = \overline{x'^2(j)} = ACF(0) \qquad (3.144)$$

Thus, the power of the total fluctuations field $x'(j)$, is given by the respective variance

$\overline{x'^2(j)}$. Given the definition of friction velocity we have:

$$u_*^2 = \overline{u'w'} = \overline{u'^2} = \overline{w'^2} \qquad (3.145)$$

u_*^2 can be regarded as the total power of the field of instantaneous velocity fluctuations. The absolute frequency f, expressed in $\sec^{-1}$, relates to the angular frequency $\omega = 2\pi/T$, expressed in rad $\sec^{-1}$, by $f = \omega/2\pi$.

The concept of angular wavenumber κ is also used in spectral analysis. The variable κ is defined as $2\pi/\lambda$, where λ is the wavelength and is expressed in rads m^{-1}. The wavenumber is expressed in $[L^{-1}]$ units. The frequency can also be dimensionless in the form $n = f\, z/\overline{u}$, where $\overline{u}$ is the mean horizontal velocity, as the wavelength for larger eddies in the surface layer relates to the distance z from the ground (Blackadar 1997). In view of the concepts discussed above, it can be noted that the TKE is produced in the range of lower wave numbers (longer wavelength and smaller absolute frequencies) and dissipated primarily in the range of large wavenumbers (greater absolute frequencies).

Turbulent eddies in the surface boundary layer are large structures, so that analysis should include measurements along several distinct points. In the most common case where measurements are done only at a single point, Taylor's hypothesis assumption is made (see Sect. 3.3).

In the case of two sensors separated by a distance r, a spatial covariance matrix $R_{ij}(x,r)$ can be established (Kaimal and Finnigan 1994):

$$R_{ij}(x,r) = \overline{u_i'(x)u_j'(x+r)} \qquad (3.146)$$

its Fourier transform is a matrix $E_{ij}(x,\kappa)$ where κ is the wavenumber vector. The $E_{ij}(x,\kappa)$ matrix contains all the information on the distribution of variability of turbulence in the entire of wavenumber space. However, generally available information about the structure of the flow is not enough to complete the matrixes $R_{ij}(x,r)$ and $E_{ij}(x,\kappa)$, and so, requires simpler analytical criteria. It is possible to resort to the concept of scalar spectral energy $E(\kappa)$, where turbulence is homogeneous in all directions. This energy is defined as the total kinetic energy, due to the kinetic energy of the flow with wavenumbers ranging from κ to $\kappa + d\kappa$, where κ is the module of vector κ (Kaimal and Finnigan 1994).

The turbulent energy spectrum is made up of the spectral ranges for storage and production of turbulent energy, inertial subrange, and dissipation range. The range for storage and production includes the group of eddies that produce TKE via tangential stresses and thermal buoyancy and consequently, storage. These eddies are mainly responsible for the turbulent transport phenomena.

The inertial subrange corresponds to a set of eddies with average frequencies and isotropic characteristics, in which there is convergence of the spectral curves, corresponding to different situations of thermal stability, to a straight line with slope -5/3 (e.g., Foken 2017). This convergence can be plotted on a graph with the variation of the natural logarithm of spectral energy density in the ordinate and the variable $\ln(\kappa)$ in the abscissa (Blackadar 1997). In this subrange, such convergence derives from the fact that kinetic energy transfer occurs in a turbulent cascade between the larger and smaller eddies. The inertial designation derives from the fact that in this intermediate frequency range, neither produce nor dissipate of kinetic energy occurs. The dissipation spectral subrange corresponds to the turbulence due to small eddies responsible for the molecular dissipation of kinetic energy into heat.

Typical length scales range for production/storage and dissipation are respectively, the Eulerian integral length scales Λ, and the Kolmogorov microscale η, of the order of 1 mm. Definitions of Eulerian integral time scales t, and length Λ, for example of the components u and w for wind velocity, are as follows (e.g. Kaimal and Finnigan 1994):

$$t_u = \int_0^\infty \left(\overline{u'(t)u'(t+\tau)}/\sigma_u^2 \right) d(\tau) \tag{3.147}$$

$$t_w = \int_0^\infty \left(\overline{w'(t)w'(t+\tau)}/\sigma_w^2 \right) d(\tau) \tag{3.148}$$

$$\Lambda_u = \bar{u}\, t_u \tag{3.149}$$

$$\Lambda_w = \bar{u} t_w \tag{3.150}$$

where τ is the time lag relative to the arbitrary point t. The Λ_u length scale is of the order of 10 to 500 m.

The Kolmogorov microscale η, with dimensions of about 1 mm is dependent on viscosity and corresponds to the range of eddies where energy dissipation occurs. Conversion of kinetic energy into heat takes place within this dissipation spectral range. The Kolmogorov microscale η, is given by:

$$\eta = \left(\frac{v^3}{\varepsilon} \right)^{1/4} \tag{3.151}$$

where v is the kinematic viscosity of air, and ε is the dissipation rate for TKE. In the superficial boundary layer, as previously mentioned above, turbulence is characterized by a wide range of frequencies related to eddies of various sizes. The energy generated in the range of absolute low frequency, relative to the energy of the eddies is transmitted by a cascading process of eddy stretching to larger wave number ranges, corresponding to the smaller eddies with isotropic turbulence (Blackadar 1997).

The influence of thermal stability in the spectral power can be graphically analyzed by curves in Fig. 3.7, where the frequency appears dimensionless as $f z/\overline{u}$ in the abscissa, and the spectral power S(f), dimensionless in the form $f S_u(f)/u_* \phi_\varepsilon^{2/3}$ is in the ordinates. The spectral power is expressed in $L^3 T^{-2}$ units. The dimensionless dissipation factor ϕ_ε, is defined by Eq. (3.96).

Figure 3.7 shows a family of curves for distinct conditions of thermal stability represented by different values of z/L (k is the von Karman constant and L is the Monin-Obhukov length).

The lower values of dimensionless frequency in the anisotropic range are more dependent on surface heating and stability conditions. On the other hand, in the inertial and dissipative subranges, the spectral curves converge towards a line with slope of about −2/3 (Fig. 3.7). This shows that at higher frequencies, the spectral intensity is no longer as dependent on conditions of thermal stability and that the smaller eddies receive energy from the inertial subrange without direct interaction with the mean flow. A convergence shown in graphs arises of the spectral curves into a line with a slope of −5/3, in which the natural log of the scalar spectral density energy in the ordinates is plotted against the ln(κ) as abscissa.

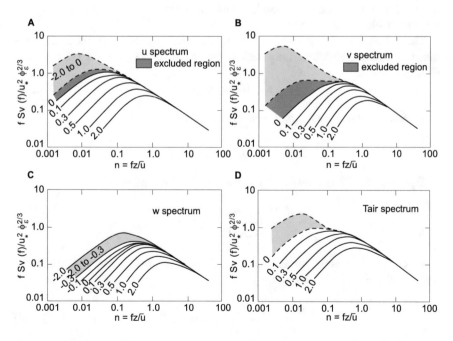

Fig. 3.7 Normalized spectra for **a** u; **b** v; **c** w; and **d** T_{air} of air velocity and temperature potential (after Kaimal and Finnigan 1994)

For negative ξ values, there is discontinuity in the so-called excluded region, which serves as a separation between spectra under stability and instability conditions and is associated with an abrupt transition in characteristic lengths of eddies (Fig. 3.7). For the spectral density curves for components u, v and w, the proposed empirical functions (Kaimal et al. 1972) are as follows:

$$\frac{f S_u(f)}{u_*^2} = \frac{102n}{(1+33n)^{5/3}} \tag{3.152}$$

$$\frac{f S_v(f)}{u_*^2} = \frac{17n}{(1+9.5n)^{5/3}} \tag{3.153}$$

$$\frac{f S_w(f)}{u_*^2} = \frac{2.1n}{(1+5.3n^{5/3})} \tag{3.154}$$

where n is the dimensionless frequency in the form $fz/\bar{u}$. To perform dimensional analysis, the u_i component is obtained from Kolmogorov's Law for the inertial sublayer (Stull 1994):

$$S_{u_i}(K) = \alpha_k \varepsilon^{2/3} K^{-5/3} \tag{3.155}$$

where K is the wavenumber module and α_k is the dimensionless Kolmogorov constant, with a value of 0.55.

The following equation shows the relationship between the spectral power function and the absolute frequencies and wavenumber module (Kaimal and Finnigan 1994):

$$K S_{u_i}(\mathcal{K}) = f S_{u_i}(f) \tag{3.156}$$

After some development, and considering that:

$$\overline{u_i'^2} = u_*^2 \tag{3.157}$$

Equation (3.156) is transformed to the following expression:

$$\frac{f S_{u_i}(f)}{u_*^2} = \frac{\alpha_k}{(2\pi k)^{2/3}} \phi_\varepsilon^{(2/3)} n^{(-2/3)} \tag{3.158}$$

If k (von Karman constant) and α_k is assigned the values 0.4 and 0.55, respectively, Eq. (3.158) becomes:

$$\frac{n S_{u_i}(n)}{u_*^2 \phi_\varepsilon^{2/3}} = \frac{n S_{u_i}(n)}{(kz\varepsilon)^{2/3}} = 0.3 n^{(-2/3)} \tag{3.159}$$

This equation is valid for the three velocity components and shows that in the inertial subrange there is convergence of the spectral curves of different velocity components. These curves correspond to different situations of thermal stability and give a straight line with a slope of $-2/3$.

Using an analogous development, a similar equation for the spectral power of the air temperature is obtained:

$$\frac{fS_T(f)}{\overline{T}_*^2 \phi_H \phi_\varepsilon^{-1/3}} = 0.4n^{-(2/3)} \tag{3.160}$$

The dimensionless stability factor for temperature ϕ_H can be determined by Eqs. (2.52) and (2.53). The configuration of the spectral curves is given in Fig. 3.7.

It can be seen from Fig. 3.7D that in the inertial subrange there is a convergence of the temperature spectral curves corresponding to different situations of thermal stability, resulting in a straight line with slope $-2/3$.

Checking for a $-2/3$ slope in the logarithmic spectral curves is then a quick way of assessing the quality of the measured data, because this slope indicates the presence of an inertial subrange typical of the spectrum of velocity and air temperature components.

Figure 3.8 shows the power spectrum measurements of instantaneous fluctuations of components u and w for velocity and air temperature T in cork oak forests (Rodrigues 2002).

3.6.7 Cospectral Analysis

In the spectral inertial subrange of the surface boundary layer, dimensional analysis shows that the cospectra, for example, of uw, wT, and wc are proportional to the wavenumber raised to the power of $-7/3$ (Wyngaard and Coté 1972). In the inertial subrange, cospectra decrease more rapidly with frequency than the power spectra. Moreover, the wavenumbers (or frequencies) corresponding to the maximum cospectral power are lower than the wavenumbers for the corresponding maximum spectra (Blackadar 1997). Mass vertical flows and energy in the surface boundary layer are thus promoted by larger eddies, and this is relevant in the application of eddy covariance measurements, discussed below.

From a similar development established for spectral analysis, the following generalized expressions for the cospectral power C_{uw} or C_{wt} (Kaimal et al. 1972) are

$$-\frac{fC_{uw}(f)}{u_*^2 G(z/L)} = 0.05n^{-4.3} \tag{3.161}$$

$$-\frac{fC_{wT}(f)}{u_*^2 T_* H(z/L)} = 0.14n^{-4.3} \tag{3.162}$$

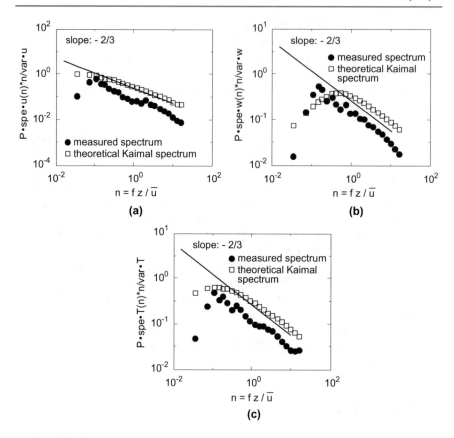

Fig. 3.8 Spectral power of components **a** u; **b** w; and **c** T, from measurements of the cork oak forest surface layer (after Rodrigues 2002)

where

$$G(z/L) = \begin{cases} 1 & -2 \leq z/L \leq 0 \\ 1 + 7.9z/L & 0 \leq z/L \leq 2 \end{cases} \tag{3.163}$$

$$H(z/L) = \begin{cases} 1 & -2 \leq z/L \leq 0 \\ 1 + 6.4z/L & 0 \leq z/L \leq 2 \end{cases} \tag{3.164}$$

Figure 3.9 shows the curves for the covariance of the product for fluctuations of uw (vertical momentum transport) and $w\theta$ (vertical transport of sensible heat).

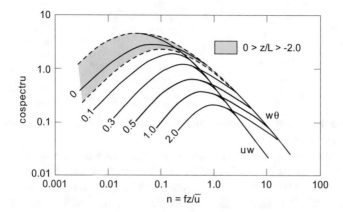

Fig. 3.9 Covariance spectral curves for uw and $w\theta$ in a flat surface showing the variation with thermal stability and frequency (after Kaimal and Finnigan 1994)

3.7 Eddy Covariance Method

Widespread application of the turbulent correlation or covariance method dates back from the last decade. This method deals with mass and energy fluxes for quantifying instantaneous fluctuations of vectorial and scalar variables (e.g., air temperature, carbon concentration, water vapor vertical component of air velocity) of turbulent eddies when they pass a measuring point. The eddy covariance method is the most accurate for calculating mass and energy fluxes and is used to calibrate other previously discussed methods such as aerodynamic (Chap. 2), Bowen methods, or the Penman–Monteith calculation for evapotranspiration described in Chap. 4.

The eddy covariance method is complex but the advent of technological tools, including improved instrumentation and software for calculations and corrections, has allowed for its extensive application in numerous natural and man-made microsystems. In this Section, only overarching principles are presented about the turbulent covariance method and instrumentation required for the flux measurements. For details about the calculations on the vertical flow of other atmospheric gases (e.g., methane) the reader is directed to Burba and Anderson (2010), Foken (2017), Papale et al. (2006), Aubinet et al. (2000), Göckede et al. (2008), Mauder and Foken (2004) and Lee et al. (2004).

Basically, the method is based on obtaining the means for covariance, as discussed above, making possible equations for the vertical atmospheric fluxes such as the following:

$$\tau = -\overline{\rho\, u'w'}\, \rho \qquad (3.165)$$

$$H = \rho \, c_p \, \overline{w'T'} \tag{3.166}$$

$$LE = \rho \, L\overline{w'q'} \tag{3.167}$$

$$C = \rho \, \overline{w'c'} \tag{3.168}$$

for the vertical momentum, sensible heat, latent heat, evaporation, and carbon dioxide, respectively.

To get the corresponding fluctuations for the above equations, suitable instrumentation and a data processing system are needed to record these turbulent fluctuations. Sonic anemometers and gas analyzers are required for measurements of fluctuations of air velocity and temperature and gases such as carbon dioxide, methane, and air humidity.

To collect information on vertical fluxes relating to the surrounding area called footprint, sensors are positioned in observation towers at a height equivalent to 1.5 times the surface elements. The sensors need to be synchronized and with sufficiently high frequency to record fluctuations of the properties of eddies able of turbulent transport. Effects of the observation tower and the instrumentation set on flow distortion as well as the detrimental effects of incorrect anemometer orientation relative to the streamlines, need to be considered. To minimize such pitfalls, the tower needs to be built with long vertical bars to support the anemometers and the measurement system, and coordinate rotation should be performed.

The general principles described make up the core of the eddy covariance method. Calculations, corrections, and data quality control are done using appropriate software packages. Among the more relevant calculations are corrections of sonic temperature due to air humidity (Schotanus et al. 1983); coordinate rotation; means calculating; WPL correction for changes in air density; data stationarity and dynamic similarity testing; evaluation of interactions between the flow and observation towers; response corrections in frequency and filtering of high-frequency peaks or spikes. Gap filling for missing data and/or replacement of poor quality data also needs to be done.

For the carbon fluxes, the main calculations are flux partitioning and night-time carbon storage. Estimations of low-frequency oscillations in the velocity field, due to non-turbulent fluctuations in means for half-hour periods, must also be carried out. The evaluation of the footprint area, which is affected by the measurements and calculations, will also be required.

The basic concepts of operating equipment for measuring fluctuations of components for wind velocity, air temperature, concentrations of water vapor, and carbon dioxide are similar. Some of the key issues regarding the eddy covariance method are:

(i) Sonic anemometers (Figs. 3.10 and 3.11) are used for recording instantaneous values of the components for flow velocity and temperature which allow the calculation of its fluctuations. Measurement of fluxes of other scalar quantities, e.g., water vapor or carbon dioxide, requires additional analyzers.

Fig. 3.10 Sonic anemometer (support in the background) and open-path carbon and water vapor analyzer installed in the field

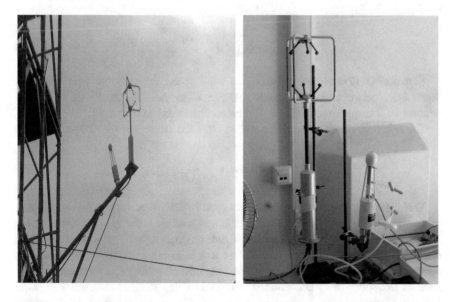

Fig. 3.11 Sonic anemometer and open-path carbon and water vapour analyser (left), with close-up view (right) installed in the field

The sonic anemometers are based on the measurement of flight time of high-frequency ultrasonic pulses between pairs of transducers. The duration of the flight time depends on the sum of the air sound speed with the air velocity along the flight path.

The measurement of air velocity depends on the size and geometry of the set of transducers. A three-dimensional anemometer is a combination of three pairs of transducers that serve to establish the magnitude and direction of the wind velocity vector. The interval for signal discharge in each pair of transducers in the three-dimensional sonic anemometers is about 10^{-3} s. The time of flight in the two opposite directions, t_1 and t_2, is measured and the following identities can be written as

$$t_1 = \frac{d_p}{c_s + V_D} \tag{3.169}$$

$$t_2 = \frac{d_p}{c_s - V_D} \tag{3.170}$$

where c_s is the velocity of sound in air, used for calculating the air temperature, V_D air velocity along the linear space between the two transducers, and d_p the length of this space or distance between the transducers. The value of d_p is typically about 0.15 m. From Eqs. (3.169) and (3.170) we get

$$V_D = 0.5 \, d_p \left(\frac{1}{t_1} - \frac{1}{t_2} \right) \tag{3.171}$$

Equation (3.171) allows recording of wind speed vector component along the linear space between the two transducers without the need for sensitivity analysis in relation to other parameters such as temperature or contaminants.

The speed of sound through air is also obtained from Eqs. (3.169) and (3.170):

$$c_s = 0.5 \, d_p \left(\frac{1}{t_1} + \frac{1}{t_2} \right) \tag{3.172}$$

The anemometer emits an ultrasound signal in both directions in the linear space of the first pair of transducers, which are stored, and air velocity is calculated using Eq. (3.171). This operation is repeated for the remaining two pairs of transducers, and the entire operation takes 2 emissions × 3 pairs ×1 millisec = 6 millisec. The results for successive emissions of sound signals are added and the means calculated. A frequency of 21 Hz is normally used for calculating atmospheric fluxes. As a result of the frequency overlaps, for spectral analysis, the limit of detectable frequencies or Nyquist frequency further described is about 10 Hz.

To minimize vibration and flow distortion, sonic anemometers need to be set up on a solid base with the larger dimension oriented in the direction of the prevailing winds. Additional factors that alter the path of the linear impulses causing measurement errors are precipitation, dew, and snow events.

(ii) Sonic anemometers enable measurement of the sonic air temperature T_S through an expression relating the speed of sound in the air with temperature, that is,

$$c_s^2 = 403T_S \qquad (3.173)$$

The value of the temperature T_S is close to the virtual temperature T_V defined (Chap. 1) by the value of the temperature of dry air with the same density as that of the humid air at the current temperature.

The contribution of the perpendicular component of wind velocity V_N to the temperature measured by the sonic anemometer is given by the following equation (Kaimal and Finnigan 1994):

$$T_S = \frac{d_p^2}{1612}\left(\frac{1}{t_1} + \frac{1}{t_2}\right)^2 + \frac{V_N^2}{403} = \frac{c_s^2 + V_N^2}{403} \qquad (3.174)$$

The effect of air humidity in the sonic temperature (Schotanus et al. 1983) for calculating the covariance $\overline{w'T'}$ of the sensible heat flux is given by

$$\overline{w'T'} = \overline{w'T_S'} - 0.51\overline{T}\overline{w'q'}\,esp \qquad (3.175)$$

giving the covariance $\overline{w'T'}$, corrected for the effects of specific air humidity q_{esp} fluctuations and V_N, from the calculation of ambient temperature using the sonic anemometer.

(iii)   Gas analyzers are used for the instantaneous measurement of gas concentrations for the corresponding covariance fluxes (such as water vapor, methane, nitrous oxide, ozone, etc.). In the case of fluctuations of water vapor and carbon analysis, the results depend on the selective absorption of radiation in the infrared range.

Analyzers can operate in two modes, open or closed path. In open-path analyzers, radiation is emitted by a source, runs an open linear circuit where it is partially absorbed by the gas, and the remaining radiation is captured by detectors at the end of the linear path. For example, water vapor and carbon dioxide absorb infrared radiation at wavelengths of 2.59 μm and 4.26 μm, respectively. Closed path analyzers operate based on the same principle of radiation absorption; the difference being that air is transported to the detection chamber through suction and tubular transport.

The open- and closed-path analyzers obey similar operating principles. In open-path analyzers radiation absorption α_i, is related to the gas concentration ρ_i, by an equation of the type:

$$\rho_i = P_{ei}f_i\left(\frac{\alpha_i}{P_{ei}}\right) \qquad (3.176)$$

where P_{ei} is the equivalent pressure of gas i, α_i is the radiation absorption by gas i, and f_i is a calibration function. If the gas mixture contains other gases that affect the way in which the ith gas absorbs radiation, then the equivalent pressure differs from the partial or total pressure of the gas.

Absorption in the infrared is due to changes in molecular vibrational and rotational states. Other factors being equal, collisions, energy transition states, and the energy absorption will increase with the pressure of the gas mixture. The absorption of radiation by a particular gas will then also depend on its concentration, as well as the concentrations of other similar gases. The function f_i is typically a polynomial function of the order of 5 or 3, for CO_2 and water vapor, respectively.

For other gases, the coefficients will be specified in the equipment calibration sheets. The availability of power and additional phenomena such as precipitation will dictate whether open- or closed-path operating systems are used. Calibration measurements with standard concentrations of gases are much faster in closed-path systems where sampling is done over a period of days, as compared with months for the open-path systems.

Closed-path analyzers operate with a time lag between the instants of suction and measurement, with losses at high frequency throughout the path for the sampled gas. Flow within the tubular system with length ranging between 3 and 16 m, can be laminar or turbulent, and airflow can be up to 50 $Lmin^{-1}$. The electrical power requirements for the closed-loop system are higher (around 30 W) than for open circuit (about 10 W). Closed-path analyzers operate in differential mode. Measurements of carbon dioxide and water vapor are based on infrared radiation absorption difference across two gas cells each with a volume of about 10.9 cm^3.

The reference cell is used for a known gas concentration of CO_2 or water vapor. The sample cell is used for a gas of unknown concentration. Infrared radiation is transmitted through both the cells, measured, and the radiation absorption rate is proportional to the gas concentration. The measuring devices have optical filters (band-pass) mounted directly on the detectors to measure the radiation wavelengths (2.59 µm for water vapor and 4.26 µm for carbon dioxide). Closed-path analyzers have detectors that are thermoelectrically cooled to about -5 °C. The gas pressure inside the cell is controlled by transducers to allow for internal pressure changes. Chemicals such as magnesium perchlorate and ascarite are typically used to purge the detectors of carbon dioxide and water vapor.

In the open system, there is no gas loss or suction in the suction tubes due to high-frequency attenuation. The flow losses are mainly due to the distance between the measurement sensors and analyzers, as well as the occurrence of rainfall, dew, or snow. In open-path systems, the WPL correction is important for density fluctuations. The calibration of CO_2 and water vapor in open-path analyzers (Fig. 3.12) is similar, although simpler than for closed path.

The basal level of the analyzer is only slightly affected by the temperature, typically 0.3 µmol/mol/°C for CO_2, and 0.02 mmol/mol/°C for H2O. This temperature drift is included in the instrument's programming at manufacture. The upper end of measurement is affected by temperature, pressure, and the stability of purge chemicals. These changes are minimal especially at ambient temperature and pressure, but in any case, the software makes the necessary corrections.

(iv) Latent heat and sensible heat fluxes, by causing expansion of the air and modifying its density change also the density of atmospheric gases. This effect

Fig. 3.12 Laboratory calibration of open-path analyzer

is particularly noticeable with smaller quantities of atmospheric gases. The sensible heat flux is not influenced by fluctuations in air density.

The effect of changes in the density of dry air $\overline{\rho}_a$, in the vertical fluxes under constant pressure, wherein the vertical flux of dry air is assumed to be zero, is given by the following expression (Webb et al. 1980):

$$\overline{\rho_a w} = \overline{\rho}_a \overline{w} + \overline{\rho'_a w'} = 0 \tag{3.177}$$

where ρ_a is the density of dry air so that Eq. (3.177) yields

$$\overline{w} = \frac{-\overline{\rho'_a w'}}{\overline{\rho}_a} \tag{3.178}$$

This results in a non-zero mean vertical velocity $\overline{w}$, whenever there are sensible or latent heat fluxes, which cause correlations of fluctuations of ρ_a with w. This velocity will induce an additional convective flux, which needs to be eliminated with the correction proposed by Webb et al. (1980) denoted by WPL, for the change in air density. The WPL correction is used in both open- and closed-path systems. In open-path systems, WPL correction for carbon dioxide flux can be expressed as (Burba and Anderson 2010)

$$F = F_{un} + \left(1.61\frac{E}{\rho_d}\right)\left(\frac{[\overline{c}]}{1 + 1.61\frac{\rho_v}{\rho_d}}\right) + \frac{H}{\rho c_p}\frac{[\overline{c}]}{T_a} \tag{3.179}$$

where F is the corrected flux, F_{un} the uncorrected flux, E evaporation, H sensible heat flux, $\overline{[c]}$ mean CO_2 concentration, ρ_V, ρ_d, and ρ water vapor density, dry air, and standard air, respectively, c_p the specific heat at constant pressure and T_a the air temperature in $^\circ K$. The expression for closed-circuit systems is slightly different.

The WPL correction is small during seasonal growth and becomes more significant outside this period (Burba and Anderson 2010). This correction applies to CO_2 fluxes, evapotranspiration, methane, and ozone, and any other trace atmospheric gas, where Eq. (3.195) should be applied. The WPL correction for carbon dioxide fluxes will be about 50%, which is much higher than that for latent heat flux (Leuning et al. 1982). In very cold environments, open-path analyzers require additional corrections not necessary e.g. in Mediterranean environments (Burba et al. 2008).

(v) An appropriate rotation of coordinates so that the u component coincides with the wind mean velocity vector, thus canceling the v and w components for correct application of the eddy covariance method. The mean values of the vertical component of wind velocity that differ from zero are due to deformation of streamlines by the ground slope, distortions caused by transducers, or tower interference. The rotation coordinates can be made before or after averaging operations (Rannik and Vesala 1999; Foken 2017) discussed in item (vi).

Coordinate rotation ensures proper positioning of the sonic anemometer in relation to streamlines, and minimizes distortions caused by the tower structure and sensors, as well as cancels the vertical and lateral air motions. Two successive rotations are required for the mean components of wind velocity or covariances that include scalar quantities. The first around the z-axis plane serves to align the u component with the x-axis to cancel component v. A second rotation along the v-axis aims at canceling w, as shown in Fig. 3.13.

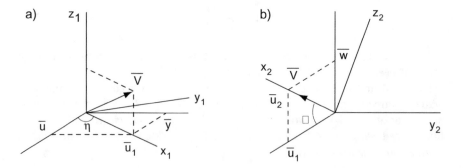

Fig. 3.13 Successive rotations of the coordinate system, **a** first rotation, and **b** second rotation (after Valente 1999)

The result of such rotations can be written in the following compact algebraic form:

$$|\overline{u_{i2}}| = [A_r][\overline{u_i}] \tag{3.180}$$

$$[\overline{su_{i2}}] = [A_r][\overline{su_i}] \tag{3.181}$$

where $\overline{u_{i2}}$ and $\overline{u_i}$ are the column matrices corresponding to the coordinates before and after the two rotations:

$$\overline{u_{i2}} = \begin{bmatrix} \overline{u_2} \\ \overline{v_2} \\ \overline{w_2} \end{bmatrix} \text{ and } \overline{u_i} = \begin{bmatrix} \overline{u} \\ \overline{v} \\ \overline{w} \end{bmatrix} \tag{3.182}$$

$\overline{s'u'_{i2}}$ and $\overline{s'u'_i}$ are the corresponding covariance matrices column that include scalar quantities:

$$\overline{s'u_{i2}} = \begin{bmatrix} \overline{s'u'_2} \\ \overline{s'v'_2} \\ \overline{s'w'_2} \end{bmatrix} \text{ and } \overline{s'u'_i} = \begin{bmatrix} \overline{s'u'} \\ \overline{s'v'} \\ \overline{s'w'} \end{bmatrix} \tag{3.183}$$

and A_r is the rotation matrix given by

$$\begin{bmatrix} cos\eta\,cos\vartheta & sin\,\eta\,cos\vartheta & sin\vartheta \\ -sin\eta & cos\eta & 0 \\ -cos\eta\,sin\vartheta & -sin\eta\,sin\vartheta & cos\vartheta \end{bmatrix} \tag{3.184}$$

The elements for A_r are obtained from the following identities:

$$\sin\eta = \frac{\overline{v}}{\sqrt{\overline{u}^2 + \overline{v}^2}}; \cos\eta = \frac{\overline{u}}{\sqrt{\overline{u}^2 + \overline{v}^2}}; \tag{3.185}$$

and

$$\sin\vartheta = \frac{\overline{w}}{\sqrt{\overline{u}^2 + \overline{v}^2 + \overline{w}^2}}; \cos\vartheta = \frac{\sqrt{\overline{u}^2 + \overline{v}^2}}{\sqrt{\overline{u}^2 + \overline{v}^2 + \overline{w}^2}}; \tag{3.186}$$

For the variance and covariance vectors, the double rotation is as follows:

$$M_2 = A_r M_0 A_r^T \tag{3.187}$$

where A_r^T is the transposed matrix of A_r and M_n matrices of (co)variance before and after the two rotations, written in the general form:

$$\begin{bmatrix} \overline{u'_n u'_n} & \overline{u'_n v'_n} & \overline{u'_n w'_n} \\ \overline{v'_n u'_n} & \overline{v'_n v'_n} & \overline{v'_n w'_n} \\ \overline{w'_n u'_n} & \overline{w'_n v'_n} & \overline{w'_n w'_n} \end{bmatrix} \tag{3.188}$$

A third rotation can be applied around the x-axis to cancel out the lateral momentum $\overline{v'w'}$. This rotation is nil in flat or slightly sloped surfaces.

(vi) The vertical turbulent flux of a quantity, $F = \overline{w'k'}$, from Eqs. (3.165) to (3.168) is given by a general expression:

$$F = \overline{w'k'} = \overline{(w - \overline{w})(k - \overline{k})} \tag{3.189}$$

Determining the means involves direct calculation of means, filtering, and removing linear trends (Aubinet et al. 2000). After data acquisition, the means of $\overline{x}_i$ are given by the statistical definition:

$$\overline{x}_i = \frac{\sum_{i=1}^{n} x_i}{n} \tag{3.190}$$

and the mean of the covariances is given by

$$\overline{x'y'} = \frac{1}{n} \sum_{k=1}^{n} \overline{x'_k y'_k} \tag{3.191}$$

Another alternative for data processing in real time is to use the running mean. Under these conditions, the running mean $\overline{x}_i$, is given by

$$\overline{x}_i = \alpha_n \overline{x}_{i-1} + (1 - \alpha_n)x_i \tag{3.192}$$

where α_n is the weighting factor given by

$$\alpha_n = \exp(-\Delta t / \tau_s) \tag{3.193}$$

In Eq. (3.193), Δt is the time interval between measurements, for example, 0.1 s. The values for τ_s (time constant) proposed vary between 200 s. and 20 min.

If $\Delta t / \tau_s \ll 1$, then Eq. (3.193) can be written as

$$\alpha_n = 1 - (\Delta t / \tau_s) \tag{3.194}$$

For removing linear trends, the mean $\overline{x}_k$ is calculated as

$$\overline{x}_k = \overline{x} + b \left(t_k - \frac{1}{n} \sum_{1}^{n} t_k \right) \tag{3.195}$$

where t_k is the time at instant k, and b is the slope of the linear trend of the sample. The slope is calculated as

$$b = \frac{\sum\limits_{k=1}^{n} x_k t_k - \frac{1}{n} \sum\limits_{k=1}^{n} x_k \sum\limits_{k=1}^{n} t_k}{\sum\limits_{k=1}^{n} t_k t_k - \frac{1}{n} \sum\limits_{k=1}^{n} t_k \sum\limits_{k=1}^{n} t_k} \tag{3.196}$$

The methods given for the calculation of the means are basically dependent on the analysis of a given finite series of measured data. The sampling time depends on the atmospheric stratification, the wind velocity, and the measurement height. Calculating the means for corresponding covariances for time series of 30 min is enough to encompass the effects of low frequency relative to all the eddies responsible for the turbulent transport (Gash and Culf 1996; Aubinet et al. 2000; Burba and Anderson 2010).

A sampling time of 30 min over the full daytime is considered as delivering no significant measurement errors. For height measurements of 2–5 m, sampling periods of 10–20 min would be required for daytime unstable stratification in summer and about 30–60 min and perhaps 120 min would be required for night-time stable stratification (Foken 2017).

(vii) Eddy covariance measurements are affected by peaks caused by environmental factors such as the footprint changes or rapid changes in turbulence or instrumentation (e.g., heavy rain falling on the sonic anemometer or an open circuit analyzer). All processes with fluctuations higher than 3.5σ are considered as spikes or high-frequency peaks. High-frequency peaks, affecting the instantaneous measurement, are removed by filtering before calculating covariance means for half-hour periods. Typically, these half-hour periods are excluded when the high-frequency peaks exceed 1% of total data (Foken 2017).

Peaks or oscillations in the time series of the mean values for half-hour periods typical for non-turbulent phenomena affect the quality of gap-filled data. The methodology used to determine these low-frequency events and consider them as outliers is based on the position of each average value of the flux over a half-hour period (e.g., the net ecosystem exchange NEE_i or carbon budget) compared to before and after adjacent periods, needed for the d_i variable (Papale et al. 2006):

$$d_i = (NEE_i - NEE_{i-1}) - (NEE_{i+1} - NEE_i) \tag{3.197}$$

where the value of NEE_i is considered a low-frequency peak if

$$d_i < Md - \left(\frac{zMAD}{0.6745}\right) \text{ or } d_i > Md - \left(\frac{zMAD}{0.6745}\right) \tag{3.198}$$

where z is a boundary value usually 4, Md is the mean of the differences, and MAD the median defined as

$$MAD = median(|d_i - Md|) \qquad (3.199)$$

These calculations are applied to periods of 13 days and separately for day and nighttime periods. In practice, these evaluation methods for high- and low-frequency peaks require computing software.

(viii)   Flow in the surface boundary layer results from eddies of different dimensions and at a given point, the eddy fluxes generate velocity and scalar concentrations at different frequencies. In the case of turbulent fluxes of a scalar k, the cospectral density C_{wk} is related to the covariance through an expression such as

$$\overline{w'k'}_{mea} = \int_0^\infty C_{wk}(f)df \qquad (3.200)$$

where f is the frequency in Hz. Atmospheric turbulence is made up of an infinite number of overlapping frequencies, and some frequency loss occurs in flux measurement. The reasons for these losses are related to the efficiency of the sensor in terms of frequency response. The main corrections for losses in frequency response are due to response time, sensor separation, or frequency attenuation in suction tubes in closed-path analyzers. In addition, factors such as losses in high frequency due to the averaging on linear circuits or volumes, high-pass filtering the lag between the sensor response time, and digital sampling also need to be considered.

All the corrections are usually included in the software associated with the sensors in the form of transfer functions, FT(f), which vary with frequency and are processed automatically, so this introductory text will only refer to its meaning. For analysis of analytical details, the reader is referred to Moore (1986), Burba and Anderson (2010), and Moncrieff et al. (1997).

Corrections for the response time compensate for sensor delay in response to rapid variations of the fluctuations contributing to the flux. The response delay is a function of the dynamic response frequency of the sensors (Moore 1986). These corrections are applied mainly to gas fluxes, and to a lesser extent for momentum fluxes, when the measurements are performed at very low heights or when the transducer response time is not sufficiently fast.

The correction for separation between the sensors corrects for the flux losses in the high-frequency domain that is because the wind velocity and scalar quantities are not sampled in the same volume (Moore 1986). In practice, although the interfering parts of devices should be kept to a minimum (Foken 2017), it is difficult to avoid some sensor separation because of their size as well as the need to minimize interference in the flow due to its volume.

This separation causes losses, as it misses mass and energy transport driven by eddies with a characteristic dimension smaller than the distance between the sensors. Thus, this correction is applied for gas fluxes, but not to the heat or momentum flux in which the sampling is done only for the sonic anemometer. The correction is lower when the cospectral shifts toward lower frequencies, due to the increase in height measurement or greater thermal instability (Laubach et al. 1994). Moore (1986) suggested that under conditions of thermal instability, the separation between the sensors should not exceed 10% of $(z-d)$. Baldocchi (1995) reported that if the ratio of the distance between the separation and $(z-d)$ is lower than 5%, the covariance error is less than 3%. These authors cite the following criteria for separation s, between sensors

$$s = (z - d)/5 \tag{3.201}$$

$$s = (z - d)/(6\pi) \tag{3.202}$$

The response correction in high frequency due to the averaging procedure in volume (in the case of closed-path systems) or in linear circuits aims to compensate for flux losses due to eddies being smaller than the sample spaces. This correction is needed for all scalar fluxes.

The response correction in high-frequency suction tubes in closed-path systems can compensate for faster dampening of fluctuations inside the tubes. This correction applies to vertical gas flux in closed-path systems. In this system, one must take into consideration that water vapor flow measurements can be affected by the internal conditions of the tube in terms of cleanliness and temperature of the walls, as well as by the temperature and humidity of the air (Moncrieff et al. 1997).

The response corrections in frequency with high-pass filtering aims to compensate for flux losses in the low-frequency range, due to the establishment of means and removal of linear trends in calculating the instantaneous fluctuations. These corrections apply to all the fluxes (Burba and Anderson 2010).

The frequency response corrections that result from different response rates among sensors is needed if, for example, a fast response sensor is coupled to a slower sensor (Moore 1986). These corrections are usually minor (Burba and Anderson 2010).

The frequency response correction, relative to digital sampling, is intended to compensate for errors arising from the discrete nature of the sampling carried out in parameters that, in fact, are continuous. This correction is established for any frequency below the critical Nyquist frequency $(1/2f)$, assessed in item xi, for avoiding overlapping frequencies, which makes it impossible to reconstruct a temporal function into spectra in the high-frequency domain.

Empirical transfer functions, FT, related to the corrections are multiplied by cospectral or spectral densities, relating to fluxes or variances and form part of the integral for absolute frequencies as follows:

$$\overline{w'k'_{mea}} = \int_0^\infty FT(f)C_{wk}(f)df \tag{3.203}$$

and so, the overall combined correction factor for all frequencies, $FC(f)$ becomes

$$FC(f) = \frac{\int_0^\infty FT(f)C_{wk}(f)df}{\int_0^\infty C_{wk}(f)FT(f)df} \tag{3.204}$$

(ix) The system for measuring fluxes with the eddy covariance method is subjected to flux losses because of the physical barrier effect from transducers and supporting structures. However, these towers and lateral arms are typically aerodynamically thin and so are responsible for little flow distortion (Wyngaard 1988; Wieringa 1980; Aubinet et al. 2000).

The velocity measurement error caused by flow distortion from the transducers and associated wakes, increases linearly with decreasing angle between the wind velocity vector direction and the linear path of the ultrasonic anemometer. This error also adds to the decrease in the ratio d_p/a, for d_p/a less than 50, where d_p is the linear space between the transducers and a is their diameter. The blocking effect at the stagnation point that arises from the impact of the streamlines on the sensors reduces or expands the vertical component w. Flow distortion contaminates the w component via the crosstalk effect, inducing horizontal components of velocity into the other two directions.

Wyngaard (1981, 1988) analyzed the effect of the flow distortion by the sensors, under conditions where the ratio between their size and the height of the measurements is less than 0.1. Under these conditions a flow U_i, before measurement is unidirectional:

$$U_i = U_i + u_i = (U_1 + u_1, u_2, u_3) \tag{3.205}$$

with index, i = 1, 2, 3, and the variables U_1, u_1, u_2, and u_3 representing the mean flow and the turbulent fluctuations, respectively. After distortion due to errors of attenuation/ amplification and crosstalk, it becomes

$$U_{ai} = U_{ai} + u_{ai} = (U_{a1} + u_{a1}, U_{a2} + u_{a2}, U_{a3} + u_{a3}) \tag{3.206}$$

and thus, has two more components for the mean velocity.

The effects of blocking errors (attenuation/amplification) and crosstalk in the fluctuations of the vertical component measured in the region of flow distortion, can according to Wyngaard (1988), be expressed as follows:

$$u_3^m = (1 + d_{33})u_3 + d_{31}u_1 + d_{32}u_2 \tag{3.207}$$

where d_{ij} are the distortion coefficients that cancel out in an area of the flow further away from the obstacle. The d_{33} coefficient is the attenuation or amplification via the blocking effect while the d_{31} and d_{32} coefficients represent the effects of crosstalk.

For scalar quantities, there is no distortion because of transducers, if the Péclet number Pe is high. This number is defined by UL/α where U is the flow velocity, L is the characteristic length, and α is the thermal diffusivity. That is, the effects of molecular diffusivity during the path time along the sensor are minimal and so the measured and real values of any scalar, c_e and $c_e{}^m$ are equal.

Equation (3.207) can be adapted to calculate the flow of a scalar quantity c:

$$\overline{(u_3 c_e)}^m = (1 + d_{33})\overline{u_3 c_e} + d_{31}\overline{u_1 c_e} + d_{32}\overline{u_2 c_e} \qquad (3.208)$$

The use of rotation coordinates, discussed above, also helps to attenuate errors arising from flow distortion. A coordinate rotation that places the mean direction of the wind velocity vector coincident with u_1 will cancel the $\overline{u_2 c_e}$ term. Equation (3.208) can then be written as follows:

$$\left(\overline{(u_3 c_e)}^m / \overline{u_3 c_e}\right) = 1 + d_{33} - (\overline{u_1 c_e})/\overline{u_3 c_e})d_{31} \qquad (3.209)$$

The values for the ratio $\overline{u_1 c_e}/\overline{u_3 c_e}$ are described by Wyngaard (1988). These authors reported that under neutral/stable conditions, the ratio is about -3, and decreases rapidly to values close to zero under instability.

According to Eq. (3.209), the effect of crosstalk is about $-3d_{31}$, under conditions of neutrality/stability and can be more pronounced than the effect of attenuation/amplification. According to Wyngaard (1988) and Baldocchi (1995), the error due to crosstalk can be eliminated by maximizing the vertical geometry of the measurement system, so that distortions are as symmetrical as possible. One way of improving this symmetry involves placing a vertical extension above the tower.

Transducers in three-dimensional sonic anemometers are set at 120° angles in a non-orthogonal symmetrical geometry, in which none of the ultrasound linear paths lie in the horizontal plane and serving to minimize disturbances in airflow. The effects of transducers and their supporting structure are corrected by anemometer software. By distributing symmetrically, the measuring system, it is possible to minimize blocking effects and losses due to stagnation (Wyngaard 1988).

(x) Three important tools for assuring data quality are analysis of stationarity, analysis of turbulent characteristics using the Monin–Obukhov (M–O) dynamics similarity (Foken and Wichura 1996; Foken 2017) and data filtering. The latter is done by assuming that friction velocity u_* (Eq. 3.145) is generally greater than 0.2 ms^{-1}, which will assure enough turbulence to promote vertical fluxes. Stationarity is needed to study the spatial variability or homogeneity of turbulent statistics, if Taylor's turbulent "freezing" holds (discussed in Sect. 3.3).

The stationarity test (Foken and Wichura 1996) involves calculating fluxes (using Eqs. 3.169 to 3.172) and divides 30 min data sets into six 5 min intervals. If the difference between the fluxes obtained for 30 min interval is less than 30% relative to the average of fluxes for all the 5 min periods, then the data is of high quality and displays stationarity. If the difference is between 30 and 50%, then the data set is considered of acceptable quality. According to these authors, the stationary test is a prerequisite for data quality.

Monin–Obukhov's theory of dynamic similarity is a key empirical tool for studying meteorological parameters and flow in the atmospheric surface layer. This similarity theory deals with all scalar quantities, variances, and linear correlations between variables.

Experiments conducted on flat terrain made it possible to confirm Monin–Obukhov hypothesis (e.g., The Kansas Field Program 1968). According to this theory, the structure of turbulence in the constant flux layer or surface layer can be described or parameterized by key variables such as height h, buoyancy g/T, kinematic tangential tension τ/ρ, and surface temperature flux $H/\rho c_p$ (Kaimal and Finnigan 1994). According to Monin–Obukhov theory, universal functions for the stability parameter ξ (equal to $(z\text{-}d)/L$) can be obtained from various atmospheric parameters and statistics (gradient, variance, and covariance) when normalized by the appropriate powers, $u_*, T_*,$ and q_* in which q is the absolute air humidity.

The relationship among the ratios $\sigma_W/u_*, \sigma_T/T_*$ and σ_q/q_* can be expressed as (Lee and Black 2003a):

$$\sigma_W/u_* = a_W[-\xi]^{1/3} \tag{3.210}$$

$$\sigma_T/T_* = a_T[-\xi]^{-1/3} \tag{3.211}$$

$$\sigma_q/q_* = a_q[-\xi]^{-1/3} \tag{3.212}$$

where σ_w, σ_u, and σ_T are the standard deviations of the vertical, longitudinal, and temperature components, respectively, and ξ is the stability parameter.

The values for a_w, a_T, and a_q are of the order of 1.9, 0.9, and 1.1, respectively. Flows can be calculated from the definitions of u_*, T_* and q_*.

A relationship proposed for the dependency of σ_W/u_*, with respect to $(z\text{-}d)/L$, (Panofsky and Dutton 1984), valid under conditions of instability, is as follows:

$$\sigma_W/u_* = 1.25[1 - 3\xi]^{1/3} \tag{3.213}$$

Compliance with the dynamic similarity relationships can be checked using the flow-variance expressions, or integral characteristics (Foken and Wichura 1996; Foken 2017):

$$\frac{\sigma_w}{u_*} = a_1[\xi]^{b1} \tag{3.214}$$

$$\frac{\sigma_u}{u_*} = a_1[\xi]^{b1} \tag{3.215}$$

	ξ	a_1	b_1	a_2	b_2
Table 3.1 Empirical constants used in Eqs. (3.214) to (3.216)					
σ_w/u_*	$-1 > \xi$	2	1/6		
	$-1 < \xi < -0.0625$	2	1/8		
σ_u/u_*	$-1 > \xi$	2.83	1/6		
	$-1 < \xi < -0.0625$	2.83	1/8		
σ_T/T_*	$-1 > \xi$			1	$-1/3$
	$-1 < \xi < -0.0625$			1	$-1/4$

$$\frac{\sigma_T}{T_*} = a_2[\xi]^{b2} \tag{3.216}$$

The empirical constants a_i and b_i are given in Table 3.1.

The quality of the data is good, with zero score, if the differences between measured and calculated values of the integral characteristics are not more than 30%.

The friction velocity threshold (typically 0.2 ms^{-1}) can be used to exclude data from the mean flux if this value is lower, because the turbulence levels are insufficient to assure a homogenous surface boundary layer and thus the validity of the measured fluxes. Typically, this threshold is obtained from values of night-time CO_2 fluxes plotted against the friction velocity values on the abscissa. The threshold is the value above which vertical flows, including respiration (Eq. 3.225) and Chap. 4 are independent of friction velocity. Friction velocity thresholds of 0.3–0.4 ms^{-1} are also common (e.g., Foken 2017).

(xi) The problem of overlapping frequencies arises when the digital sampling rate of a continuous signal is insufficient to quantify all frequencies of interest present in the continuous signal. In this case, there is an overlapping of the higher frequencies over the lower ones. Sampling for analysis of data in digital format is made at equal intervals and so the sampling range needs to be defined.

If the sampling points are too close, then the information logged will be highly correlated and excessive, but on the other hand if sampling is too slow, it will be difficult to reconstruct the time function particularly in the higher frequency domain. In such cases, the wave function obtained from the sampling points will be of a lower frequency than the function characterized causing overlap of the sinusoids of higher and lower frequency. This overlap is known as aliasing (Fig. 3.14).

The sampling theorem specifies the minimum rate, or the largest interval, necessary for spectral characterization up to a specified frequency, n_{max}. It is expressed as follows: a continuous signal which does not contain significant components at frequencies above n_{max} hertz may, in principle, be recovered from its sampled version, if this sampling interval is less than $1/(2n_{max})$ seconds.

Fig. 3.14 Schematic
illustrating the aliasing
problem (from Bendat and
Piersol 1971)

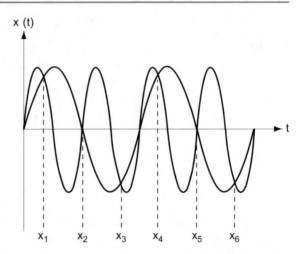

The critical frequency for sampling:

$$f_c = 2n_{max} \tag{3.217}$$

is the Nyquist frequency. The Nyquist sampling criterion states that two is the minimum number of samples required per period of the maximum frequency, n_{max}, in the continuous signal.

According to Bendat and Piersol (1971), the problem of frequency overlap can be quantified from the following identity, where $t = 1/(2f_c)$:

$$\cos(2\pi nt) = \cos\left(2\pi(2if_c \pm j)\frac{1}{2f_c}\right) = \cos\frac{\pi n}{f_c} \tag{3.218}$$

where i is an integer positive index. For any frequency j within the interval $0 \leq j \leq f_c$, there is an overlap between j and an infinite number of frequencies written in the general form:

$$(2f_c \pm j), (4f_c \pm j), \ldots, (2if_c \pm j), \ldots \tag{3.219}$$

Solving the problem involves including the respective corrective transfer function in the product for transfer functions (Moore 1986), and using sampling rates in keeping with the Nyquist criterion.

(xii)   Flux measurements should not be made too close to the top of canopy surfaces, in a way that the measured values correspond only to points very close to the nearest rough elements, being affected by roughness, nor should these be made too far above, so that the data does not only relate to the canopies, being also contaminated by advective effects from adjacent surfaces.

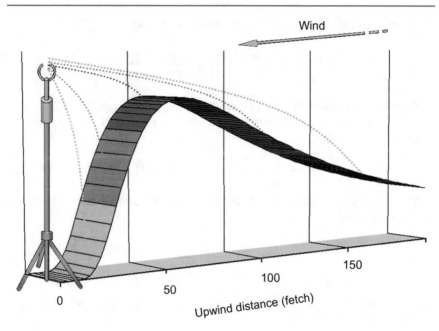

Fig. 3.15 Area upwind from the observation tower (after Burba and Anderson 2010)

The area abridged by measurements with instrumentation in the observation tower, known as a footprint or fetch length, is upwind. The defining criterion is the direction of predominant winds, so that the measured fluxes correspond as accurately as possible to the vertical mass and energy fluxes generated in this area (Fig. 3.15).

Fetch must be considered to determine the best possible location and height of the measurement tower. As a rule of thumb, the ratio of the fetch and measurement height should be about 100 (Oke 1992).

The relationship between the measurement heights and the footprint involves a given height, wind profile velocity, and surface roughness, calculating the pattern of contributions over the upwind area for the measured fluxes. If F is the flux of a scalar quantity measured in a horizontal point x, located at coordinate 0 at height z, the cumulative contribution F_x of a source area extending from the observation point to the upward distance $x > 0$, is given by (Schuepp et al. 1990):

$$F_x = F \, exp(-(z - d)u_c/(u_* kx)) \tag{3.220}$$

where k is the von Karman constant equal to 0.41, and u_c wind velocity assumed constant, defined as the mean wind speed between the surface and the observation height, z.

Given a logarithmic wind profile, the constant velocity u_c, is given by

$$u_c = \int\limits_{d+z_{0M}}^{z} u(z)dz/ \int\limits_{d+z_{0M}}^{z} z = \frac{u_*[ln((z-d)/z_{0M}) - 1 + z_{0M}/(z-d)]}{k[1 - z_{0M}/(z-d)]} \quad (3.221)$$

From the cumulative flux at a point $(0, z)$, given by Eq. (3.220), the upwind contributions from a source located at a distance x, is obtained by differentiation in abscissa:

$$(1/F)dF_x/dx = b((z-d)u_c/(u_*kx^2))F_x/F \quad (3.222)$$

The abscissa corresponding to the point of maximum contribution x_{max}, for the flux, is as follows:

$$x_{max} = \frac{u_c(z-d)}{u_* 2k} \quad (3.223)$$

and the corresponding ordinate, or the maximum relative contribution to the total measured flux is given by

$$\left(\frac{1}{F}\frac{dF_x}{dx}\right)_{max} = \frac{4u_*k}{u(z-d)}(exp(-2)) \quad (3.224)$$

The models described are valid under conditions of thermal neutrality and logarithmic profiles. Obstacles and deviations from the logarithmic profile due to greater surface roughness will tend to increase the effect of sources near the measurement tower.

If other conditions are held constant, increasing measurement height will lead to an increase in the footprint distance upward from the location(s) of the point(s) of the maximum contribution, whereas the magnitude of this contribution decreases. The distance upward (or footprint area) covered by the observation tower increases substantially with the height of the measurements, as does the area with null contribution adjacent to the tower. Footprint size increases with the height of measurements and decreases with surface roughness and thermal stability.

The effect of surface roughness on the footprint is more significant for measurements made at a lower height than for these made at greater heights. This marked effect is due to a greater contribution to fluxes from the area adjacent to the measurement tower. In practical terms, this means that surface roughness should be considered in selecting the height of measurements, with respect to the location of the tower and sensors.

Changes in atmospheric stability can increase the footprint by several orders of magnitude. Under conditions of atmospheric instability, the area encompassed by the measurements is reduced. The distance from the maximum contribution, x_{max}, is reduced by 57% if $(z-d/L)$ is equal to -0.84 (Schuepp et al. 1990). Thus, the flux data under highly unstable conditions should either be corrected or eliminated because much of the flux comes from the perturbed area near the observation tower. Clearly, the influence of the instability on measurements is greater if measurement height is lower or if the surface roughness is higher.

(xiii) Gap-filled data sets obtained over 30 min intervals sometimes have missing data or gaps that require filling. This is often the case with the measurement of vertical fluxes of carbon dioxide and water vapor. The need for gap filling may result from problems with the measurement instruments and this can lead to poor quality data.

Basically, the gap-filling processes can be done by interpolation, whereby the missing data are calculated from those obtained under the similar micrometeorological conditions. These conditions relate to the air temperature and relative humidity, and solar radiation. The gap filling is usually done using appropriate software. It can be necessary to resort to appropriate time windows of about 7 to 15 days before and after obtaining the means for the missing data fluxes and interpolate from data corresponding to equivalent values of the microclimate variables. As a last resort, averaged climate data for the days with missing data can be used. In the case of carbon dioxide fluxes in plant ecosystems, it will likewise be necessary to partition fluxes of the carbon budget, which were measured into the gross (photosynthesis) and respiration.

Basically, the calculations for respiration are made from night-time data (therefore no photosynthesis) defined by a criterion in which the radiation is, for example, less than 20 Wm^{-2}. The algorithm described by Reichstein et al. (2005) for the calculation of ecosystem respiration is based on non-linear regression as follows:

$$R(T) = R_{ref} e^{E_0 \left(\frac{1}{T_{ref} - T_0} - \frac{1}{T - T_0} \right)}$$

(3.225)

where the total respiration $R(T)$ is a function of the air or ground temperature T, of a reference temperature T_{ref} of 10 °C and of a regression constant T_o equal to −46.03 °C. E_o is the activation energy and R_{ref} the respiration at the reference temperature. From the measured night-time carbon budget values, and assuming equality of respiration, we can estimate the R_{ref} and T_{ref} parameters over successive 10 days intervals. The values for these parameters at the specified time intervals are stored and used for calculating the daytime respiration as a function of the air and soil temperature.

An additional correction for systems in the biosphere relates to night-time carbon storage term, S_c, expressed by an equation as follows:

$$S_c = \frac{P}{RT} \int_0^h \frac{\Delta[c]}{\Delta t} dz \approx \frac{P}{RT} h[c]_t$$

(3.226)

where P is the atmospheric pressure, R the constant for ideal gases, T the air temperature, $[c]$ the concentration of CO_2, h the height of the surface elements (e.g., trees) and $[c]_t$ the CO_2 concentration at the top of the surface elements. In stable night periods, the storage and advection terms, in the equations for budgets relative to vertical flows scalar and vectorial quantities, are not negligible and eventually cannot be measured with enough accuracy (Foken 2017).

Calculation of fluxes on calm nights, with turbulence attenuated in the surface boundary layer, is done by applying the friction velocity rule, mentioned above, with the exclusion of fluxes for the half-hour periods in which the friction velocity is lower than the 0.2 ms^{-1} threshold. These fluxes are replaced by gap filling, based on values of fluxes corresponding to similar meteorological conditions. On calm nights, with thermal stability, there is static accumulation of carbon dioxide (e.g., Eq. 3.226) which is removed when morning turbulence resumes. In this case, the stored amount of CO_2 should be added to the morning vertical flux. If in this situation the correction by the described friction velocity is applied, then the night-time flux accumulation would be accounted twice. Thus, friction velocity correction should be applied only if the stored CO_2 overnight is removed by lateral advection, and thus not detected at sunrise due to the removal.

(xiv)   A general process for calculating the vertical carbon fluxes in a typical biosphere, such as a forested site, thus begins with calculating the mean covariance over 30 min intervals., followed by rotation of coordinates so that the component u coincides with the mean local velocity vector, thus nullifying the v and w components, and removal of linear trends. Next, the WPL correction is applied to account for changes in air density and the Schotanus correction for the influence of air humidity on sonic temperature and the correction for overnight storage of CO_2.

Data filtering should then be applied for vertical fluxes to eliminate those corresponding to vertical velocities with deviations from zero mean higher than 0.35 m; the occurrence of high-frequency peaks in the data sets above 1%, occasional occurrence of undulating/wave phenomena, using the absolute median deviation around the median described above, and frictional velocities below a threshold of 0.2 ms^{-1}. Data with these characteristics is of bad quality. The data selected by this filtering is submitted to the stationarity test (item **x**) assigning ratings of 0, 1, and 2. If the difference between the mean covariance for 30 min intervals as compared with 5 min is less than 30% then a 0 rating is applied. In a similar way, if the difference is between 30% and 50% and more than 50% ratings of 1 and 2, respectively are applied.

Data submitted to the dynamic similarity test to the vertical wind component (Table 3.1) are considered of good quality if the differences between the measured and calculated values for the integral characteristics are not higher than 30%. The final data classification will be 0, 1, or 2 if stationary test and dynamic similarity test give a zero result, if one or both give a result of 1 and if at least one result is 2, respectively.

Fig. 3.16 A 30 m observation tower on a flat surface

According to Mauder and Foken (2004), the value 0 is indicative of good data quality, which can be used in fundamental research, and 1 is indicative of data that can be used for long-term observations. A ranking 2 suggests that this data is of poor quality and should be disregarded. These bad quality data, together with data removed in the first filtering, should be replaced by gap-filling processes.

An instrumented observation tower of about 30 m in height, located on flat terrain with homogeneous cover within a radius of several hundred meters, (Fig. 3.16) should assure the collection of good quality data thereby reducing problems associated with day or night-time advection.

References

Aubinet, M., Grelle, A., Ibrom, A., Rannik, Ü., Moncrieff, J., Foken, T., et al. (2000). Estimates of the annual net carbon and water exchange of forests: The Euroflux methodology. *Advances in Ecological Research, 30,* 113–175.

Baldocchi, D. D. (1995). Instrumentation III (Open and Closed Path CO_2 and Water Vapor Sensors), Lecture 12. In *Advanced Short Course on Biometeorology and Micrometeorology,* Università di Sassari, Italia.

Bendat, J., & Piersol, A. J. (1971). *Random data: Analysis and measurement procedures.* Wiley, 407 pp.

Blackadar, A. K. (1997). *Turbulence and diffusion in the atmosphere.* Springer, 185 pp.

Burba, G., & Anderson, D. (2010). *A Brief Practical Guide to Eddy Covariance Flux Measurements.* LI-COR Biosciences, 213 pp.

Burba, G., McDermit, D. K., Grelle, A., Anderson, D. J., & e Xu, L. (2008). Addressing the influence of instrument surface heat exchange on the measurements of CO_2 flux from open-path gas analyzers. *Global Change Biology, 14*(8), 1854–1876.

Businger, J. A., Wingaard, J. C., Izumi, Y., & Bradley, E. F. (1971). Flux profile relationships in the Atmospheric surface layer. *Journal of the Atmospheric Sciences, 28,* 181–189.

Chapra, S. C., & Canale, R. P. (1989). *Numerical methods for engineers,* 2nd edn, 813 pp.

Foken, T. (2017). *Micrometeorology,* 2nd edn, Springer, Berlin, 362 pp.

Foken, T., & Wichura, B. (1996). Tools for quality assessing of surface-based flux measurements. *Agricultural and Forest Meteorology, 78,* 83–105.

Fox, R. W., & McDonald A. T. (1985). *Introduction to fluid mechanics*. Wiley, 742 pp.

Gash, J. H. C., & Gulf, A. D. (1996). Applying a linear detrend to eddy correlation data in real time. *Boundary Layer Meteorology, 79,* 301–306.

Göckede, M., Foken, T., Aubinet, M., Aurela, M., Banza, J., Bernhofer, C., et al. (2008). Quality control of CarboEurope flux data. Part 1. Coupling footprint Analyses with flux Data quality assessment to evaluate sites in forest ecosystems. *Biogeosciences, 5,* 433–450.

Kaimal, J. C., & Finnigan, J. J. (1994). *Atmospheric boundary layer flows* (p. 289). Their Structure and Measurement: Oxford University Press.

Kaimal, J. C., Wyngaard, J. C., Izumi, Y., & Coté, O. R. (1972). Spectral characteristics of surface layer turbulence. *Quarterly Journal of the Royal Meteorological Society, 98,* 563–589.

Laubach, J., Raschendorfer, M., Kreilein, H., & Gravenhorst, G. (1994). Determination of heat and water vapour fluxes above a spruce forest by eddy correlation. *Agricultural and Forest Meteorology, 71,* 373–401.

Lee, X., Massman, W., & Law, B. (2004). *Handbook of micrometeorology*, Atmospheric and oceanographic sciences library, Vol. 29, Kluwer Academic Publishers, 251 pp.

Leuning, R., Denmead, O. T., Lang, R. G., & Ohtaki, E. (1982). Effects of heat and water vapor transport on eddy covariance measurement of CO_2 fluxes. *Boundary Layer Meteorology, 23,* 209–222.

Lynn, P. A., & Fuerst, W. (1998). *Introductory signal processing with computer applications*, 2nd ed, Wiley, 479 pp.

Mauder, M., & Foken, T. (2004). Quality control of eddy covariance measurements. CarboEurope-IP Task 1.2.2 (C:0,1,2).

Moncrieff, J. B., Massheder, J. M., de Bruin, H., Helbers, J., Friborg, T., Heuinkveld, B., Kabat, P., Scott, S., Soegaard, H., & Verhoef, A. (1997). A system to measure surface fluxes of momentum, sensible heat, water vapour and carbon Dioxide. *Journal of Hydrology* 188–189 and 589–611.

Moore, C. J. (1986). Frequency response corrections for eddy correlation systems. *Boundary Layer Meteorology, 37,* 17–35.

Morrison, N. (1994). *Introduction to fourier analysis*. Wiley, 562 pp.

Oke, T. R. (1992). *Boundary layer climates*, 2nd ed., Routledge, 435 pp.

Panofsky, H. A., & Dutton, J. A. (1984). *Atmospheric turbulence* (p. 397). Models and Methods for Engineering Applications: Wiley.

Papale, D., Reichstein, M., Aubinet, M., Canfora, E., Bernhofer, C., Kutsch, W., et al. (2006). Towards a standardized processing of net ecosystem exchange measured with eddy covariance technique: Algorithms and uncertainty estimation. *Biogeosciences, 3,* 571–583.

Reichstein, M., Falge, E., Baldocchi, D., Papale, D., Aubinet, M., & Berbigier, P. et al. (2005). On the separation of net rcosystem exchange into assimilation and ecosystem respiration: Review and improved algorithm. *Global Change Biology, 11*(9), 1424–1439.

Rodrigues, A. M. (2002). *Fluxos de Momento, Massa e Energia na Camada Limite Atmosférica em Montado de Sobro*. Ph.D. Thesis (Environment, Energy profile) Instituto Superior Técnico, U.T.L., Lisbon, 235 pp. [in portuguese].

Schotanus, E. K., Nieuwstadt, F. T. M., & de Bruin, H. A. R. (1983). Temperature measurement with a Sonic Anemometer and its application to heat and moisture flux. *Boundary-Layer Meteorology, 26,* 81–93.

Schuepp, P. H., Leclerc, M. Y., Macpherson, J. I., & Desjardins, R. L. (1990). Footprint prediction of scalar fluxes from analytical solutions of the diffusion equation. *Boundary Layer Meteorology, 50,* 355–373.

Shaw, R. (1995a). Statistical description of turbulence, Lecture 7. In *Advanced Short Course on Biometeorology and Micrometeorology*, Università di Sassari, Italia.

Shaw, R. (1995b). *Turbulent kinetic energy and atmospheric stability*, Lecture 8. In *Advanced Short Course on Biometeorology and Micrometeorology*. Università di Sassari, Italia.

Shaw, R. (1995c). *Surface layer similarity*. Lecture 10. In *Advanced Short Course on Biometeorology and Micrometeorology*. Università di Sassari, Italia.

Shaw, R. (1995d). *Instrumentation I (Response Characteristics, Vector Wind Sensors)*. Lecture 13. In *Advanced Short Course on Biometeorology and Micrometeorology*. Università di Sassari, Italia.

Stearns, S. D., & Hush, D. R. (1990). *Digital signal analysis*, 2nd edn. Prentice-Hall International, Inc., 441 pp.

Stull, R. S. (1994). *An introduction to boundary layer meteorology*. Kluwer Academic Publishers, 666 pp.

Tennekes, H., & Lumley, J. L. (1980). *A first course in turbulence*. MIT Press, 300 pp.

Valente, F. M. R. T. (1999). *Intercepção da Precipitação em Povoamentos Esparsos. Modelação do Processo e Características Aerodinâmicas dos Cobertos Molhados*. Ph.D. Thesis, Universidade Técnica de Lisboa, Instituto Superior de Agronomia, Lisbon, (in portuguese).

Webb, E. K., Pearman, G. I., & Leuning, R. (1980). Correction of flux measurements for density effects due to heat and water vapor transfer. *Quarterly Journal of the Royal Meteorological Society, 106*, 85–100.

Willis, G. E., & e Deardorff, J. W. (1976). On the use of Taylor's translation hypothesis for diffusion in the mixed layer. *Quaterly Journal of the Royal Meteorological Society, 102*, 817–822.

Wyngaard, J. C. (1981). The effects of probe-induced flow distortion on atmospheric turbulence measurements. *Journal of Applied Meteorology, 20*, 784–794.

Wyngaard, J. C. (1988). Flow distortion effects on scalar flux measurements in the surface layer: Implications for sensor design. *Boundary Layer Meteorology, 42*, 19–26.

Wyngaard, J. C., & Coté, O. R. (1972). The budgets of turbulent kinetic energy and temperature variance in the atmospheric surface layer. *Journal of the Atmospheric Science, 28*, 190–201.

Wyngaard, J. C., Coté, O. R., & Izumi, Y. (1971). Local free convection, similarity, and the budgets of shear stress and heat flux. *Journal of the Atmospheric Sciences, 28*, 1171–1182.

Exchange of Energy and Mass Over Forest Canopies

4

Abstract

This chapter aimed to characterize the aerodynamic mass and energy fluxes over forest canopies. These canopies are rough surfaces wherein flow-gradient assumptions may not be valid, leading to exponential wind profiles in trunk spaces and understory, superimposed by logarithmic profiles above the top of canopies. This leads to airflow regimes wherein canopy stomatal resistance is dependent on tree physiological factors and of short and long-term phenomena, in interaction with aerodynamic variables. Forest canopies are also very prone to intermittent events, such as gusts or ejections, and wake formation in trees downstream, adding components to the usual kinetic energy equations and influencing turbulence spectral dynamics and particle transport. The analysis of drag processes in canopies, in wind tunnel and field, allows also to characterize phenomena such as tree bending and pulling. The evapotranspiration regime of these canopies is also coupled with the moisture of soil and atmosphere, in comparison with lower canopies, with transient flow mechanisms particularly relevant in trunk and understory spaces. Finally, forest canopies are relevant in carbon balance dynamics, especially related to the net ecosystem exchange, on a micro or global scale, and these dynamics reflect the influence of physical and biological factors in carbon sinking.

4.1 Introduction

The magnitude of atmospheric mass and energy fluxes over forest canopies, for example, in relation to water loss, soil erosion, or the microclimate, depends on processes at the level of the biosphere, within the control volume between the ground surface and the top of the surface boundary layer. The latter is divided, as referred to in Chap. 2, into the inertial and roughness sublayers.

© The Author(s), under exclusive license to Springer Nature Switzerland AG 2021   105
A. Rodrigues et al., *Fundamental Principles of Environmental Physics*,
https://doi.org/10.1007/978-3-030-69025-0_4

Forest canopies are combinations of vegetation of considerable height and rough surfaces, adjacent to the atmosphere, as compared to lower and smoother vegetable canopies such as pastures or crops. The physical environment of forest stands is determined by factors such as the penetration of solar radiation and the extent of airflow within canopies. The configuration of the canopy dictates the development of air velocity fields and radiation transmission regimes. Branches absorb the linear momentum of flow through drag, while simultaneously absorbing and dispersing solar radiation, minimizing its transmission to lower levels. In this way, the surface boundary layer and the air layer within the canopy are aerodynamically coupled.

Air circulation regulates the microenvironment and plant growth via the exchange of carbon dioxide, heat, and water vapor on leaf surfaces and through diffusion of heat and mass between the area inside the canopy and the air layer above it. Heat and mass exchanges at the level of the foliage occur by molecular diffusion coupled with higher exchange resistance at the molecular boundary layers. Turbulent exchanges of heat and mass occur outside the molecular boundary layer, adjacent to the foliage surfaces.

Airflow also induces direct mechanical action on plants and other obstacles as well as dispersion of particles, microorganisms, fungal spores, and pollen released by plants (Shaw 1995). The forces exerted may be enough to damage branches and trunks or even uproot trees. Such processes are mainly due to intermittent turbulent phenomena. The dispersion rate of bacterial or fungal diseases is also dependent on the velocity of the wind fields, like wise determined by the vegetative configuration.

The Penman–Monteith equation , based on the big-leaf principle and the Bowen ratio , can be used to quantify vertical fluxes of mass and energy in forests, as an aerodynamic and turbulent covariance method (Jarvis and McNoughton 1986; Kelliher et al. 1990). The Penman–Monteith equation quantifies vertical flux of water vapor and patterns of the evapotranspiration regime. This equation includes aerodynamic and canopy resistance, which relate to the canopy physiology and configuration. Characterization of flow within forest canopies, also addresses topics such as parameterization of the rough sublayer and the interaction between carbon and water vapor fluxes, which are of great importance, e.g., in the Mediterranean region.

4.2 Aerodynamic Characterization and Stability in the Rough Sublayer

The roughness sublayer (Fig. 2.1) is in the atmospheric zone, adjacent to the canopy surface including the understory, where the dynamics of circulation are influenced by the spacing between the elements with the development of circulation wakes downwind.

In the canopy roughness sublayer, the main flux-gradients principles are not normally followed, because anomalous phenomena occur. These run in the opposite direction to the gradients and are mainly due to the influence of coherent, intermittent processes for turbulent transport in this sublayer (Kaimal and Finnigan 1994). The occurrence of these counter-gradients causes a cyclical variation in the energy from sensible and latent heat stored in the air within the canopy. Under the counter-gradient process, air parcels from the trunk space penetrate through the crown into the atmosphere or a gust can enter the atmosphere (Foken 2017). The cycle begins with a calm period when the atmosphere of the rough sublayer accumulates substantial amounts of sensible and latent heat as well as carbon dioxide (Denmead and Bradley 1985). Turbulent eddies then form, and the warmer atmosphere is replaced with cooler, drier air with lower carbon content. This process lasts for about 30 s, after which the air is heated and humidified until the occurrence of another discrete phenomenon.

In a forest stand the maximum air temperature happens in the upper crown at around 1–2 pm. Below the crown, the daytime temperatures are lower. During the daytime unstable stratification above the canopy prevails. At night time, the minimum temperature associated with radiative cooling occurs at the top crown level. The minimum temperature in the forest ground occurs a bit later due to the downward flow of cool air from the crown tops. During night time, stable stratification above forest stands is common (Foken 2017).

Mean velocity and instantaneous fluctuations are strongly attenuated within the canopies. The mean flow profile is exponential under the canopy, in contrast with the logarithmic profile above it (Fig. 4.2). A second velocity maximum occurs at the lower level where the space among trunks is more open than within the crowns.

Airflow exerts drag forces on natural or modified terrain surfaces and bodies. In agroforests surfaces, drag is caused by the canopy and is imparted by its elements onto the velocity field. This drag is made up of skin friction and form drag forces (Shaw 1995). The latter, above the canopy, where momentum absorption takes place, are stronger under turbulent flow conditions. Over and within agroforest canopies, examples of a wide range of existing individual and grouped elements include foliage, trees, crops, forest stands, or even water streams.

Those objects present irregular boundaries of retarded air, forming cascades of eddies in leeward surfaces. The drag on leaves can be measured in a wind tunnel under a steady and controlled airflow, e.g., with leaf metallic replicas, allowing to evaluate how the drag coefficient varies with wind speed and direction. The drag on real leaves increases due to the influence of cuticles and hair, insofar that for aluminum specimens, covered with real leaves, that increase was of about 20% at wind velocities of 1.5 ms^{-1} and 50% at 0.5 ms^{-1}, due to higher relevance of skin friction at lower wind speeds.

Wind tunnel studies are theoretically restricted to laminar flow, which is usually common under these experimental conditions, while in natural conditions turbulent flow prevails with associated increases in momentum transfer, especially if the vegetal elements are shaken by airflow. In this context, Monteith and Unsworth (2013), referred that in real turbulent conditions aerodynamic resistances to momentum transfer are downgraded by an order of magnitude as much as 1.5.

Leaves in real vegetal canopies are seldom isolated and thus the drag coefficient of foliage will depend on a set of factors including foliage density and wind speed. In agroforest canopies, most leaves, branches, and needles are exposed to turbulent flows in the interacting wakes of windward elements, insofar that a shelter effect, defined as the ratio between ratios of real drag coefficients was proposed and the coefficient measured for isolated elements. Values representative of shelter coefficients ranges from 1.2 to 1.5 for shoots of apple trees to 3.5 for pine forests (Monteith and Unsworth 2013).

Drag coefficients of leaf specimens and aerodynamic resistance to momentum transfer can be obtained from shear stress, derived, e.g. from energy obtained from velocity values and fluctuations according to calculations described in Chaps. 2 and 3, and Annex 2. Results from tunnel experiments described in Monteith and Unsworth (2013) showed that when leaves were oriented in the direction of the airstream the drag is minimal due mainly to skin friction . If the airstream forces are exerted over concave or convex surfaces, form drag will be much higher than skin friction. In this case, a representative equation of combination of form drag and skin friction will be the following:

$$c_d = c_f + nU^{-0.5} \tag{4.1}$$

where c_d is the total drag coefficient which can be envisaged as the sum of a form drag component c_f with a term representative of a frictional component $nU^{-0.5}$ with n being a constant. The dimensionless form drag coefficient c_d is isotropic. The tangential stresses near the canopy top, are necessary for the concentration of kinetic energy, and allow for high-intensity turbulent regimes within the canopy.

From wind tunnel laboratory results that drag force F on individual trees can be related to wind speed as follows:

$$F = 0.5 c_d \rho U^2 h^2 \tag{4.2}$$

where h is the tree height, c_d the drag coefficient, and U the average wind speed to which an individual tree is exposed. In Eq. (4.2), estimating the drag applied to trees, the biometrical variable is height instead of the perpendicular area to wind flow as it is commonly used with other elements. A specific empirical estimation of drag with British grown Sitka spruce (*Picea sitchensis*) reflecting the effect of streamlines was the following:

$$c_d = a \left[\frac{M}{h^3}\right]^{0.67} exp\left(-bU^2\right) \tag{4.3}$$

where M is the branch mass (kg), U the average wind speed (ms^{-1}), and a and b are constants with values 0.71 and 9.8×10^{-4}, respectively. Applying Eq. 4.3 to experimental results it was shown that for trees with 15 m height and a mass branch of 50 kg, the drag coefficient is reduced by half when wind velocity increased from 1 to 27 ms^{-1}. Monteith and Unsworth (2013) cited experimental results showing that, for conifer trees exposed to a wind tunnel, the relation between drag force and windspeed was affected by the increase of streamlining by individual leaves and whole branches, reflecting a decrease in the effective cross section of the crowns. In the same way, the effects of higher streamlining in small trees and in the top of large trees were reflected in a lower increase rate of drag with air velocity, by comparison with rigid objects.

The elements within the canopy are also sources of small-scale horizontal turbulence, due to wakes formed downwind, not directly exposed to flow. Independently of turbulence and of the shelter effects, the influence of wind speed on drag is very complex, because of interactions of aerodynamic forces on leaf elements and resistant elastic forces. With increasing air velocity, leaf elements are shaken with the formation of elastic forces and an increase of momentum transfer. So, drag arises due to a lack of equilibrium between aerodynamic and these elastic forces.

In field conditions, the drag caused by the complex airflow velocity field causes turning moments and forces acting in the bottom of stems, pulling the plants. In these plants, rotation moments are thereafter formed by the mass and soil attached to root systems. The forces generated by wind fields are opposed by resistance tensions in root systems and by soil shear resistance. In canopies, energy dissipation mechanisms and interference among adjacent plants can damp mechanical oscillations. Plants will fail when the resulting bending moment , in the bottom zone of contact of stems with soil surfaces, overwhelms the strong resistance of stems and soil root systems (Monteith and Unsworth 2013). Individual plant failure to drag forces develops as the stem bow or root system rotates and is pulled out from the soil.

As trees develop, they gradually adjust to dominant airflows by building a root, stem, and branch systems that can withstand wind forces along their lifetimes. The likelihood of stem buckling is higher under conditions such as weakening of lower parts of the plant system by diseases, decreasing soil shearing strength close to roots by water accumulation; weakening of higher parts of plants by excessive water interception or excessive nitrogen fertilization. Isolated trees, due, e.g., to harvesting or mortality, are also less resistant to external forces. Practices aiming to reduce stem lengths, and thus bow tendencies, can be handicapped by greater force transmission to roots, which can result in higher root lodging. Results with wheat, reported in Ennos (1991), and cited by Monteith and Unsworth (2013), showed that the pulling resistance of stems was 30% higher than that of root systems, insofar that the latter would collapse firstly. In a tunnel laboratory, this author found that the moment stresses needed to brake stems was around 0.2 Nm. The axial pulling resistance of roots was found to change with distinct soil shear strengths.

In rice plants, Tani (1963) found that in laboratory wind tunnel plant stems broke under moment forces of around 0.2 Nm, which were higher by a fourfold order of magnitude than the equivalent rotational forces recorded on the field. This difference in rotary moments was attributed to factors such as: (i) the much larger forces exerted in the field under strong turbulent gust events when the instantaneous wind speed can be larger than the average by threefold; (ii) the resonance between the natural period of oscillation of plants and the dominant period of turbulent eddies over the canopy and eventual diseases in plants grown in the field. The collapse of rice plants in the field due to drag forces was shown to occur when plants achieved 0.84 m height and windspeed exceeded 20 ms^{-1}.

Results of wind tunnel simulations with tree canopies showed that the mean drag on sheltered tree specimens was only about 6–8% of that of isolated specimens subjected to airflow. This result was interpreted as consequent to the exponential decay of wind speed from the top to the bottom in specimen canopies, with change the air velocity field, by comparison with a uniform air velocity which remains almost unchanged with exposure to individual specimens.

Foken (2008) suggests typical values for roughness height for sublayer, Z_*, over the canopy top of about two or three times the tree height. Fazu and Schwerdtfeger (1989), report that the interface, z_*, between the roughness and inertial sublayers, is about 50–100 times the height of the momentum roughness length, z_{0M}, where $z_* = Z_* - d$. Garratt (1980) gives experimental values for the ratio z_{*H}/z_0, of the order of 100 for the temperature profile and 35 and 150 for the velocity profile for dense and less dense tree canopies, respectively. For z_{0M} values of the order of 5–10% of the height of the canopy, h, this author suggested that for dense canopies the z_{*M}/z_0 ratio can be about 10. Monteith and Unsworth (1991) suggest a z_* of about 10 z_0, considering z_0 of the order of 0.1 h. Mihailovic et al. (1999) report that in tree canopies, the length of the roughness sublayer ranges from $d + 10z_{0M}$ to $h + 20z_{0M}$. They reported that with z_{0M} of the order of 0.1 h, the height of the rough sublayer will vary one- or twofold with the canopy height.

Cellier and Brunet (1992) suggest the following expressions, corrected for dimensionless gradients (Eqs. 2.37 and 2.38, in Chap. 2):

$$\frac{d(u)}{d(z)}\frac{k(z-d)}{u_*} = \phi_M^*\left(\xi, \left(\frac{z-d}{Z_*-d}\right)\right) \approx \varphi_M(\xi)\varphi_M^*\left(\frac{z-d}{Z_*-d}\right) \tag{4.4}$$

$$\frac{d(T)}{d(z)}\frac{k(z-d)}{T_*} = \phi^*H\left(\xi, \left(\frac{z-d}{Z_*-d}\right)\right) \approx \phi_H(\xi)\phi_M^*\left(\frac{z-d}{Z_*-d}\right) \tag{4.5}$$

in which ϕ_M^* is given by Garrat (1994):

$$\phi_M^*\left(\frac{z-d}{z_*-d}\right) = \exp\left(0.7\left(1 - \frac{z-d}{Z_*}\right)\right) \tag{4.6}$$

where ϕ_M and ϕ_H are functions given by Eqs. (2.52) and (2.53) in Chap. 2.

Potential temperature profiles (Fig. 4.1) within the canopy indicate that thermal stability within the atmosphere varies significantly in the vertical direction (Kaimal and Finnigan 1994). During the day, in the lower inner zone, the Richardson gradient number, Ri_g (Eq. 2.45) is positive, indicating stability and the Richardson flux number Ri_f (Eq. 2.46) is negative, indicating instability. This situation is due to intermittent incursions of large eddies, which cause a positive heat flux (negative Ri_f) interrupted by longer periods of thermal stability (positive Ri_g).

In hardwood stands in temperate zones, the intermittent low-frequency phenomena, that last about 25 s, can increase pressure by about 2.5 Pa (Shaw et al. 1990). During transitional periods of instability, periodic waves tend to form in a similar way to the flow over isolated hills (Chap. 5) with a natural frequency of the order of Brunt–Väisälä, N_{BV}, given by the expression:

$$N_{BV} = \left(\frac{g}{\theta}\frac{\partial\theta}{\partial z}\right)^{0.5} \tag{4.7}$$

At night, trunk spaces are unstable, while the upper levels are thermally stable with little turbulence, as shown in Fig. 4.1 (Kaimal and Finnigan 1994). This leads to the formation of dew at the top of the crown due to radiative cooling, while the environment surrounding the trunks is dry because of turbulent mixing. At higher levels corresponding to d height (Chap. 2) where fluxes vary according to the gradient direction, the Ri_f parameter can be used to characterize thermal stability.

4.3   Turbulent Transport of Kinetic Energy

Turbulent flow in plant environments is associated with processes governing momentum, heat, and mass exchanges between the atmosphere and forest canopies (Raupach and Thom 1981). These exchange processes regulate microclimate in canopies and have a special impact on carbon dioxide and water vapor fluxes,

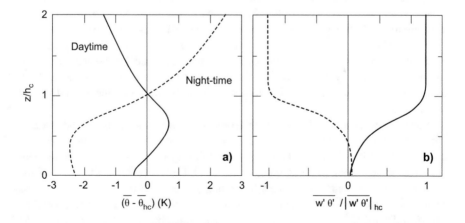

Fig. 4.1 Typical profiles within and above the forest canopy in day and night-time conditions for **a** potential temperature, and **b** sensible heat, (h_c = canopy height) (after Kaimal and Finnigan 1994)

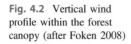

Fig. 4.2 Vertical wind profile within the forest canopy (after Foken 2008)

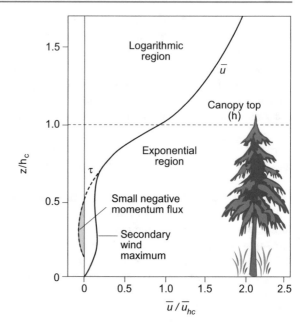

linked to photosynthesis and evapotranspiration. The impact of this dynamic is very strong in issues related, e.g., with agroforestry, carbon and nitrogen cycles, or climate change, justifying the continuous development of efforts in research and knowledge applications.

Over forest canopies, as opposed to low vegetation, turbulent transport is mainly caused by instantaneous short-lived high-intensity phenomena (seconds to minutes). These are driven by large eddies, called coherent structures , which are independent of the mean profiles. Such phenomena have a ramp or saw tooth configurations and display some regularity over space and time.

These phenomena are designated as ascending or descending interactions, ejections, and gusts. The predominant downward vertical transport of linear momentum is mainly carried out by the ejection phenomena (u' < 0 and w' > 0) and gusts (u' > 0 and w' < 0). Upward vertical transport is less representative and mainly due to fast interactions.

Lee and Black's (1993a) work on temperate coniferous forests indicated that ejections and gusts accounted for 72% of the phenomena. Moreover, more than half of the vertical flux momentum occurred during only 9.6% of the total measurement time. In softwood stands , Green et al. (1995) reported that 40% of the total vertical–horizontal transfer of horizontal momentum took place during less than 10% of the time measurements. For a softwood stand, Denmead and Bradley (1985) indicated that the largest part of turbulent transport of heat and mass was due to descending eddies from a height comparable to that of the trees, with an average duration of 30 s, transporting cold, dry air that penetrated the canopy at 3 min intervals. Another indicator of the high regularity of intermittent phenomena over canopies is the correlation between the vertical and horizontal velocities:

$$R_{uw} = \overline{u'w'}/\sigma_u\sigma_w \qquad (4.8)$$

so that at the interface between the canopy and the adjacent atmosphere is of the order of -0.5, significantly greater than the typical value for a flat surface, -0.3, or at a point located well above the interface (Shaw 1995).

Rapid turbulent phenomena will promote air exchange at different temperatures, regulating the energy exchanges within the canopy. The retardatory effect of the airflow is more intense in zones with higher canopy biomass, whereas areas with lower biomass density will have a secondary wind speed maximum (Fig. 4.2). This velocity profile will affect vertical exchanges within the canopy.

Figure 4.3 shows vertical wind velocity profiles for forest canopy modeled in wind tunnel , for pine in two locations (Bordeaux and Uriarra) and for eucalypt (Moga).

The intensity of turbulence i_{ui}, defined by the ratio between the standard deviation of velocity fluctuations and the mean horizontal velocity (Eq. 3.13), is related to kinetic energy and is used to quantify levels of fluctuations related to the mean flow. The turbulence intensity increases with the density of the canopy (Raupach and Thom 1981).

Similarly, Shaw et al. (1988) reported that in a hardwood forest (Camp Borden, Canada) turbulence intensity in the atmospheric layer above the canopies, decreased with a lower leaf area index and was higher under conditions of thermal instability.

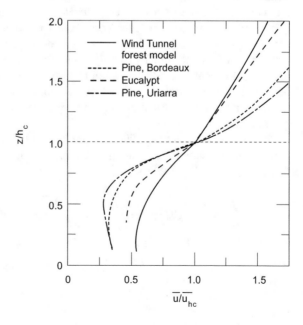

Fig. 4.3 Representative diagram of the mean velocity profile of air, dimensionless at canopy height (after Kaimal and Finnigan 1994)

These authors indicate turbulent intensity values from 0.2 to 0.45 for hardwood stands within the limits ($i_u < 0.5$) using Taylor's hypothesis (turbulent freezing) for measurements at a single point. The turbulent intensities are ordered as follows:

$$i_w < i_v < i_u \tag{4.9}$$

The turbulent kinetic energy budget in the forest roughness sublayer requires terms in relation to the formation of turbulent wakes downwind from the surface elements. This turbulence associated with the development of wakes can be significant, about twice the magnitude of turbulence from tangential interaction with the mean velocity field (Raupach and Shaw 1982).

Coppin et al. (1986) consider that wake turbulence production, P_w, within canopies, can be quantified as

$$P_w = \left\langle \bar{u} \frac{\partial \left(\overline{u'w'} \right)}{\partial z} \right\rangle \tag{4.10}$$

wherein angle brackets represent the horizontal average operator. This equation is obtained from spatial averaging of the turbulent kinetic energy (Eqs. 3.88 and 3.89) and accounts to produce turbulent kinetic energy (TKE) from the mean kinetic energy (MKE). The energy converted by wake production from MKE to TKE is equally done on the work done on the mean flow by canopy elements producing form drag (Raupach and Thom 1981).

Leclerc et al. (1990), based on results from the hardwood canopy atmospheric surface layer at Camp Borden, Canada, suggest that conditions of thermal stability are more conducive to kinetic energy development resulting from wake turbulence (Eq. 4.10). The wake turbulence corresponds to eddies with a maximum length scale that must be smaller than those relative to momentum transport. Thus, the absorption of momentum by aerodynamic drag on the foliage is linked to an accelerated rate of viscous dissipation of turbulent kinetic energy (Baldocchi and Mayers 1988). The wake effect induced by forest canopies draws kinetic energy from average mean flow and to large intermittent descending eddies (Raupach and Thom 1981).

Data from Leclerc et al. (1990) are presented in Fig. 4.4 that show the effect of conditions of instability and thermal neutrality in the terms for the kinetic energy budget equation (Eq. 3.89):

$$\frac{\partial \bar{e}}{\partial t} = \underset{\text{I}}{\frac{g}{\theta_v} \left(\overline{u'_3 \theta'_v} \right)} - \underset{\text{III}}{\frac{\partial \left(\overline{u'_3 e} \right)}{\partial x_3}} - \underset{\text{IV}}{\left(\frac{1}{\bar{\rho}} \right) \frac{\partial \left(\overline{u'_3 p'} \right)}{\partial x_3}} - \underset{\text{V}}{\overline{u'_1 u'_3} \frac{\partial \overline{u_1}}{\partial x_3}} - \underset{\text{VI}}{} \underset{\text{VII}}{\epsilon} \tag{3.89}$$

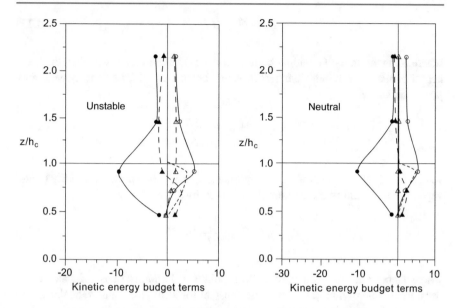

Fig. 4.4 Vertical profiles of dimensionless kinetic energy budget under conditions of thermal instability and neutrality (O mechanical production; Δ buoyancy production; ◆ turbulent transport; – wake production; • residual term) (after Leclerc et al. 1990)

According to these authors, term III for buoyancy production is more important under unstable conditions, in which case it is of the same order of magnitude as term VI for mechanical kinetic energy production. Term VI is positive resulting from the negative sign for downward vertical momentum flux, $\overline{u'_1 u'_3}$. Under conditions of instability and thermal neutrality, the turbulent kinetic energy flux over the canopy rises in the air layer above the canopy.

Figure 4.4 shows the importance of the negative residual component of the kinetic energy budget. This component is defined as the sum of the terms V for transport via pressure and VII for viscous dissipation. The relevance of wake fine turbulence within the canopy is shown in Fig. 4.4 for unstable and neutral conditions. This calculation on wake turbulence was carried out by temporal average.

4.4   Evaluation of the Vertical Flows of Heat and Mass Using the Bowen Ratio Method

The Bowen ratio for measurement of vertical heat and mass fluxes is derived from the energy budget of the surface. This budget, expressing the average fluxes per unit area, in simplified form, can be summarized as (Monteith and Unsworth 1991)

$$R_n - G = H + LE \tag{4.11}$$

where R_n is the radiative budget components of small and large wavelength, G the heat flux in soil, H the sensible heat flux, and latent heat flux, LE. The Bowen ratio, β, is defined by

$$\beta = \frac{H}{LE} = \frac{\rho\, c_p \Delta T}{\rho\, L\, \Delta q} \tag{4.12}$$

where L is the latent heat of water vaporization, and q the specific humidity (water vapor mass per unit mass of humid air) given by

$$q = \frac{\chi}{\rho} = \frac{M_w e}{RT} \Big/ \rho \tag{4.13}$$

where T and χ are the temperature and absolute air humidity (mass of water vapor per unit volume of humid air), R is the constant for ideal gases, and ρ the density of humid air. Considering that this density is the weighted sum of the density of its components, we have

$$\rho = (M_w e + M_A(p - e))/RT \tag{4.14}$$

where p is the total air pressure, and e the partial pressure of water vapor in humid air, Eq. (4.13) can be written as

$$q = \frac{\chi}{\rho} = \frac{M_w e}{M_w e + M_A(p - e)} \tag{4.15}$$

where $\varepsilon = 0.622$ is the molecular ratio between the molecular weight of water vapor M_w and of air M_a

$$\Delta q = \frac{\Delta \chi}{\rho} = \frac{\varepsilon\,\Delta e}{\varepsilon e + (p - e)} \approx \frac{\varepsilon \Delta e}{p} \tag{4.16}$$

Equation (4.12) can be written

$$\beta = \frac{c_p \Delta T}{L\Delta q} = \frac{p c_p \Delta T}{L\varepsilon \Delta e} = \gamma \frac{\Delta T}{\Delta e} \tag{4.17}$$

where γ is the psychometric constant (equal to $c_p p/L\varepsilon$) and the ratio $(\partial T/\partial e)$ is obtained from temperature and vapor pressure data at various levels. The latent heat flux, LE, can then be calculated from Eq. (4.18) as follows:

$$LE = \frac{(R_n - G)}{1 + \beta} \tag{4.18}$$

and for sensible heat H:

$$H = \beta \frac{R_n - G}{1 + \beta} \tag{4.19}$$

it can be demonstrated that the flux for any gas, F_c, such as carbon dioxide is given (Monteith and Unsworth 1991) by

$$F_c = \frac{R_n - G}{\partial C/\partial T_e} / \rho c_p \tag{4.20}$$

where T_e is the equivalent temperature, defined as

$$T_e = T + (e/\gamma) \tag{4.21}$$

that represents the temperature of a volume of air after adiabatic condensation, without heat loss, of the total of water vapor contained in the air, initially saturated (Annex A2).

Equations (4.18)–(4.20) indicate that to calculate the latent, sensible heat, and gas fluxes, it is necessary to have measurements or estimates of the net radiation, heat flux in soil, temperature data, vapor pressure, and gas concentration at various levels. The values of means over 30 min periods for the various parameters are adequate.

The Bowen ratio method assumes steady-state conditions for the radiative field and wind velocity as well as conditions of steady vertical flux. This method is not constrained by stability conditions, as it only requires similarity between K_H and K_V and not with K_M (Oke 1992; Monteith and Unsworth 2013). The Bowen ratio method does not require correction factors, when $(R_n - G)$ tends to zero, for example, at night or under conditions where net radiation is low. Typical values for the Bowen ratio are 0.1 for tropical oceans, 0.1–0.3 for tropical forests, 0.4–0.8 for temperate forests and pastures, 2–6 for semi-arid areas, and more than 10 for the deserts (Oke 1992). In Portugal, Bowen ratio values for cork oak woodlands , grown under the Mediterranean climate and measured in summer, were about 2.4 (Rodrigues 2002).

4.5 Evaluation of Evapotranspiration and Energy Coupling Using the "Big-Leaf" Approach

An alternative way to calculate the latent heat is to use the Penman–Monteith equation . This equation introduces a parameter, r_c, describing canopy resistance . It assumes that the canopy behaves as a thin layer of vegetation, like a big leaf, with the physiological properties of amphistomatous (with stomata on both sides) leaf, and the stomatal resistance is like that of the canopy. The stomata and canopy resistance parameters, r_c, have similar roles in the water vapor transfer processes over crops and leaves. The latent heat flux is then given by

$$LE = \frac{\Delta(R_n - G) + \rho c_p \{e_s(T(z)) - e(z)\}/r_{aM}}{\Delta + (\gamma(r_c + r_{aM})/r_{aM})} \tag{4.22}$$

The Δ parameter corresponds to the rate of change of saturated vapor pressure, e_s, at air temperature, T. For air temperatures below 40 °C, the rate is as follows:

$$D = LM_w e_s(T)/(RT^2) \tag{4.23}$$

According to Tan and Black (1976), canopy resistance is a function of stomatal resistance, leaf area index, and resistance to water vapor diffusion through the air volume of crowns. Some representative values for r_c (s/m) are zero in water, 70 in low grasses, 50 in agricultural crops, and 80–150 in forests (Oke 1992). The Amazonian forest resistance values ranging between 100 and 1000 s/m were suggested by Shuttleworth et al. (1984). In temperate coniferous forests, Stewart and Thom (1973) and Lee and Black (1993b) reported r_c, values ranging between 100 and 400 s/m, and between 150 and 450 s/m, respectively. In Portuguese Cork Oak Woodland, measured values were about 320 s/m (Rodrigues 2002).

Equation (4.22) considers canopy as a big leaf allowing the calculation of turbulent flux of latent heat, by combining the surface energy budget with the mass flow equation using resistance parameters, in analogy with the electrical circuit concept. Among the weaknesses of the Penman–Monteith equation is (Baldocchi 1994): (i) use of canopy resistance concept, which is dependent on various individual factors and (ii) difficulty in characterizing sparse canopies as it assumes horizontally homogeneous surfaces.

To analyze the available energy contributions, atmospheric moisture deficit and canopy resistance as inputs for the evapotranspiration process, Jarvis and McNoughton (1986) reformulated Eq. (4.22) as follows:

$$LE = \Omega \frac{\Delta(R_n - G)}{\Delta + \gamma} + (1 - \Omega) \frac{L\{e_s(T(z)) - e(z)\}}{R_a + R_c + R_b} \tag{4.24}$$

where the terms Ω and R_b correspond to the decoupling coefficient and the laminar resistance, respectively, at the level of the leaf surfaces.

Equation (4.24) introduces the concepts of equilibrium evapotranspiration LE_{eq}, imposed evapotranspiration LE_i, and decoupling coefficient Ω. Equilibrium evapotranspiration, LE_{eq}, is defined as

$$LE_{eq} = \frac{\Delta}{\Delta + \gamma}(R_n - G) \qquad (4.25)$$

The potential evapotranspiration condition LE_{pot}, relates to the maximum evaporation, in the extreme case of a uniformly humid surface, which may be free water or vegetation with high water availability. With the gradual saturation of the atmosphere adjacent to the surface vapor pressure deficit and evaporation will decrease reaching the equilibrium evaporation. The equilibrium evaporation is, therefore, an extreme situation corresponding to the rate of free evaporation on a surface, after saturation of the adjacent atmosphere (Cunha 1977).

Equation (4.25) corresponds to the first term of the right side of Eq. (4.24) when $\Omega = 1$. The term ΩLE_{eq} is the evapotranspiration rate that would occur if the energy balance of a surface was determined by the diabatic radiative term of the Penman–Monteith equation , in the absence of any relationship to the atmospheric conditions. However, the Δ parameter in the equation, related to air temperature, imparts some dependency of LE_{eq} on the atmosphere.

The Priestley–Taylor equation relates to the potential and equilibrium evaporations as follows (Vogt and Jaeger 1990)

$$LE_{pot} = 1.26 LE_{eq} \qquad (4.26)$$

showing that the evaporation potential is greater than the equilibrium evaporation. The proportionality constant in Eq. (4.26) depends on how the boundary layer height evolves throughout the day (Chap. 1), and how to air dry is transported in this layer by downward movements of air masses, trapped by the top inversion layer.

The vertical flux of water vapor is about 24–36% of LE_{eq} (Balddochi et al. 1997) for forest canopy under dry conditions. Homogeneous forest canopies, with closed covers and large water availability, can transpire at a rate of 1.26 times the equilibrium evapotranspiration .

A further quantity $(1 - \Omega)LE_i$, the imposed evaporation, representing the second term on the right side of Eq. (4.24), relates to the evaporation rate that would occur if the energy budget of a surface was dominated by an adiabatic term. This term increases with the value of $\rho c_p (e_s(T(z)-e(z))/r_a$, integrating the right-side numerator of Eq. (4.24), when the value of r_a is very low in strong winds and rough forest canopies.

An electrical analogy can be applied to the second term on the right side of Eq. (4.24), where the vapor pressure deficit is the potential difference, and the electrical resistance is the sum of the different resistances to vapor diffusion.

The decoupling factor Ω, which is the separating factor in Eq. (4.24) between the balance and imposed evaporation, is defined as

$$\Omega = (\Delta/\gamma + 1)/(\Delta/\gamma + 1 + (r_c/r_a)) \tag{4.27}$$

Equation (4.27) shows the preponderant role of canopy resistance in determining Ω, for forest-type canopies, characterized by high r_c and low r_a values.

The Ω factor is associated with the analysis of the relative change of canopy resistance with the relative variation of water vapor flux (Jarvis and McNoughton 1986). This factor is about one in smooth and well-watered surfaces, with evapotranspiration rates of about LE_{eq}. The Ω factor has a value close to zero for surfaces with greater aerodynamic roughness, where the evapotranspiration rates are coupled to the atmospheric vapor pressure deficit.

Typical values for Ω are 0.1 and 0.2 for forests (strong coupling) and 0.8 and 0.9 for lower canopies (Monteith and Unsworth 1991). In the absence of precipitation and surface dryness, the lower canopies with higher aerodynamic resistance are weakly coupled to atmospheric dynamics. Thus, the respective evaporation regimes will be more dependent on either available energy or net radiation, rather than the vapor pressure deficit. In contrast, high canopies are strongly coupled to thick atmospheric layers, so that the latent heat flux will be more closely associated with changes in atmospheric vapor pressure deficit.

The mean annual Ω, obtained by the inverse Penman–Monteith equation for eucalypt stands in Portugal, was about 0.26 to 0.11. These values obtained between 2004 and 2005 correspond to conditions of extreme drought when there was about a 50% reduction in precipitation relative to the long-term average of 709 mm (Rodrigues et al. 2011). Initially, in 2004, the evapotranspiration of about 723 mm was more dependent on net radiation, whereas, during the second phase in 2005, the evapotranspiration decreased dramatically to 392 mm due to stronger stomatal control, under conditions of extreme water stress, and imposed evapotranspiration . Values for the decoupling factor were about 0.18 for cork oak stands in Portugal (Rodrigues 2002).

If E_c and E_f are the evapotranspiration for the lower canopy and forest, and T_c and T are the temperatures of the canopy and the air, the evapotranspiration regimes of the two cover types can be given by the following simplified equations (Oke 1992):

$$E_c = (e_s(T_c) - e(T))/(r_c + r_a) \tag{4.28}$$

$$E_f = (e_s(T_c) - e(T))/(r_c + r_a) \approx (e_s(T) - e(T))/r_c \tag{4.29}$$

In Eq. (4.29), the term for aerodynamic resistance is negligible and the assumption is made that the leaves are small and with rough surfaces, so that $T_c \approx T$.

During active daytime transpiration, the surface temperature of lower canopies is higher than the air temperature, thus the numerator of Eq. (4.28) will be greater than that of Eq. (4.29). Moreover, the total resistance of lower canopies is smaller than that of the forest canopies with higher canopy resistance, r_c, (Oke 1992). Thus, the denominator of Eq. (4.28) is lower than for Eq. (4.29) and under dry conditions, $E_c > E_f$. Equation (4.28) shows that the evaporative potential of the lower canopies depends on canopy temperature which is a function of the absorbed solar radiation. Equation (4.29) couples the evaporative potential of forest canopies to the air temperature and atmospheric vapor pressure deficit. Higher roughness of forest canopies implies that the direct flux of latent heat is controlled primarily by physiological factors and air humidity.

The equilibrium evapotranspiration rates also depend on flow intensity in the boundary layer adjacent to the forest canopy. The soil–plant–atmosphere dynamics over canopies are disturbed by intermittent turbulent phenomena, but the intervals between these phenomena are not long enough to affect equilibrium evapotranspiration rates (Kelliher et al. 1990).

The stomatal resistance increases and according to Eq. (4.27), Ω decreases because of water deficiency in soil (Monteith and Unsworth 2013; Balddochi et al. 1997). Water equilibrium is regulated by stomatal activity, through mechanisms such as the release of abscisic acid from roots, which promote stomata closure (Balddochi et al. 1997). The soil water content will depend on soil texture and capacity to retain or release water, and the possibilities for roots to tap deep into the soil. Stewart and de Bruin (1985) describe the importance of soil moisture in the dynamics of canopy resistance in a pine forest shown in Figs. 4.5 and 4.6.

The leaf area index is another important factor regulating canopy resistance and transpiration processes. Blanken et al. (1997) report major changes in the coupling coefficient in a boreal forest between the periods before and after foliage formation. Before leaf formation, the coupling coefficient Ω was 0.08, which according to Eq. (4.27), indicates high canopy resistance and low aerodynamic resistance. These factors are representative of strong coupling between the rough canopy and the surface layer. Following foliage growth, the coupling coefficient Ω was 0.31, that is, lower coupling to the surface layer because of an aerodynamically smoother leaf layer with a higher r_a value.

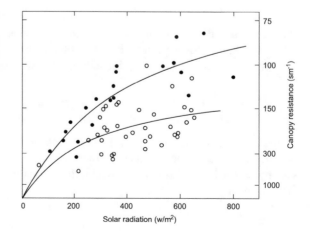

Fig. 4.5 Variation of canopy resistance as a function of solar radiation in a pine forest ecosystem with different soils (• wet soils, o dry soils) (after Stewart and de Bruin 1985)

Fig. 4.6 Variation of canopy resistance as a function of the specific humidity deficit in a pine forest ecosystem with different soils (• wet soils, o dry soils) (after Stewart and de Bruin 1985)

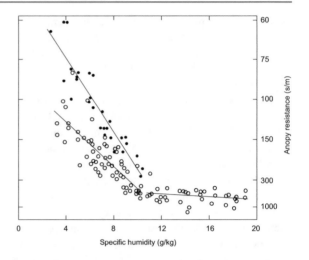

Baldocchi and Meyers (1991) mentioned that the dominant timescale for turbulent fluxes in a moderately open deciduous forest canopy ranged between 200 and 300 s corresponding to an eddy frequency under which large-scale vortices swept into the trunk space from above, inducing an ejection of air from the canopy. During the quiescent transitory periods between events, air humidity, and respired carbon dioxide accumulated inside the canopy atmosphere, insofar that it could be considered that evaporation from forest soil/litter surfaces and understory, was a non-steady-state process quasi-close coupled with the above canopy saturation deficit.

The following equation interprets how soil moisture evapotranspiration from soil/litter surfaces in forest understory, during quiescent periods, influence saturation deficit $D(t)$ decreases with time:

$$\frac{dD(t)}{dt} = \frac{\Delta(R_n - G) - (\Delta + \gamma)LE(t)}{\rho h c_p} \tag{4.30}$$

with h being the height of control volume for evapotranspiration calculation. Equation (4.30) allows by differentiation obtaining the time variation of evapotranspiration in the understory (Monteith and Unsworth 2013):

$$LE(t) = LE(0)\exp\left(-\frac{t}{\tau}\right) + \frac{\Delta}{\Delta + \gamma}(R_n - G)\left[1 - \exp\left(\frac{-t}{\tau}\right)\right] \tag{4.31}$$

being $LE(0)$ the latent heat at the beginning of the transitory period when the saturation deficit in the understory strata is equal to that of the air above the canopy. From Eq. (4.25) defining equilibrium evapotranspiration, Eq. (4.31) can be written as

$$LE(t) = LE(0)\exp\left(-\frac{t}{\tau}\right) + LE_{eq}\left[1 - \exp(\frac{-t}{\tau})\right]LE \tag{4.32}$$

With the time constant τ (Eq. A1.3) expressed by

$$\tau = r_a h \frac{\Delta + \gamma\left(1 + \frac{r_g}{r_{aV}}\right)}{\Delta + \gamma} \tag{4.33}$$

with r_g and r_{aV} being the soil surface resistance and the boundary layer resistance for water transfer from the soil surface to the top of canopy understory strata , respectively.

From Eq. (4.32), it can be concluded that the evaporation rate in the canopy understory strata is dependent on the saturation deficit of the air above the canopies tending to equilibrium evapotranspiration when instant t is much higher than the time constant τ. The values of boundary layer resistance for water transfer, r_{aV}, range between 50 and 100 sm^{-1}. The values of soil surface resistance, r_g, are about 0 for wet soils and range between 500 and 3000 sm^{-1} for dry soils.

Equation (4.30) shows that for dry soils surface resistances and time constants are higher, the latter ranging between 1500 and 5000 s, and evapotranspiration in canopy understory strata tend to be dominated by saturation deficit of air above canopies. This is because turbulent eddies with time scales of about 200–300 s, lower than those of dry soils, can renew air within bellow-strata before full soil equilibrium evaporation is achieved. On the other hand, for wet soils, faster evaporation rates corresponding to time scales ranging between 100 and 200 s, allow achieving evapotranspiration equilibrium before the start of large eddies carrying air from heights above canopies (Monteith and Unsworth 2013). Those considerations allow inferring the complexity of the physical and biological factors concerning the processes of control exerted by the different layers of forest canopies in the total water uses and regimes.

Baldocchi et al. (1997) also reported that in a rough pine forest with reduced foliage, the canopy resistance was an order of magnitude higher than aerodynamic resistance. These authors also indicated that the Bowen ratio can be greater than 1 when the stomatal resistance is much higher than the aerodynamic resistance. Similarly, Lindroth (1985) found a fourfold variation in the Bowen from 0.5 to 2 for pine forest in Jädraas, Sweden, showing high seasonal variability due to simultaneous variations in parameters such as canopy resistance and latent heat flux, necessary for the adaptation of forests to the immediate environment.

Overall, rougher higher vegetation with canopy resistance values well above aerodynamic resistance will tend to have high H/LE ratios (Baldocchi et al. 1997). In contrast, under the same conditions, low flat vegetation will tend to have low H/LE ratios.

The diurnal variation r_c associated with the influence of various factors has an important bearing on stomata functioning. Blanken et al. (1997) in a hardwood boreal forest reported a low canopy resistance early in the morning followed by a steady increase throughout the day. In the morning, under conditions of low vapor pressure deficit and high internal water content, the stomata open allowing for photosynthesis to occur with minimal water loss. Under conditions of high net radiation, the increase in the vapor pressure deficit leads to an increase in stomatal resistance until midafternoon.

In temperate softwood forests, Gash and Stewart (1975) reported a pattern of variation along the day, with increasing canopy resistance between 100 and 300 s/m. Lee and Black (1993b) as well as Stewart and de Bruin (1985), for softwood stands indicate that canopy resistance increases throughout the afternoon, due to an increase in vapor pressure deficit and/or a decrease in net radiation. In coniferous stands, Oke (1992) also referred to daily increases of canopy resistance from 100 to 350 s/m, due to an increase in atmospheric vapor pressure deficit.

Factors such as vapor pressure deficit, net radiation, and air moisture influence canopy resistance variations mainly in the short term (Baldocchi et al. 1997; Kelliher et al. 1995). According to these authors, the short-term factors interact with long-term factors intrinsic to biogeochemical processes and collectively influence stomatal activity, photosynthesis, and evapotranspiration processes. In well-drained soils, for example, conditions of modest rainfall and short growing season lead to low rates of decomposition and mineralization of organic matter, low nutrient availability, low growth rates, nutrition, and leaf development. This provides a setting for low leaf area, and large stomatal resistance, thereby limiting photosynthesis and transpiration.

4.6 Carbon Sequestration in Forests

Emissions of carbon dioxide and other greenhouse gases, such as methane and nitrous oxide, have increased since the nineteenth century. Indeed, atmospheric CO_2 content has risen from 280 ppm in the early phase of the Industrial Revolution to 370 ppm at present. According to the Intergovernmental Panel on Climate Change (IPCC) in the absence of precautionary measures, including afforestation, there will be an increase in atmospheric CO_2 concentration to 500 ppm by 2050, which will increase further throughout the remainder of the twenty-first century (e.g., Rodrigues and Oliveira 2006). The Kyoto Protocol defined a target of 5% for reduction of CO_2 emissions by industrialized countries during the period 2008–2012. A reasonable goal using measures proposed in the Kyoto Protocol would be the stabilization of atmospheric CO_2 concentrations at 550 ppm throughout the twenty-first century. This would correspond to an overall air temperature increase of 2–3 °C above the current level.

Forests contribute greatly toward carbon sequestration or sinking. Strategies for mitigating global warming should have long-term goals, bearing in mind that the residence time of GHG in the atmosphere is 50 to 100 years. The theoretical contribution of the long-term Portuguese forestry sector in reducing GHGs in terms of carbon sequestration under the National Plan on Climate Change should be around 1.6 Mt/year corresponding to a total of afforestation of 500,000 to 600,000 ha plus 4 Mt/year, corresponding to forest management activities in existing forest areas.

The 1990s debate on climate change has since led to greater dissemination of scientific information on the importance of forests. This has mainly resulted from the widespread application of the eddy covariance method, spurred by research programs like CARBOEUROPE between 2003 and 2008 (involving about 70 partners in Europe and 30 entities both within and outside Europe) as well as parallel programs, such as Ameriflux implemented in North American, among others. The global scientific network FLUXNET, involving partners of earlier scientific programs, was also created. The outcome was an extended knowledge base on the main factors that contribute to seasonal and annual variations of carbon balance components with latitude (Falge et al. 2002).

The carbon balance termed net ecosystem exchange (NEE), and gross primary assimilation (GPP) in forest ecosystems are related to the biology of plants and their physical environments such as the leaf area index, the temporal dynamics of the microclimate variables, and the duration of growing season, temperature, and soil moisture. NEE represents the net carbon dioxide measurable flux between a given soil–plant ecosystem and the atmosphere.

GPP is the carbon dioxide uptake, resultant from the gross primary productivity or gross photosynthesis by vegetal components of ecosystems such as trees and understory species, e.g., bushes or grasses. Gross assimilation of carbon is particularly dependent on the intercepted solar radiation, mostly in the range of photosynthetic active radiation (PAR) and the total ecosystem respiration (TER) is mainly associated with the air and soil temperatures (Carrara et al. 2004; Reichstein et al. 2002).

In continental European forest stands TER increases with latitude. TER relates to the additional flux of respired carbon dioxide with two components which are autotrophic and heterotrophic respiration.

In this context, net ecosystem exchange can be defined as the difference between GPP and TER, following a criterion wherein a loss of carbon to the atmosphere is negative. Micrometeorological and edaphic variables are the main influencers embedded within those processes. Evapotranspiration flux variations along with several time scales and their interactions with, e.g., carbon dioxide exchange dynamics can also be evaluated.

The soil respiration can be measured in the field with analyzers (Fig. 4.7). Climate variability is one of the key features of climate change. In the case of the Mediterranean areas, there is an increased tendency for droughts which cause substantial reductions in both the NEE and GPP (Ciais et al. 2005; Granier et al. 2007; Pereira et al. 2007). In these regions, droughts are largely responsible for the interannual variability of carbon fixation/sequestration, particularly due to stomatal control of evapotranspiration (Tenhunen et al. 1985; Pereira et al. 1986) as well as gas exchanges at the leaf level and photosynthesis. In northern Europe and North America, these interactive effects between droughts and carbon sequestration are not as pronounced.

Fig. 4.7 Analyzer for
measuring soil respiration

Eddy covariance measurements carried out in Portugal on intensive eucalypt
stands managed as coppice (Rodrigues et al. 2011) over an 8-year period were
used to evaluate the impact of two dramatic events in relation to NEE and GPP .
These were the occurrence of intense and prolonged drought in 2004 and 2005
(mentioned above) and tree felling in 2006 when trees with an average height of
20 m reached the end of a 12-year productive cycle (Fig. 4.8).

The eddy covariance instrumentation was installed at the top of a 33 m tower.
The effects of drought were seen mainly in 2005, the year in which the total NEE
(357 gCcm^{-2}), GPP (1255.1 gCcm^{-2}), and TER (898.48 gCcm^{-2}) decreased
drastically relative to the typical year of 2002 (865 gCcm^{-2}, 2206 gCcm^{-2}, and
1340 gCcm^{-2}, respectively) by means of tight stomatal control, a typical
regime-imposed evapotranspiration . The effect of this drought was similarly felt in
a holm oak forest (Pereira et al. 2007) in which the values of NEE , GPP , and TER
, measured by eddy covariance method with a closed-path system, were 38 gCcm^{-2},

Fig. 4.8 Eucalypt trees in the
Setúbal Peninsula, managed
as coppice for pulp and paper
production, with an average
height of 20 m

548 gCcm^{-2}, and 510 gCcm^{-2} during the hydrological year 2004/2005. The corresponding values were 140 gCcm^{-2}, 1400 gCcm^{-2} e 1260 gCcm^{-2} during the hydrological year 2005/2006, when precipitation was 685 mm.

This pattern of annual carbon sequestration in Mediterranean regions shows seasonal variation, corresponding to a year of physiological activity. In summer, stomata close in response to severe water stress due to high solar radiation (above 2000 MJm^{-2}), high relative humidity and air temperature, and overall tight coupling with atmospheric conditions. Monthly accumulated values of NEE and GPP were dependent on photosynthetically active radiation and atmospheric vapor pressure deficit.

After felling (Fig. 4.9), eucalypt stems sprouting from the stump gave rise to new plants that emitted carbon over a period of seven months. Thereafter, the coppice recovered its carbon sink capacity. Total NEE was 200 gCcm^{-2} in 2008, 209 gCcm^{-2} in 2009, and 96 gCcm^{-2} in 2010. This partial recovery of the ability to capture carbon was simultaneous with excess stems cutting on stumps in the winter and excess heterotrophic respiration was due to biomass abandoned on the ground. In 2010, the average height of the trees was about 7 m (Fig. 4.10).

On the other hand, the seasonal variation regime for carbon fixation, after cutting, was opposite to the normal regime with maximum absorption in the summer, under conditions of higher water stress, and without absorption during winter. This was mainly due to a functionally mature root system, typical of trees about 20 m in height which provided the young plant canopy with an adequate water supply for the low biomass content, compared to the mature forest (Rodrigues et al. 2011). However, during the first two years, there was no carbon fixation in the winter due to the greater sensitivity of young plants, to dew and to the excess shoots cutting. In 2010, the seasonal trend of carbon fixation reversed, returning to the seasonal absorption pattern, typical of mature forest.

During the 2000–2010 decade, the average annual solar radiation exceeded 6000 MJm^{-2} in the eucalypt plantation in southern Portugal. This was much higher than the annual average of 3700 MJm^{-2} typically seen for native Scots pines , for example, with 80-year-old trees in Brasschaat in Belgium in a temperate zone. For the same period, the average annual solar radiation was 5300 MJm^{-2} for dense holm forests in Puechabon with 50-year-old trees in the French Mediterranean region (Pita et al. 2013), this being lower than the values for the eucalypt stand.

Fig. 4.9 Felled eucalypt stands immediately after cutting

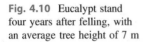

The water vapor pressure deficit in the atmosphere, for the three different forest covers, was highest in June (5 hPa) for a Scots pine stand , in July (19 hPa) for holm oak , and in August (24 hPa) for eucalypt. The typical annual GPP for pine cover was about 1200 $gCcm^{-2}$ (Carrara et al. 2003) and for the holm oak forest cover in Puechabon was 1300 $gCcm^{-2}$ (Allard et al. 2008). These values relating to different parts of the European continent were lower than the GPP for eucalypt, 2206 $gCcm^{-2}$. Clearly, the main reason for this difference, despite the more intensive farming system, is the availability of solar radiation. The average annual NEE (1999–2002) for the Scots pine cover was 120 $gCcm^{-2}$, with a 2-year period (1999 and 2000) as a carbon sink. The corresponding NEE holm oak in France (2001–2006) was about 300 $gCcm^{-2}$. NEE values of these two forest types are lower than NEE values for eucalypt (865 $gCcm^{-2}$ in a normal year and 357 $gCcm^{-2}$ in a dry year).

The fixation monthly variation pattern for the Scots pine canopy was different from that for eucalypt, with carbon uptake only during the summer and autumn and physiological activity being lower during the remainder of the year. In contrast, the pattern of carbon fixation over a similar period for the holm oak canopy stretched over the whole year without any interruption in the summer due to water stress. The decoupling coefficient was lowest for the eucalypt canopy, where the annual mean was about 0.1 in 2005, when climatic factors inducing water stress (global solar radiation and vapor pressure deficit) were more severe.

These three case studies of forest canopies in Europe serve to evaluate differences in patterns and variability of carbon sequestration regimes over annual and seasonal periods. Interactions between fluxes of carbon and water vapor and between physical and physiological variables were also found to be different.

An application of carbon sink capacity by forest canopies is the use of short rotation coppice (SRC) for biomass production for energy. This cluster should be evaluated under global forecasting scenarios, that indicate by the year 2050 energy demand will be about 1041 EJ. In the same year the potential for total biomass production, without affecting food production is estimated at 1135 EJ (Ladanai and Vinterbäck 2009).

Fig. 4.11 Poplars managed under SRC in central Portugal

The SRC can be grown in abandoned or contaminated agricultural land, so as not to compete with crops in fertile soils with less energy-intensive management. Basically, this type of culture is based on the intensive exploitation of forest species, such as eucalypt, poplar (Fig. 4.11), or willow for coppicing in short production cycles. In Europe, the cycles are about three years with a density of 6000–8000 trees ha^{-1} whereas the American system involves 5-year cycles and density of 1100–1600 ha^{-1}.

These lignocellulosic woody crops, together with other types of biomass, can contribute to the replacement of fuels, such as coal, in the production of electricity and liquid fuels. The yields obtained with poplars cultivated at a density of 5000–8000 trees ha^{-1} and 3-year production cycles are in the range of 8–20 tMS ha^{-1} yr^{-1} (4–10 tC ha^{-1} yr^{-1}). Under the same conditions for eucalypt SRC cultivations, the yields range from 4 to 20 tC ha^{-1} yr^{-1}.

CO_2 emissions (4 gCO_2/MJ) from burning wood-based biofuel compare favorably with that for coal (96.8 gCO_2/MJ). Lifecycle analysis shows that the average ratio of energy output and fossil energy input to produce energy is about 36 for the SRC cultivations and 0.9 for coal (Djomo et al. 2011).

References

Allard, V., Ourcival, S. R., Joffre, R., e Rocheteau, A. (2008). Seasonal and annual variation of carbon exchange in an evergreen forest in southern france. *Global Change Biology, 14,* 714–725.

Balddochi, D. D. (1994). A comparative study of mass and energy exchange over a close C_3 (wheat) and open C_4 (corn) canopy: I. the partitioning of available energy into latent and sensible heat exchange. *Agricultural and Forest Meteorology, 67,* 191–220.

Baldocchi, D. D., & Meyers, T. P. (1988). Turbulence structure in a deciduous forest. *Boundary Layer Meteorology, 43,* 345–364.

Baldocchi, D., & Meyers, T. (1991). Trace gas exchange above the floor of a deciduous forest: 1. Evaporation and CO_2 efflux. *Journal of Geophysical Research, 96,* 7271–7285.

Baldocchi, D. D., Vogel, A. C., & Hall, B. (1997). Seasonal variation of energy and water vapor exchange rates above and below a boreal jack pine forest canopy. *Journal of Geophysical Research, 102,* 28939–28951.

Blanken, P. D., Black, T. A., Yang, P. C., Neumann, H. H., Nesic, Z., Staebler, R., et al. (1997). Energy balance and canopy conductance of a boreal aspen forest: partitioning overstory and understory components. *Journal of Geophysical Research, 102,* 28915–28927.

Carrara, A., Janssens, I., Yuste, J., & Ceulemans, R. (2004). Seasonal changes in photosynthesis, respiration and NEE of a mixed temperate forest. *Agricultural and Forest Meteorology, 126,* 15–31.

Carrara, A., Kowalski, A. S., Neirynck, J., Janssens, I. A., Yuste, J. C., & e Ceulemans, R. (2003). Net ecosystem CO_2 exchange of mixed forest in Belgium over 5 years. *Agricultural and Forest Meteorology, 119,* 209–227.

Cellier, P., & Brunet, Y. (1992). Flux-gradient relationships above tall plant canopies. *Agricultural and Forest Meteorology, 58,* 93–117.

Ciais, P., Reichstein, M., Viovy, N., Granier, A., Ogee, J., Allard, V., et al. (2005). Europe-wide reduction in primary productivity caused by the heat and drought in 2003. *Nature, 437,* 529–533.

Coppin, P. A., Raupach, M. R., & Legg, B. J. (1986). Experiments on scalar dispersion within a plant model canopy, Part II: An elevated plane source. *Boundary Layer Meteorology, 35,* 167–191.

Cunha, F. R. (1977). *Meteorologia Geral e Agrícola.* Mesologia e Meteorologia Agrícolas, Instituto Superior de Agronomia, Lisboa, (in portuguese).

Denmead, O. T., & Bradley, E. F. (1985). Flux-gradient relationships in a forest canopy, pp. 421–442. In B. A Hutchinson, B. B. Hichs (eds.). *The Forest-Atmosphere Interaction.* Proceedings of the Forest Environmental Measurements Conference, Reidel Publishing Company, p. 850.

Djomo, S. N., Kasmioui, O., & Ceulemans, R. (2011). Energy and greenhouse gas balance of bioenergy production from poplar and willow: A review. *GCB Bioenergy, 3,* 181–197.

Ennos, A. R. (1991). The mechanics of anchorage in wheat (Triticum aestivum L.). *Journal of Experimental Botany, 42,* 1607–1613.

Falge, E., Baldocchi, D., Tenhunen, J., Aubinet, M., Bakwin, P., Berbigier, P., Bernhofer, C., Burba, G., Clement, R., Davis, K. J., Elbers, J. A., Goldstein, A. H., Grelle, A., Granier, A., Gumundsson, J., Hollinger, D., Kowalski, A. S., Katul, G., Law, B. E., Malhi, Y., Meyers, T., Monson, R. K., Munger, J. W., Oechel, W., Paw Uv., Tha, K., Pilegaard, K., Rannik, Ü., Rebmann, C., Andrew Suyker, A., Valentini, R., Wilson, K., & Wofsy, S. (2002). Seasonality of ecosystem respiration and gross primary production as derived from FLUXNET measurements. *Agricultural and Forest Meteorology, 113,* 53–74.

Fazu, C., & Schwerdtfeger, P. (1989). Flux-gradient relationships for momentum and heat over a rough natural surface. *Quarterly Journal of the Royal Meteorological Society, 115,* 335–352.

Foken, T. (2008). *Micrometeorology* (p. 306). Berlin: Springer Verlag.

Foken, T. (2017). *Micrometeorology* (2nd ed.) (p. 362). Springer Verlag, Berlin.

Garrat, J. R. (1980). Surface influence upon vertical profiles in the atmospheric near-surface layer. *Quarterly Journal of the Royal Meteorological Society, 106,* 803–819.

Garrat, J.R. (1994) *The atmospheric boundary layer.* Cambridge University Press, p. 316.

Gash, J. H. C., & Stewart, J. B. (1975). The average surface resistance of a pine forest derived from bowen ratio measurements. *Boundary Layer Meteorology, 8,* 453–464.

Granier, A., Reichstein, M., Breda, N., Janssens, I. A., Falge, E., Ciais, P., et al. (2007). Evidence for soil water control on carbon and water dynamics in european forests during the extremely dry year: 2003. *Agricultural and Forest Meteorology, 143,* 123–145.

Green, S. R., Grace, J., & Hutchings, N. J. (1995). Observations of turbulent air flow in three stands of widely spaced sitka spruce. *Agricultural and Forest Meteorology, 74,* 205–225.

Jarvis, P. G., & McNaughton, K. J. (1986). Stomatal control of transpiration: Scaling up from leaf to region. *Advances in Ecological Research, 15,* 1–48.

Kaimal, J. C., & Finnigan, J. J. (1994). *Atmospheric Boundary Layer Flows* (p. 289). Their Structure and Measurement: Oxford University Press.

Kelliher, F. M., Leuning, R., Raupach, M. R., & Shulze, E. D. (1995). Maximum conductance for evaporation from global vegetation types. *Agricultural and Forest Meteorology, 73*, 1–16.

Kelliher, F. M., Whitehead, D., Mcaneney, K. J., & Judd, M. J. (1990). Partitioning evapotranspiration into tree and understorey components in two young Pinus radiata D. Don Stands. *Agricultural and Forest Meteorology, 50*, 211–227.

Ladanai, S., & Vinterbäck, J. (2009). Global potential of sustainable biomass for energy, Report 013. SLU, Uppsala, p. 32.

Leclerc, M. Y., Beissner, K. C., Shaw, R. H., Den, H. G., & Neumann, H. H. (1990). The influence of atmospheric stability on the budgets of reynolds stress and turbulent kinetic energy within and above a deciduous forest. *Journal of Applied Meteorology, 29*, 916–933.

Lee, X., & Black, T. A. (1993a). Atmospheric turbulence within and above a Douglas fir stand. Part I: statistical properties of the velocity field. *Boundary Layer Meteorology, 64*, 149–174.

Lee, X., & Black, T. A. (1993b). Atmospheric turbulence within and above a douglas fir stand. Part II: eddy fluxes of sensible heat and water vapour. *Boundary Layer Meteorology, 64*, 369–389.

Lindroth, A. (1985). Seasonal and diurnal variation of energy budget components in coniferous forests. *Journal of Hydrology, 82*, 1–15.

Mihailovic, D. T., Branislava, L., Rajkovic, B., & Arsenic, I. (1999). A roughness sublayer profile above a non-uniform surface. *Boundary Layer Meteorology, 93*, 425–451.

Monteith, J. L., & Unsworth, M. H. (1991). *Principles of environmental physics* (2nd Ed.) (p. 291), Edward Arnold.

Monteith, J. L., & Unsworth, M. H. (2013). *Principles of environmental physics* (4th ed., p. 403). Oxford: Academic Press.

Oke, T. R. (1992). *Boundary layer climates* (2nd ed.), 435. Routledge.

Pereira, J. S., Tenhunen, J. D., Lange, O. L., Beyschlag, W., Meyer, A., e David, M. M. (1986). Seasonal and diurnal patterns in leaf gas exchange of eucalyptus globulus labill. Trees growing in Portugal. *Canadian Journal of Forest Research, 16*, 177–184.

Pereira, J. S., Mateus, J. A., Aires, L. M., Pita, G., Pio, C., David, J. S., et al. (2007). Net ecosystem carbon exchange in three contrasting mediterranean ecosystems—the effect of drought. *Biogeosciences, 4*, 791–802.

Pita, G., Gielen, B., Zona, D., Rodrigues, A., Rambal, S., Janssens, I. A., & Ceulemans, R. (2013). Carbon and water vapor fluxes over four forests in two contrasting climatic zones. *Agricultural and Forest Meteorology, 180*, 211–224.

Raupach, M. R., & Shaw, R. H. (1982). Averaging procedures for flow within vegetation canopies. *Boundary Layer Meteorology, 22*, 79–90.

Raupach, M. R., & Thom, A. S. (1981). Turbulence in and above plant canopies. *Annual Review of Fluid Mechanics, 13*, 97–129.

Reichstein, M., Tenhunen, J. D., Roupsard, O., Ourcival, J. M., Rambal, S., Miglietta, F., Peressotti, A., Pecchiari, M., Tirone e, G., Valentini, R. (2002). Severe drought effects on ecosystem CO_2 and H_2O fluxes at three mediterranean evergreen sites: Revision of current hypotheses? *Global Change Biology, 8*(10), 999–1017.

Rodrigues, A. M. (2002). *Fluxos de Momento, Massa e Energia na Camada Limite Atmosférica em Montado de Sobro.* Ph.D. Thesis (Environment, Energy profile) Instituto Superior Técnico, U.T.L., Lisbon, 235 pp. [in portuguese].

Rodrigues, A. M., Oliveira, H. (2006). Sequestro de Carbono, Tendências Globais e Perspetivas do Sector Florestal Português. *Ingenium, II Série, 92*, 68–71. [in portuguese]

Rodrigues, A., Pita, G., Mateus, J., Kurz-Besson, C., Casquilho, M., Cerasoli, S., et al. (2011). Eight years of continuous carbon fluxes measurements in a portuguese eucalypt stand under two main events: drought and felling. *Agricultural and Forest Meteorology, 151*, 493–507.

Shaw, R. C (1995). *Statistical description of turbulence.* Lecture 9, in: *Advanced Short Course on Biometeorology and Micrometeorology.* Università di Sassari, Italia.

Shaw, R. H., Hartog, G. D., & Neumann, H. H. (1988). Influence of foliar density and thermal stability on profiles of reynolds stress and turbulence intensity in a deciduous forest. *Boundary Layer Meteorology, 45,* 391–409.

Shaw, R. H., U. Paw, K. T., Zhang, X. J., Gao, W., Hartog, G., & Den Neumann, H. H. (1990). Retrieval of turbulent pressure fluctuations at the ground surface beneath a forest. boundary layer. *Meteorology 50,* 319–338.

Shuttlewortth, W. J., Gash, J. H. C., Lloyd, C. R., Moore, C., Roberts, J., Filho, A. O. M., Fisch, G., Filho V de, P.S., Ribeiro, M. N. G., Molion, L. C. B., Sá, L. D. A., Nobre, J. C. A., Cabral, O. M. R., Patel, S. R., & Moraes, J.C. (1984). Eddy correlation measurements of energy partition for amazonian forest. *Quarterly Journal of the Royal Meteorological Society, 110,* 1143–1162.

Stewart, J. B., & Thom, A. S. (1973). Energy budgets in pine forest. *Quarterly Journal of the Royal Meteorological Society, 99,* 154–170.

Stewart, J. B., & de Bruin, H. A. R. (1985). *Preliminary study of dependence of surface conductance of thetford forests on environmental conditions*, pp. 91–104. In B. A. Hutchinson, B. B. Hichs (Eds.). *The Forest-Atmosphere Interaction.* Reidel Publishing Company

Tan, C. S., & Black, T. A. (1976). Factors affecting the canopy resistance of a Douglas-Fir forest. *Boundary Layer Meteorology, 10,* 475–488.

Tani, N. (1963). The wind over the cultivated field. *Bulletin of the National Institute of Agricultural, Science,* Tokyo A 10, 99.

Tenhunen, J. D., Lange, O. L., Harley, P. C., Beyschlag, W., Meyer, A., e David, M. M. (1985). Limitations due to water stress of leaf net photosynthesis of Quercus Coccifers in the Portuguese evergreen scrub. *Oecologia, 67,* 23–30.

Vogt, R., & Jaeger, L. (1990). Evaporation from a pine forest-using the aerodynamic method and bowen ratio method. *Agriculture and Forest Meteorology, 50,* 39–54.

Flow Over Modified Surfaces

5

Abstract

In this chapter, an assessment of airflow over modified surfaces was made, mainly corresponding to urban areas, hills, or of transition between surfaces with different roughness, different land uses, and ambient temperatures. This surface heterogeneity that leads to a development of an internal boundary layer, including sublayers, with a determined height and influence length or fetches were analyzed. In the same way, the variation of surface temperatures leads to internal thermal boundary layers with estimable heights. Airflow patterns over isolated arrayed building elements were assessed with turbulent changes in vertical velocity profiles, typical wakes or cavities, or horizontal wrapping horseshoe vortices, in the context of urban heat island or stratification effects, with implications in items as distinct as atmospheric thermal regimes or pollutant dispersion. The main patterns of airflow over isolated or grouped hills were discussed, with an analysis of air circulation around and over hills under different stability conditions, characterized through Froude number variation. Finally, a discussion was carried out of the dynamics of specific atmospheric case studies, such as katabatic winds or Foehn and Bora-type descending flow in hills under inversion conditions, with potential relevant theoretical extrapolations.

5.1 Introduction

In the previous chapters, atmospheric fluxes in the surface layer over relatively homogeneous and flat surfaces were described. Such idealized surfaces occur in seas, plains, ice, and forests. However, in many real situations, surface characteristics vary over distances as short as 100 m. Among examples of surface heterogeneity are coastlines, valleys, hills, and transition areas between rural and urban areas, as well as between different vegetation types. Flow over these obstacles changes due to factors such as surface roughness, temperature, humidity, and altitude, which may act in concert. The combined action of these factors is unpredictable and nonlinear (Arya 1988).

Cold air flows from source areas as open hilltops, forest slopes, and other inclined surface areas are the most common micrometeorological phenomena. Basically, these processes can be envisaged as flows of compacted air parcels which can be interrupted or damped by obstacles. These flows show several levels of risk of frost deposition, e.g., in cold air nights with strong cooling by longwave radiation due to cloudless skies (Foken 2017).

This chapter covers some fundamentals on the interactions between transport of air mass in movement and urban canopies (in analogy with forest canopies) or small topography changes in the realm of environmental physics. For further details, the reader is referred to Arya (1988), Stull (1994), Oke (1992), Garratt (1994), and Foken (2017).

5.2   Internal Boundary Layer

Modification of surface roughness affects the boundary layer structure, with the formation of layers for each boundary surface. Changes in the surface roughness are very common due, e.g., to deforestation or building development (e.g., Foken 2017) and localized profiles for velocity, temperature, and air humidity depend on the specificities of the surfaces. Under different thermal stability conditions, the development of a discontinuous layer of air, including sublayers with different thicknesses caused by horizontal advection in contiguous surfaces, occurs (Fig. 5.1) (Foken 2008; Stull 1994). This layer is termed as an internal boundary layer. Relevant factors affecting this development are roughness lengths and friction velocity, both upwind and downwind from the zone where roughness changes. The roughness change is felt at a downwind distance of up to about 300 z_o, with consequent modification of the velocity profile, with z_o being the aerodynamic roughness length.

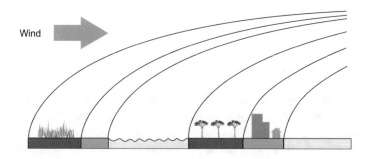

Fig. 5.1 Internal boundary layers flow over different land use types (after Stull 1994)

The atmosphere above the internal boundary layer is not modified by the characteristics of the surfaces upstream of the initial surface roughness change. The height of the internal boundary layer, above a given point, depends on the horizontal distance from this point to that of the initial transition in surface roughness, with the length of this influence length defined as fetch.

Atmospheric flow from a flat to a rough surface causes air to decelerate inside the internal boundary layer, with the formation of horizontal convergence and upward movement above the border between the two surfaces. Opposite effects take place when flow moves from a rough to a smooth surface (Stull 1994). These vertical movements interact with other convective movements, influencing the transport of pollutants, for example, in transition zones between cities and their surroundings. The differences in the flow regimes between smooth and rough surfaces dictate the need for energy-generating wind turbines to be mounted on smooth surfaces (Fig. 5.2).

The height of the internal boundary layer, δ, is calculated as a function of the fetch x (Stull 1994)

$$\frac{\delta}{z_{o1}} = c \left(\frac{x}{z_{o2}}\right)^d \tag{5.1}$$

where z_{o1} and z_{o2} are the aerodynamic roughness lengths upwind and downwind from the contact border between two distinct adjacent surfaces. The power d is about 0.8 under thermal neutrality, being slightly lower (0.6–0.7) under thermal

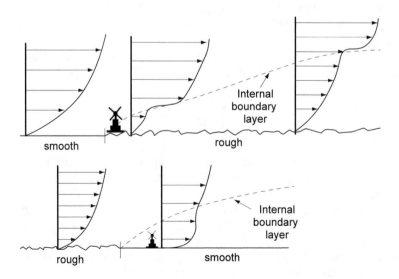

Fig. 5.2 Wind turbines under two conditions within the internal boundary layer (adapt. Foken 2008)

stability and a bit higher (0.8–1) under thermal instability conditions. The parameter c, is about 0.2–0.8, being higher under thermal instability and lower under thermal stability conditions.

The value of c can be also obtained as a function of the two roughness lengths (Stull 1994)

$$c = 0.75 + 0.03 \ln \left(\frac{z_{o2}}{z_{o1}} \right) \tag{5.2}$$

Another equation for estimating the height of the internal boundary layer is as follows (Raabe 1983):

$$\delta = 0.3x^{0.5} \tag{5.3}$$

The various sublayers of the internal boundary layer mix at heights ranging between 30 to 100 m above the surface, with a mean flow due to the overall influence of the various surfaces (Foken 2008). This homogenization of the internal boundary layer is related to average characteristic length C_x of surface roughness higher than 1 km, and air mixing height of about $C_x/200$ (Mahrt 1996). In a similar way, the thermal internal boundary layer also stratifies when in contact with surfaces with abrupt changes in temperatures, coupled with changes in sensible heat flow. These surface temperature changes can be due to factors such as different land uses or variations in air humidity. Because evaporation needs energy, which can be provided by a downward sensible heat flux, an increase in evaporation from soil can be possible under thermal neutrality. High wind velocities and soil dehydration are factors contributing to soil erosion, e.g., in sandy soils. In this context, windbreaks can be used in areas with high wind velocities for reducing soil erosion potential (Foken 2017).

An expression for estimating the height of the thermal boundary layer, δ_T, over a temperature gradient is as follows (Raynor et al. 1975):

$$\delta_T = \frac{u_*}{u} \left[\frac{x(\Delta\theta)}{|\partial T/\partial z|} \right]^{1/2} \tag{5.4}$$

where $\Delta\theta$ is the horizontal potential temperature variation. The vertical temperature gradient is measured upwind of the point for surface temperature variation. All other parameters in Eq. (5.4) are measured at a reference level (Foken 2008).

Airflow from a warm to a cold surface induces the formation of a stable thermal boundary layer. After crossing the temperature transition point, turbulence decreases sharply contributing to the homogenization of the cold air. Under conditions of thermal stability, the following expression can be used to estimate the depth of the thermal boundary layer, δ_{Test}, (Garrat 1987):

$$\delta_{Test} = 0.014\overline{u}\left[\frac{x\overline{\theta}}{g\overline{\Delta\theta}}\right]^{1/2} \tag{5.5}$$

where $\overline{u}$ is the mean velocity flow and $\overline{\Delta\theta}$ the difference between the air temperature, prior to the development of the layer, and the temperature of the cooler surface, and x the fetch.

The internal thermal boundary layers can form in winter when atmospheric flow moves from cold land surfaces to warmer water surfaces or in summer when the air over cold-water surfaces is transported to warmer land surfaces. Under these conditions, the height of the convective boundary layer increases with distance from the separation zone and is characterized by strong turbulence which can be suppressed by downward air mass from low-pressure zones (Chap. 1).

5.3   General Characterization of the Urban Boundary Layer

Urban areas encompassing residential, commercial, and industrial areas show distinct aerodynamic, radiative, and climatic characteristics as compared with surrounding non-urban areas. The changes are mainly confined to the urban boundary layer, although pollutant urban plumes can stretch across dozens of kilometers into surrounding areas (Arya 1988). The urban boundary layer, with a height of about 1000 m, is a mesoscale atmospheric boundary layer phenomenon and its characteristics are dictated by the features of the urban canopy. Urban canopies located below building tops are characterized by microscale processes that occur at the street level among buildings (Oke 1992). Because the buildings and streets in cities have high heat capacities, urban air cools more slowly at night than surrounding areas, leading to the formation of urban heat islands (Foken 2017).

The urban boundary layer is influenced by the heat island effect, which is due to air temperatures higher than in adjacent non-urban areas, as well as higher surface roughness. In adjacent buildings areas, temperatures at the ground surface and air are normally warmer than the temperatures of free surface due to thermal losses from buildings and to sheltering effect against winds.

Radiation effects over urban canopies include the decrease in solar radiation in shaded areas and local increase in radiation reflected from building surfaces such as walls exposed to sun. In areas next to buildings, there is also a reduction in radiative cooling, due to the lower emission of ascending long-wavelength radiation from surfaces, associated with the smaller form factor in relation to the sky (Chap. 1), and to a greater emission of descending long-wavelength radiation from buildings surfaces and heated homes (Oke 1992).

In the absence of topography, buildings with highly variable geometric distributions impart roughness on the urban canopy. The roughness length, z_o, is of the order of centimeters in urban areas with low houses and of the order of meters in more modern areas with tall buildings.

Under mild winds and clear skies, the atmospheric flow due to urban island heating prevails to dominate over the effects of increased roughness. Thermal induced circulation vortices are therefore formed, rising towards the tallest buildings, and descending with subsidence motion over the lower buildings. During the day, this circulation can reach the top inversion of the atmospheric boundary layer which can then acquire a dome shape. In this dome, accumulation of emissions such as dust, smoke, particles, and haze, takes place under conditions of little or no wind.

In the daytime mixed layer, vertical profiles of variables such as temperature, air velocity or specific humidity vary considerably among surface locations, following the rule of vertical homogeneity up to the inversion at about 1000 m (Arya 1988). Vertical fluxes of sensible heat and water vapor were reported by Ching (1985) to vary two to fourfold, as did Bowen ratios. Maximum Bowen ratios of 1.8 and 0.2 were found in areas with tall buildings and in non-urban areas and lower housing, respectively.

At night due to strong stability, the urban boundary layer decreases to a height of a few hundred meters and is higher in zones with taller buildings compared to the non-urban areas. This reduction in height is due to the stability of the nighttime airflow, which suppresses vertical turbulent mixing (Arya 1988).

In mid-sized urban areas, the combined heat island effect and higher roughness can disrupt the inversion of the surface night layer, by changing the velocity and temperature vertical profiles of the boundary layer moving towards the inner-city canopy. A similar situation may occur in temperate zones during sunrise in winter, in which, the initial thermal stability can attenuate vertical turbulent mixing due to warmer urban areas (Oke and East 1971). When moving over urban areas, the lower air layer becomes unstable, while the layers above remain neutral or slightly stable. The atmosphere above the top of the inversion maintains the initial characteristics of the surrounding boundary layer of flat non-urban areas.

The daytime variations in atmospheric stability of the urban boundary layer are much smaller than for the boundary layer above the surrounding non-urban areas, despite daytime variations in height (Arya 1988). This urban boundary layer is mixed intensively by convective processes throughout the daytime cycle, in contrast with the boundary layer in surrounding non-urban areas. This accounts for the higher intensity of nighttime surface winds in the urban surface.

5.4   Flow in Urban Areas

Man-made obstacles to the atmospheric flow have in general simple well-defined geometries arising from combinations of cubic or hemispherical shapes. Flow through these obstacles has three main features that distinguish it from flow on flat surfaces: the accelerated flow resulting from the concentration of streamlines, separation and recirculation, and higher eddy concentration (Rohatgi and Nelson

1994). The atmospheric flow in inner urban canopies is complex and its study is applied mainly to specific situations which are not directly generalized (Arya 1988). Most of the studies carried out to date are done in a wind tunnel using models of individual buildings including prototypes.

The areas of accelerated flow correspond to higher mean wind velocities. Separation zones correspond to lower mean velocity zones and greater turbulence, and the flow in the center of the vortices is intense, strongly turbulent, and periodic in nature. As an example of a practical application, above mentioned, the localization of wind turbines requires information in relation to areas where accelerated flow predominates or where wind velocity is not significantly diminished. In this context, street and buildings microenvironments are also important for the safety and well-being of inhabitants.

The pattern of airflow over a rectangular isolated building with a flat roof positioned perpendicular to the wind (with its upwind side at 90° to the flow) has four zones namely, background, displacement, cavity, and wake (Fig. 5.3). In the displacement zones, flow over an edge at the back of an obstacle leads to a wake flow in an opposite direction.

The displacement separation on an obstacle edge (e.g., a building) is stationary, independent of the Reynolds number (for $Re > 10^4$), and occurring at a fixed location, thereby facilitating experimental modeling studies in terms of scaling the obstacles that make up the urban canopy (Arya 1988). The displacement zone (Fig. 5.3) corresponding to an unperturbed flow, wraps around the building, the wake, and the cavity region. The streamlines in this zone are diverted upwind and laterally, wrapping around the developing wake (Arya 1988). After passing the building, the separation zone follows a downward direction, attenuates, and disappears downstream.

After contact with the building wall, the streamlines flow upwards to the top, downward, or laterally around the building (Fig. 5.4). The maximum pressure occurs in the central area of the windward wall, where the air velocity reaches a stagnation point, and the pressure decreases according to Bernoulli's equation (Annex A2). If the building is rectangular, there is flow separation at the top and sides. These areas are subject to suction and flow reversal, causing recirculation of

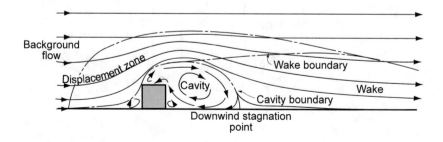

Fig. 5.3 Schematic of flow areas around a two-dimensional building with sharp edges (after Arya 1988)

flow in the cavity area. This recirculation of air along the building walls causes a circulation pattern in the horizontal plane of the cavity zone, in which, the streamlines form a horseshoe pattern running in opposite directions.

Streamlines impact building walls at the downwind stagnation point and recirculate in the cavity region (Fig. 5.3). Around tall buildings, the air diverges at the contact point where it is deflected downwards, and longitudinal and lateral recirculation occurs at higher velocity (Fig. 5.4). Ratios between the wind velocities adjacent to the building relative to an open space are indicated in Fig. 5.4. These ratios are about 2.5–3. The cavity region is characterized by lower mean velocity recirculation, but greater turbulence (Oke 1992). This region can accumulate high concentrations of contaminants despite the occurrence of turbulent diffusion phenomena at higher levels (Arya 1988).

Dimensions of the cavity region depend on the form ratios of the building configuration (Width/Height, W/H, or Length/Height, L/H) or flow characteristics such as the ratio between boundary layer depth of the predominant flow and the building height. The maximum cavity length can vary from H (building height) to $15H$ (Hosker 1984), while the maximum depth can vary from H (for $L/H < 1$) to $15H$ (for $W/H > 100$ and $L/H < 1$). The width/height ratio also influences the mean velocity flow as it decreases further downwind for wide buildings. Wind tunnel studies (Meroney 1977) show that at a downwind distance of about 10 times the building height with a ratio W/H of 3, velocity decreased by 11%, whereas for a W/H ratio of 1, the decrease was 5%.

In the separation zone immediately above the building, streamlines concentrate, and higher velocity air currents form above the wake zone. Below this higher velocity zone, velocity is much lower, and some flow recirculates (Fig. 5.5). Further downwind, the impact of the building on the flow decreases, begins to normalize, and reattaches to the surrounding atmosphere.

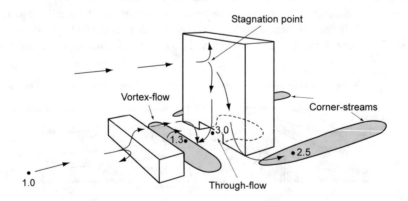

Fig. 5.4 Schematic of flow patterns induced by a tall building with prismatic contours. Numbers refer to the ratio of the air velocities adjacent to the building and open space (after Oke 1992)

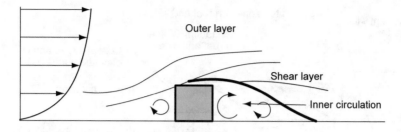

Fig. 5.5 Schematic of atmospheric flow around two-dimensional obstacle (after Rohatgi and Nelson 1994)

The upwind velocity profile is logarithmic and distorts on the roof, with partial reversal in direction. The terminal downwind profile on the building (Fig. 5.6) does not have interactive effects and is not as sloped as the upwind profiles due to the residual effect of wakes promoting greater transfer of vertical momentum (Oke 1992).

Additional building configuration and orientation induce changes in the general flow pattern. For example, if the same cubic shape is oriented diagonally to the wind, there are two walls, windward and leeward oriented obliquely to the flow. This tends to reduce the strength of the suction zones, particularly on the roof. If the roof is sloped, there will be flow separation at its crest, along with a symmetrical lateral downwind horseshoe flow (Fig. 5.7). For a slope greater than 20°, the windward walls will be subject to increased pressure forces, whereas the leeward wall will be subject to greater suction (Oke 1992).

The downwind flow involving the cavity forms wakes (Figs. 5.3 and 5.5). Wakes correspond to the whole flow region downwind of the obstacle and cavity area, disturbed by an interaction with the flow (Fig. 5.7). The wakes, with characteristic dimensions dependent on the size ratios of the buildings, form when the flow is disturbed by bluff bodies, which are obstacles with flat surfaces perpendicular to it,

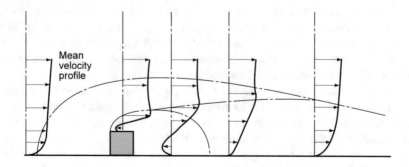

Fig. 5.6 Mean velocity profiles at various locations around a rectangular building oriented perpendicular to the flow (after Oke 1992)

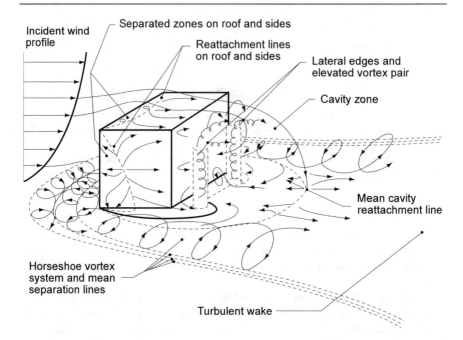

Fig. 5.7 Schematic of separated flow zones in the next wake zone through a sharp-edged three-dimensional building (after Arya 1988)

or by smooth obstacles. The wakes can be divided into a near and far wake region. The former occurs before the streamlines detach from the surface (Fig. 5.7), caused by disturbances of the tangential separation of air layers, and by eddies formed in the boundary layer of building edges.

After separation, the fluid layers become unstable and turbulent, eventually leading to periodic large Von Karman eddies which grow in the displacement zone downwind along with horseshoe-shaped eddies, as mentioned above. These two types of large eddies cause oscillations in the wake boundary, while smaller eddies transport linear momentum across streamlines (Arya 1988). As a result of these processes downwind turbulence increases over a length of an order of magnitude of several building heights, along with a decrease of mean flow velocity. The far wake zone is downstream of the near wake, with a larger characteristic dimension. Turbulence decays exponentially with distance until disappearing downstream of the obstacle.

Studies in wind tunnel indicate that the flow perturbations caused by buildings extend downwind up to 10 to 20 times the height of buildings and in other directions up to 2 to 3 times the height (Arya 1988). The wakes formed in contact with the edges of the separation zone are very different when the flow is oblique relative to the building (47° angle) and can stretch across distances of about 80 times the building height. When the flow is perpendicular to the building, the corresponding distance is about 13 times the building height (Rohatgi and Nelson 1994).

The airflow pattern for a cluster of buildings depends on the ratio H_a/W_s, where H_a is the mean building height and W_s is the along-wind spacing between buildings. When the flow is perpendicular to the largest building dimension, with the buildings reasonably spaced ($H_a/W_s < 0.3$ for row buildings), its pattern is like that of an isolated building (Fig. 5.8a). When the spacing is closer (H_a/W_s of about 0.65 for row buildings) the wake from each building interferes with that of the adjacent one, making the overall pattern more complex (Fig. 5.8b) (Oke 1992). At closer spacing, the main flow skims over the building tops, forming vortices in the cavity region, often a street (Fig. 5.8c). The succeeding buildings reinforce the eddy dynamics by the downwind deflection in spaces between them.

If the wind is oriented obliquely to the streets, then airflow follows a corkscrew-type motion along them. If the wind is oriented in the street direction there is a strong concentration of streamlines, causing higher flow velocities (Oke 1992).

The situation is different if a tall building emerges above the general urban canopy. The wind impacts against the windward surface of the tallest building, causing a stagnation edge in the center, at about three-quarters of the building height. From the stagnation edge, the air flows upwind to the top generating an eddy downwind to the base of the building that causes turbulence at the lower level at the rear of the building.

(a)

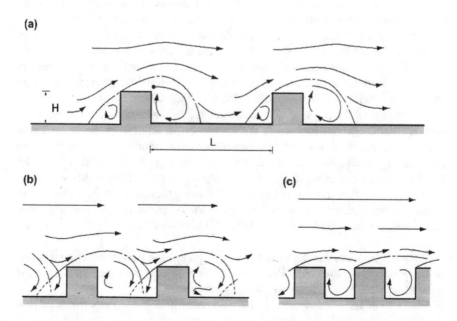

(b)   **(c)**

Fig. 5.8 Schematic of air circulation for different configurations of prismatic buildings, oriented perpendicular to the mean flow (after Oke 1992)

There is also lateral air movement which flow along the building sides in a symmetrical horseshoe-shaped flow. Eventually, these lateral flows along various buildings merge, with a prevalence of turbulence from the tallest building. If the buildings have empty spaces at their base (e.g., building raised on pillars or if there is a walkway) there will be a concentration of streamlines with a crossflow in their spaces. In this context, instead of providing shelter, taller buildings promote the deflection of faster moving upper air down to the ground level. These low-level winds can have velocities about three times higher (Fig. 5.4) than on a flat open space (Oke 1992).

In clusters of buildings that vary considerably in shape and configuration, air circulation regimes are complex and asymmetrical, depending on buildings heights, distances between them, and the characteristics of mean airflow. Low buildings located upwind of tall buildings affect the configuration of symmetrical horseshoe-shaped eddies, which tend to wrap around downstream buildings, while the cavity and wake regions of the taller buildings are only slightly affected (Arya 1988). In this case, taller buildings impose a downwind movement to the mean flow, generating various types of air displacement (Fig. 5.4). On the contrary, small buildings are negatively affected by recirculation inherent to the cavity and wake regions if they are located downwind from taller ones. (Hosker 1984).

For urban hemispherical geometries (Fig. 5.9), horseshoe-shaped eddies dominate the flow, and wakes persist over long distances. Studies in wind tunnels have shown that the vortex lines approaching the hemispheric obstacle are rotated towards a longitudinal direction and stretched downwind as they overpass the obstacle (Hansen and Cermak 1975). As the wakes immerse in a boundary layer with tangential stresses, their mean velocity increases vertically. Downwind eddies to the hemisphere move in an oscillating pattern and cannot be measured at a single point (Fig. 5.9). Downstream the hemisphere obstacle, a huge vertical momentum transport occurs between the top of the boundary layer and the ground, causing an increase in longitudinal mean velocity (Rohatgi and Nelson 1994).

Knowledge of the physical and environmental behavior of wind in urban areas with tall buildings, can be used in protecting against adverse effects of air circulation, with resultant savings in maintenance and improve the well-being of pedestrians and city dwellers as well as dispersion of pollutants (Oke 1992). The buildings are programmed to withstand loads due to high-velocity airflow both in terms of mean flow and turbulence. In designing buildings, it is necessary to consider characteristics such as resistance loads, compactness, porosity, roof inclination, pressure distribution, as well as the location of flow separation areas and suction effects.

Figure. 5.10 illustrates the velocity profile upwind and downwind of a prismatic building with two symmetrical sloped roofs. The area just above the roof has greater turbulence, similarly, to flow over the hills, described below.

Winds over urban canopies are also associated with the various forms of precipitation, related moisture regimes, and consequent surface wear. The location and configurations of arrays of urban buildings also affect the radiation fields in terms of sunny and shade spaces, thermal load accumulation.

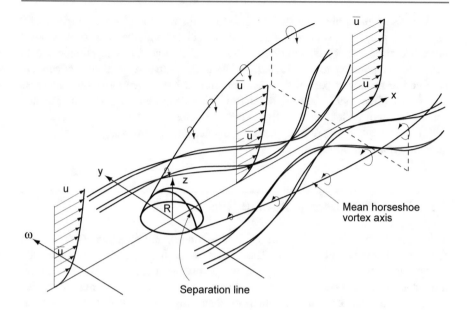

Fig. 5.9 Schematic of air circulation where the mean flow is oriented perpendicular to a hemispherical building (after Rohatgi and Nelson 1994)

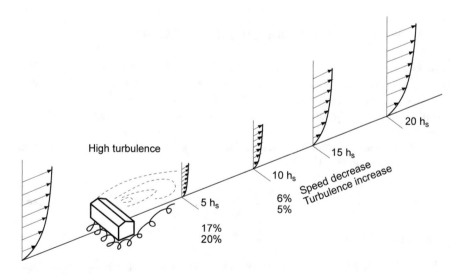

Fig. 5.10 Schematic of air circulation regime when the mean flow is oriented perpendicular to a prismatic building with two symmetrical sloping roofs (after Rohatgi and Nelson 1994)

Aspects of urban planning such as the spatial distribution of streets, green areas, pavements, playgrounds, or parking lots, are associated with the range of previously described factors. Clearly, buildings need to have large bases in comparison to height, corresponding to a ratio *W/H* of about 2, so that recirculation effects are confined to the upper part of the base and should have open spaces for air cross circulation (Oke 1992). Atmospheric flow also determines the distribution and accumulation of pollutants, e.g., from motor vehicles so that open spaces and street width are adequate to promote good air displacement, mitigating the poor recirculation typical of narrow streets.

5.5 Flow Over Gently Sloping Hills

This topic deals with the effects of small hills, ridges, and escarpments on flow in the atmospheric boundary layer. The effects of high mountain chains go well beyond the atmospheric boundary layer, involving mesoscale motions. Flows around hills and gentle ridges are sensitive to the dominant wind regime, unlike flows around urban canopies in which mechanically and thermally generated turbulence is mainly responsible for fluctuations in the mean flow (Arya 1988). The interaction between these obstacles and the wind regime must consider conditions for thermal stability or instability.

5.5.1 Flow Under Different Conditions of Stability

The effects of topography on the dominant flow are affected by thermal stratification. The Froude number, F_r, is the dimensionless ratio often used for quantification of stratification of atmospheric flow around topography. Physically, the Froude number represents the ratio of inertia to gravity forces determining the flow over topography.

The traditional definition applied in fluid mechanics for flow in an open system is (Fox and McDonald 1985)

$$F_r = \frac{\bar{u}}{\sqrt{gL}} \tag{5.6}$$

Equation (5.6) can be analyzed as follows:

(i)   if $F_r < 1$, the flow is subcritical. In this case, predominates the denominator representing the velocity of the perturbation wave flow, so that perturbations can move in a direction independent of the prevalent current. For example, the waves can travel upstream, while the current can move downstream.

(ii) if $F_r = 1$, the flow is critical.

(iii)   if $F_r > 1$, the flow is supercritical. In this case, the current velocity is higher than the velocity of the waves, so that these are dragged in the same direction.

To apply the Froude number to airflow around or over topography, it is assumed that under thermal stability, layers of disturbed air oscillate vertically in the Brunt–Väisälä, N_{BV}, frequency

$$N_{BV} = \left(-\frac{g}{\theta_0} \frac{\partial \bar{\theta}}{\partial z} \right)^{1/2} \tag{5.7}$$

where $\bar{\theta}$ is the mean air temperature potential and θ_0 the temperature potential at height z_0. The frequency defined in Eq. (5.7) is the natural frequency of the internal gravitational waves, or downwind waves.

The oscillation of an air mass with this frequency, in an air mass with a mean velocity $\bar{u}$, induces a wave in this flow with a natural wavelength λ_{BV}:

$$\lambda_{BV} = \frac{2\pi\bar{u}}{N_{BV}} \tag{5.8}$$

The Froude number for flow over topography with a characteristic length C, can be expressed as (Oke 1992)

$$F_r = \frac{\pi\bar{u}}{N_{BV}C} \tag{5.9}$$

The Froude number defined by Eq. (5.9) represents the ratio between inertia and buoyancy forces. It is about the same order of magnitude as the inverse square of the Richardson mass number (Arya 1988)

$$F_r \approx R_{im}^{-2} \tag{5.10}$$

Thermal stratification influences flow over topography in a significant way. This influence in flow patterns is qualitatively like that under neutrality conditions, although with different intensity. Other effects, such as the generation of gravity waves, vortices formation, hydraulic jumps, the inability of low-level fluid to go over hills, and upstream blocking can only occur in stable stratified flows over or around topography (Arya 1988).

If $F_r \ll 1$, the stratification is strong, while for $F_r \gg 1$, conditions are near neutral. For strictly neutral conditions, $F_r = \infty$ corresponding to a case wherein the Froude number no longer represents the flow dynamics. A variety of airflow conditions can be envisaged for an isolated hill, depending on the Froude number (Stull 1994). For Froude numbers about 0.1, corresponding to thermal stability, air surrounds the obstacle horizontally without vertical flow (Fig. 5.11). Upwind of the hill, some blocking and atmospheric stagnation can also occur. The flow blocked by topography has a preponderant influence on the local microclimate and on the dispersion of pollutants emitted upwind (Arya 1988).

For winds with higher velocity, or lower thermal stability (F_r about 0.4), some air rises the hill (Fig. 5.11), while airflow at lower levels diverge moving around obstacles symmetrically and horizontally. It is, therefore, possible to distinguish between streamlines of airflow around topography and of rising airflow (Snyder et al. 1985). The height H_s, of this separation streamline, can be estimated assuming that the kinetic energy of a fluid layer, following a streamline, is equal to the potential energy of the stratification. The overall equation of the process is given by (Sheppard 1956)

$$\frac{1}{2}\rho u_o^2(H_s) = g \int_{H_s}^{H_h} (H_h - z)\left(-\frac{\partial \rho}{\partial z}\right)\partial z \tag{5.11}$$

where H_h is the height of the hill. For the case of stratified flow, with a constant density gradient, the height H_s is given by (Hunt and Snyder 1980)

$$H_s = H_h(1 - F_r) \tag{5.12}$$

The wavelength of the flow over the hill is much smaller than its height, and thus induces the formation of gravity waves, downwind from the hill, which is separated from the surface of the topographic obstacle. This flow separation will occur above the air that does not oscillate and surrounds the hill. The fraction of the air column, with height equal to the hill height, flowing over the top of the hill, is of the order of magnitude of the Froude number (Stull 1994)

$$Fr = \frac{z_{h0}}{z_{hill}} \tag{5.13}$$

where z_{ho} is the height of the air column rising above the hill that separates downwind from the obstacle surface and z_{hill} is the total height of the hill.

Experimental results bear out the separation streamline principle (Snyder et al. 1985). Regardless of the thermal stratification, the angle of the flow orientation relative to a hill and the upwind hill slope are variables that can also be decisive in allowing the air to rise or surround the topographical obstacle (Arya 1988).

Figure. 5.11 represents the effects of thermal stability on the flow on an isolated hill. For a unit Froude number, the stability is weaker, and the winds are stronger, so that the natural wavelength matches the height of the hill. Large amplitude gravity waves or mountain waves are generated by this natural resonance with the possibility of recirculation near the ground close to the wave crests. In this situation, surface air stagnates at periodic intervals downwind of the hill, and reverse flow can

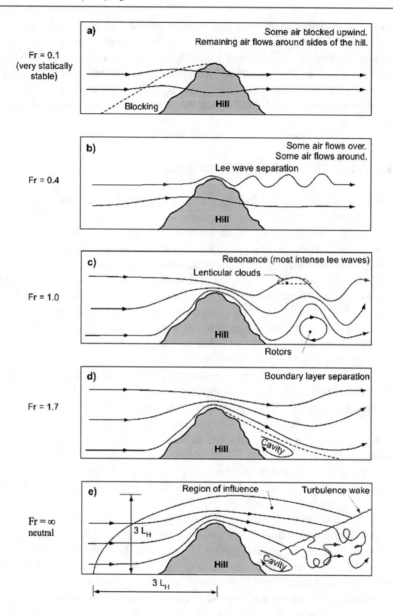

Fig. 5.11 Schematic of air circulation regimes over an isolated hill for different Froude numbers (after Stull 1994)

occur at the surface under the eddies. Under moist conditions, stationary lenticular clouds can form along the crests, and clouds can form downwind at the top of the recirculation zone (Stull 1994).

For even stronger winds with low thermal stability (F_r about 1.7), the natural wavelength is longer than the hill height. This causes separation of the downwind boundary layer of the hill, delivering a cavity zone, in the opposite direction to the main flow. The drag force associated with the flow over a hill is maximum for the resonance state ($F_r = 1$). This force is lower if F_r is lower than 1 and for the separation boundary layer caused by natural waves of greater wavelength ($F_r > 1$) (Stull 1994).

5.5.2 Flow Under Neutral Conditions

The Froude number tends to infinity under neutral stability, and so cannot be used for the characterization of the atmospheric flow. Over smooth topography, flow separation and the formation of cavity areas are like those described above for urban canopies. These processes are lighter and may even be absent (Taylor et al. 1987). Under neutrality, the streamlines are disturbed upstream and above the hill at about threefold the hill height. Beyond this zone, the flow is not affected by the hill. At the hilltop the streamlines converged, causing the wind to accelerate. Below the downwind hill slope, strong winds cause a cavity area related to the separation of the boundary layer and to the beginning of turbulent wakes. Their height is about the same as the height of the hill, but increases in size along decreasing with turbulence intensity, increasing downwards.

In general, airflow accelerates as it moves upstream over the top of the hills. The acceleration factor A_f, or fractional speedup is defined as the ratio $\Delta\overline{u}(x,z)/\overline{u}_0(z)$, where $\Delta\overline{u}(x,z)$ is the difference between the mean horizontal streamwise velocity at height z and $\overline{u}_0(z)$. The $\overline{u}_0(z)$ is a reference velocity, at a downwind distance where it is not influenced by the hill (Kaimal & Finnigan 1994). This A_f factor decreases with an increase in height (Fig. 5.12b) and varies between 1 and 2.5, depending on hill shape, slope, and the ratio between the width and height. The magnitude of speedup in hilltop is relevant for topics as diverse as wind power facilities or estimation of wind load in buildings.

In the case of flow around a cone-shaped hill, rather than accelerating over the top, the flow passes laterally around the surface (Fig. 5.12c) which remains unchanged relatively to the flat surface (Rohatgi and Nelson 1994).

The highest rates of acceleration are seen around lightly sloped three-dimensional hills (Arya 1988). The flow moving at oblique angles to the hills is characterized by lateral vortices and a helical turbulent flow wake.

The mean hill width $L_{1/2}$, is defined as the horizontal distance from the hill peak to a point corresponding to half the peak height (Fig. 5.12a). For slightly sloping hills, wind acceleration can be about twice the ratio of the hill height and $L_{1/2}$. This ratio is about 1.6 for isolated hills (Taylor and Teunissen 1987). For very slightly sloping hills (height 100 m and $L_{1/2}$ about 250 m), wind acceleration above the top will be about 60% compared to normal conditions.

Such data is relevant to the installation of wind turbines. The height at maximum acceleration, Δz_{max}, can be estimated as follows (Stull 1994):

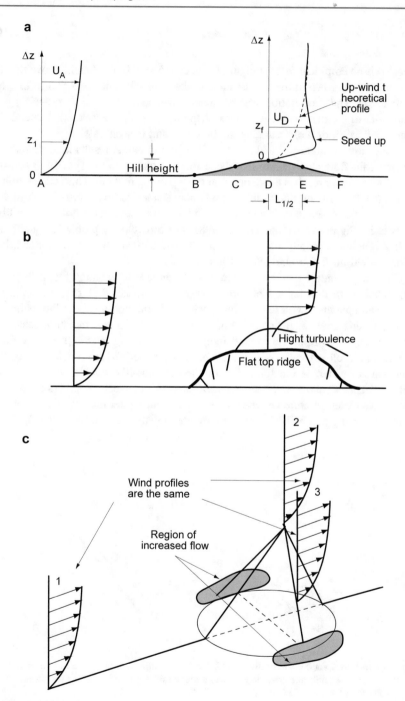

Fig. 5.12 Schematic of air circulation over isolated hills with different profiles. **a** gently sloping hill, **b** flat-topped hill, and **c** theoretical conical hill after Stull (1994); **b** and **c** after Rohatgi and Nelson (1994)

$$\Delta z_{max} \cong z_o e^{\left(2k^2 L_{1/2}\right)^b} \tag{5.14}$$

where b is an empirical factor ranging between 0.5 and 1. The height above the top of a slight hill, wherein the acceleration peaks is of the order of 2.5–5 m (Taylor et al. 1987). The acceleration will be greater over natural concave obstacles, perpendicular to the main flow. A convex shape decreases the acceleration due to the lateral deflection of the circulating air (Rohatgi and Nelson 1994).

In the case of hill arrays and gentle valleys, wind tunnel results indicate that for the first pair, flow changes are like those for an isolated pair (Arya 1988) and the acceleration factor reaches a maximum at the top of the first hill. This acceleration is attenuated downstream due to interactions between successive pairs of valleys and hills. Thus, the acceleration factor value on the top of successive downstream hills tends to be slightly above one (Arya 1998). The top velocity profile is logarithmic with a roughness length higher (Fig. 5.13a and b) than that of the undisturbed velocity field on flat ground (Stull 1994).

Cold air generation and drainage in valleys can lead to radiation fog, if the air is cooled below the dew point. The formation of fog generates the strongest cooling at its top, and thus the lowest temperatures are now found not in valleys but mainly in slopes. In this context, it happens commonly that in the low radiation season, a reduced longwave from the ground induces lifting of fog, and delivering diurnal low stratus in closed valleys. A correlated process, derived from cold air flowing over warmer water, is the formation of the cumulus-like clouds, with the same meters' deepness, above water surfaces. Because this convective process happens at differences of temperature between the water and air higher than 10 °C, fog occurs mainly in spring in small lakes when they are already warm (Foken 2017).

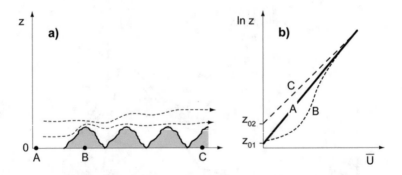

Fig. 5.13 a Flow over a chain of hills, and **b** Velocity profile A, with acceleration over the first hill B, and a new equilibrium condition C, over a rough hill chain surface with a roughness length z_{02}, greater than for a flat surface (after Stull 1994)

5.6 Katabatic Winds Under Stability Conditions

Katabatic winds form when cold, dense air accelerates downstream by gravity over a gentle slope, and are commonly found in the stable boundary layer, up to 10–15 m high. These winds are caused by an inverse thermal structure formed during periods of mild winds at meso and synoptic scales. The katabatic winds occur particularly on calm nights on down slopes under regimes of thermal stability and radiation cooling (Oke 1992).

Even gentle sloping (0.001 to 0.01) across large areas can cause katabatic winds of about 1–2 m/s (Stull 1994). These winds occur over most surfaces, with the general exception of large lakes and oceans.

For this type of flow, the velocity profile is characterized by friction drag generated at the surface with an intermediate layer where the velocity peaks horizontally, along with gravity and increased advection. At a higher level above the gravitational flow, shear stress increase mixing and turbulence (Fig. 5.14). As the air layer descends the horizontal velocity tends to increase.

The virtual potential temperature of katabatic flow is lower on the slope surface and increases progressively with height (Stull 2000). In the absence of an environment mean flow, katabatic winds may reach speeds of 0.5–3.5 m/s (Stull 1994). When the mean velocity of wind flow increases, gravitational flow velocity can increase if oriented in the direction of the wind. In contrast, it will decrease if oriented in the opposite direction, with an eventual suppression of katabatic wind. (Heilman and Dobosy 1985).

A general equation aligned with the downward slope direction, for characterizing the dynamics of katabatic winds is (Stull 2000)

$$\frac{\partial u_d}{\partial t} + u_d \frac{\partial u_d}{\partial x} + v \frac{\partial u_d}{\partial y} = g \frac{\Delta\theta}{\theta} sin(\alpha) + f_c V - C_D \frac{u_d^2}{h} \qquad (5.15)$$

$$\text{I} \qquad \text{II} \qquad \text{III} \qquad \text{IV} \qquad \text{V} \qquad \text{VI}$$

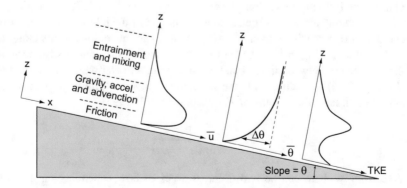

Fig. 5.14 Schematic showing katabatic winds along a hillside with slope α (after Stull 1994)

where u_d is the mean wind velocity along the slope, h is the height of the cold air layer, α the slope angle, C_D is the joint drag coefficient caused by slope surface and by the lower velocity airflow layer and $\Delta\theta$ is the potential temperature difference between the ambient air and air in the katabatic flow. The term I represents acceleration, II and III, the slope and longitudinal advection transport, term IV the buoyancy, term V the Coriolis force (at mid-latitude about 10^{-4} s^{-1}), and VI the turbulence drag.

Initially, the katabatic wind is mainly influenced by buoyancy (IV) and advection (II) (Stull 2000). At this stage, the mean velocity u_d is given by

$$u_d = \left[|g| \frac{\Delta\theta}{\theta} x sin(\alpha) \right]^{(1/2)} \tag{5.16}$$

and after a 2nd stage, an equilibrium velocity, u_{eq}, is attained where the gravity (term IV) is compensated by frictional drag (term VI)

$$u_{eq} = \left[g \frac{\Delta\theta}{\theta} \frac{h}{C_D} sin(\alpha) \right]^{(1/2)} \tag{5.17}$$

5.7 Descending Winds Under Inversion

Fohen and Bora-type winds are phenomena that can occur in the atmosphere under conditions of thermal stability (cold air under warm air layer). These winds typically occur during the winter in temperate zones, under conditions when a synoptic prevailing wind impacts on isolated or arrayed hills (Stull 2000).

If the height of the hill is higher than the height of the upwind cold air, then the cold air is retained upwind and does not move uphill. Warm wind above the cold layer, can jump the hilltop and descend adiabatically, without heat exchange with the exterior ambient, along the downstream slope. This results in mild winds called Foehn, typically found in several places in Europe (Fig. 5.15a). If the height z_i of the cold ascending air layer is greater than the height of the hilltop z_{hill}, then a Bora phenomenon occurs wherein very fast cold winds can move downwards along the lee side of the hill (Fig. 5.15b). Under this phenomenom, the wind firstly accelerate in the contraction between the hill and the overlying air layer with a pressure drop, to P_1 (Fig. 5.16) according with the Bernoulli principle. Within the contraction height, a modified Froude number F_r^* becomes (Stull 1994)

$$F_r^* = \frac{\bar{u}}{N_{BV}(z_i - z_{hill})} \tag{5.18}$$

Accordingly, there can be two different types of flow depending on the wind velocity. For low wind velocities, ($F_r^* << 1$) the acceleration over the hill can move the inversion downward. When the winds are weak, flow separation occurs

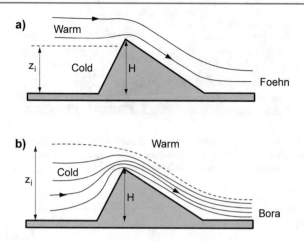

Fig. 5.15 Schematic of descending winds under inversion conditions (cold air under warm air layer). **a** Fohen wind, and **b** cold Bora wind. The solid lines represent streamlines, and the dotted lines indicate the separation between the layers of warm and cold air (after Stull 2000)

downstream along the hill due to the Bernoulli effect with static pressure drop, concentration of streamlines, and increase velocity.

In stronger winds ($F_r^* = 1$) with a strong inversion top, the mixed layer is transported downward with greater acceleration in a narrow layer of higher velocity known as downslope windstorm (USA) or Bora (Balkan Peninsula) (Fig. 5.15b). These gusts can reach speeds of 50 m/s and last for 4–6 days (Stull 1994).

The downstream transport of warm air along the slope requires energy to overcome buoyancy forces, causing a slight deceleration at the bottom relative to the hilltop, also reducing the severity and destructive effect of downslope winds. Katabatic and Bora phenomena correspond to cold downslope winds, although driven by different processes. The main difference between them, is that Bora depends on strong winds with higher dimensional scales, associated with low-pressure zones. On the other hand, katabatic winds are associated with local thermal stratification in high-pressure zones with mild winds.

In Bora-type flow over hills, the hydrostatic pressure in the warm air layer, P_2, is lower than the pressure at lower levels P_1 and pressure at ground surface P_{sfc} at heights below (Fig. 5.16).

Assuming that pressure P_2 is constant along dashed streamline that separates warm and cold air, it follows that, along the streamline in the location of strong Bora effect, the lower pressure P_2 is getting closer to the ground in comparison with a top hill so that pressure at the surface downstream of the hill is equal to the pressure at the top. Compaction of streamlines in cold air layer moving downward in the hill also contributes to a declining pressure, according to the Bernoulli effect. Further downstream, the surface pressure is again normal, at level P_1 (Fig. 5.16), so that, according to the same effect, horizontal pressure increases, simultaneously with a decrease in wind velocity. Under these conditions, the inversion top warm air rises to the initial height, greater than the hill height, under a swift turbulent jump termed as hydraulic jump (Fig. 5.16).

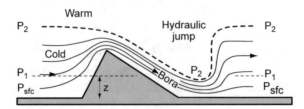

Fig. 5.16 Schematic showing development of hydraulic jump in Bora-type flow where P_2 is the pressure constant along the dashed streamline, P_1 is the hydrostatic pressure at the full line at z height, and P_{sfc} is the pressure at the ground surface (after Stull 2000)

References

Arya, S.P. (1988). *Introduction to Micrometeorology*. International Geophysics Series 42. Academic Press, p. 307.

Ching, J. K. S. (1985). Urban-scale variations of turbulence parameters and fluxes. *Boundary Layer Meteorology, 33,* 335–361.

Foken, T. (2008). *Micrometeorology* (p. 306). Berlin: Springer Verlag.

Foken, T. (2017). *Micrometeorology* (2nd ed.). Berlin: Springer Verlag, p. 362.

Fox, R. W., & McDonald A. T., (1985). *Introduction to fluid mechanics*. John Wiley & Sons, P. 742.

Garrat, J. R. (1987). The stably stratified internal boundary layer for steady and diurnally varying offshore flow. *Boundary Layer Meteorology, 38,* 369–394.

Garrat, J. R. (1994). *The atmospheric boundary layer*. Cambridge University Press, p. 316.

Hansen, C. A., & Cermak, J. E. (1975). Vortex containing wakes of surface obstacle, Report CER 75-76ACH-JEC16, Colorado State University, Fort Colins, CO.

Heilman, W., & Dobosy, R. (1985). A nocturnal atmospheric drainage flow simulation investigating the application of one-dimensional modelling and current turbulence schemes. *Journal of Applied Meteorology, 24,* 924–936.

Hosker, R. P. (1984). Flow and diffusion near obstacles, pp. 241–326. In D. Randerson (ED.), *Atmospheric science and power production* (pp. 241–326). Tech. Info. Cent., U. S. Dept. Energy, Oak Ridge, Tennessee.

Hunt, J. C. R., & Snyder, W. H. (1980). Experiments on stability and neutrally stratified flow over a model three-dimensional hill. *Journal of Fluid Mechanics, 96,* 671–704.

Kaimal, J. C., & Finnigan, J. J. (1994). *Atmospheric boundary layer flows* (p. 289). Their Structure and Measurement: Oxford University Press.

Mahrt, L. (1996). The bulk aerodynamic formulation over heterogeneous surfaces. *Boundary Layer Meteorology, 78,* 87–119.

Meroney, R. N. (1977). *Wind in the perturbed environment: its influence on WECS, American wind energy association*. Boulder Colorado: Spring Conference, May 11–14.

Oke, T. R. (1992). *Boundary layer climates* (2nd ed.). Routledge, p. 435.

Oke, T. R., & East, C. (1971). The urban boundary layer in montreal. *Boundary Layer Meteorology, 1,* 411–437.

Raabe, A, (1983). On the relation between the drag coefficient and fetch above the sea in the case of the off-shore wind in the near shore zone. *Journal of Meteorology, 41,* 251–261.

Raynor, G. S., Michael, P., Brown, R. M., & Sethu Raman, S. (1975). Studies of atmospheric diffusion form a nearshore oceanic site. *Journal of Applied Climatology and Meteorology, 78,* 351–382.

Rohatgi, J. S., & Nelson V. (1994). Wind characteristics, an analysis for the generation of wind power. Alternative Energy Institute, West Texas A&M University, USA, p. 239.

Sheppard, P. A. (1956). Airflow over mountains. *Quarterly Journal of the Royal Meteorological Society, 82,* 528–529.

Snyder, W. H., Thompson, R. S., Eskridge, R. E., Lawson, R. E., Castro, I. P., Lee, J. L., et al. (1985). The structure of strongly stratified flow over hills: dividing-streamline concept. *J. Of Fluid Mechanics, 152,* 249–288.

Stull, R. S. (1994). *An introduction to boundary layer meteorology.* Kluwer Academic Publishers, p. 666.

Stull, R. (2000). *Meteorology for scientist and engineers* (2nd ed., p. 502). Thomson Learning: Brooks/Cole.

Taylor, P. A., & Teunisson, H. W. (1987). The Askervein project: overview and background data. *Boundary Layer Meteorology, 39,* 15–39.

Taylor, P. A., Mason, P. J., & Bradley, E. F. (1987). Boundary layer flow over low hills (a review). *Boundary Layer Meteorology, 39,* 107–132.

Heat and Mass Transfer Processes

6

Abstract

Heat and mass transfer principles, coupled with the modulation of energy and mass budgets, were covered in this chapter. Transfer processes regulate a spectrum of phenomena, such as variations of carbon and water exchanges or sediment transport. The main topics related to heat conduction were Fourier's law and the thermal exchanges in the upper layers of soils, considering their influence in stationary and transient processes. Natural and forced convection processes were assessed through empirical parameters, ratios, and equations reflecting the interactions between fluid mechanics, in air and water, and the transfer of heat, particles, and molecules. Empirical tools for convection analysis considered the prevalence of viscous and inertial forces, allowing also estimation of ratio between depths of thermal and momentum boundary layers in objects. The fundamentals of radiation emissivity, transmissivity, absorption, net radiation, and radiative spatial relationships were also considered. The theoretical foundations and environmental relevance of mass transfer of gases and particles and the importance of Brownian, viscous, or Newtonian drag were also developed. The associated processes of creeping, jumping, impaction or wet transfer, dependent on the balance of forces such as friction, inertia, or gravity through the wind field and transport of sediments in watercourses were highlighted.

6.1 Conduction

6.1.1 General Principles

Conduction is a key mechanism of heat transfer in environment microsystems. The biosphere is a heat sink during heating periods (daytime, summer) and a source of heat during the colder periods (night, winter). Thus, surface organic and inorganic materials act as transient thermal reservoirs which attenuate the atmosphere climate near the ground. In plants and cold-blooded animals, internal energy production is low, and the organism's function as passive conductors. Warm-blooded animals generate large amounts of internal energy and can, in principle, provide thermal energy to the environment (Lee 1978).

Heat transmission mechanisms involving conduction, convection, and radiation are key physical processes for the stationary and transient surface temperature regimes. These processes ensure that physical bodies at different temperatures transfer heat from the hot to the colder body. These bodies at different temperatures promote energy transfer from molecules with higher to lower energy levels.

Conduction is the primary way of heat propagation within a body through internal molecular motion in solids and fluids at rest. Gaseous thermal conduction involves molecular kinetic energy so that in a given volume at high temperature its molecules are at higher velocity/speeds than a volume of gas at lower temperatures. Molecules tend to move between these volumes at different temperatures and transfer kinetic energy and momentum between them. The mechanism of thermal energy conduction in liquids is qualitatively similar (Holman 1983), except that the constituent molecules are closer to each other and the molecular force fields exert a greater influence on energy exchange during a collision, allowing a greater heat transfer rate. In good solid conductors, the heat conduction occurs primarily by transport by free electrons moving between zones at different temperatures. Solids that are good heat conductors are generally good electrical conductors. Thermal conduction in solids via vibration of the structure, typical of insulating solids, is not as efficient as energy transmission by electrons in motion (Holman 1983).

Heat conduction tends to evenly distribute the temperature inside a body. A temperature gradient in the body leads to an energy transfer from the high to the lower temperature regions. The process can be modeled using Fourier's Law (Eq. 6.2) based on the empirical observation of a one-dimensional stationary heat flux through a solid body. In general terms, the heat transfer rate in a given direction per unit area, via solid conduction is proportional to the temperature gradient in the same direction

$$\frac{q}{A} \approx \frac{\partial T}{\partial x} \qquad (6.1)$$

and introducing a proportionality constant k

$$q = -kA\frac{\partial T}{\partial x} \qquad (6.2)$$

where q is the heat transfer rate, A the area perpendicular to the heat transfer direction, and $\partial T/\partial x$ is the temperature gradient in this direction. The positive constant k is thermal conductivity with $Wm^{-1} K^{-1}$ units, and the heat flux is expressed in Watts. Thermal conductivity varies with temperature and its negative

Table 6.1 Thermal conductivity for some materials (adapt. Holman 1983)

Material	Thermal conductivity ($Wm^{-1} K^{-1}$)
Copper	385
Silver	410
Iron	73
Steel	43
Quartz	41.6
Sandstone	1.83
Oak	0.17
Mercury	8.21
Water	0.56
Hydrogen	0.18
Air	0.024
Saturated water vapor	0.0206

sign indicates that the heat flows towards decreasing temperature. Table 6.1 shows typical values for various materials.

Equation (6.2) can be generalized to a three-dimensional state, per unit area

$$q = -k \left(\frac{\partial T}{\partial x} \vec{e}_x + \frac{\partial T}{\partial y} \vec{e}_y + \frac{\partial T}{\partial z} \vec{e}_z \right) = -k \nabla T \qquad (6.3)$$

where ∇T is the spatial temperature gradient. Equation (6.3) is enough to characterize the heat conduction under steady-state conditions defined by a constant temperature at any point in time.

A general equation for analyzing non-stationarity or transient state conditions in which the temperature of the body is variable with time, or if there are heat sources or sinks within it, should include these heat transmittance conditions.

Considering an infinitesimal cubic element, the energy balance needs to take into account that the sum of the energy conducted to the inside through the left side, $-kA\frac{\partial T}{\partial x}$, with the heat generated inside the element $\dot{q}Adx$, equals the sum of the internal energy change, $\rho c A \frac{\partial T}{\partial \tau} dx$ with the energy conducted through the right side

$$-kA \left(\frac{\partial T}{\partial x} \right)_{x+ds} = -A \left[k \frac{\partial T}{\partial x} + \frac{\partial}{\partial x} \left(k \frac{\partial T}{\partial x} \right) dx \right] \qquad (6.4)$$

and so,

$$-kA \frac{\partial T}{\partial x} + \dot{q}Adx = \rho c A \frac{\partial T}{\partial \tau} dx - A \left[k \frac{\partial T}{\partial x} + \frac{\partial}{\partial x} \left(k \frac{\partial T}{\partial x} \right) dx \right] \qquad (6.5)$$

or

$$\left[\frac{\partial}{\partial x}\left(k\frac{\partial T}{\partial x}\right)+\dot{q}\right]=\rho c\,\frac{\partial T}{\partial \tau} \tag{6.6}$$

where $\dot{q}$ is the heat generated per unit volume, c the specific heat of the material, and ρ the density.

Equation (6.6) is the one-dimensional heat equation, which can be generalized to a three-dimensional scale

$$\frac{\partial}{\partial x}\left(k\frac{\partial T}{\partial x}\right)+\frac{\partial}{\partial y}\left(k\frac{\partial T}{\partial y}\right)+\frac{\partial}{\partial z}\left(k\frac{\partial T}{\partial z}\right)+q=\rho c\,\frac{\partial T}{\partial \tau} \tag{6.7}$$

If the thermal conductivity is constant, the general equation for heat becomes

$$\left(\frac{\partial^2 T}{\partial x^2}\right)+\left(\frac{\partial^2 T}{\partial y^2}\right)+\left(\frac{\partial^2 T}{\partial z^2}\right)+\frac{q}{k}=\frac{1}{\alpha}\frac{\partial T}{\partial \tau} \tag{6.8}$$

where the variable $\alpha = k/(\rho c)$ is called thermal diffusivity of the material and the denominator ρc, is its thermal capacity. A low thermal capacity represents reduced energy storage for an increase of the material temperature, and of the transfer to the exterior. The α value increases with thermal conductivity and decreases with increasing the thermal capacity.

6.1.2 Heat Conduction in the Soil

Soil surface temperatures vary over space and time. An ideal surface has a uniform temperature which varies only over time in response to transient energy fluxes. At a given point, the surface temperature is a function of the energy budget, dependent on the radiative balance, atmospheric exchange processes near the surface, type of canopy, and soil top layer thermal properties (Arya 1988).

Other factors that determine soil temperature are the latitude, time of the year, and time of day. Figures 6.1 and 6.2 show seasonal and daily vertical variability of soil temperature.

A difference between the surface temperature and air temperature adjacent to the surface can be established. The latter is measured by automatic weather stations at heights of about 1–2 m.

The study of vertical heat flow in soil under stationary or transient conditions, when $\partial T/\partial t \neq 0$, is an application of heat conduction principles in Eqs. (6.6) and (6.8). These equations highlight the need to measure thermal diffusivity. Soil thermal diffusivity measures its ability to promote the diffusion of thermal effects controlling the velocity at which heatwaves move vertically and the depth at which the effect of surface temperature variations is felt.

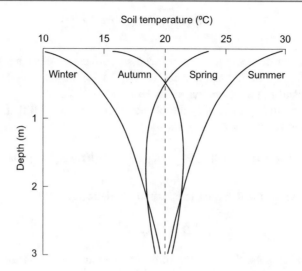

Fig. 6.1 Diagram of seasonal variation of soil temperature (after Hillel 1982)

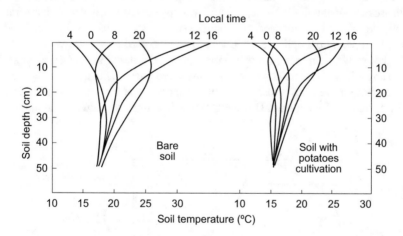

Fig. 6.2 Representative profiles of soil temperature variation with depth. The top values show the hours for different temperature variation curves (after Monteith and Unsworth 1991)

In cultivated soil, compaction, texture, composition, and water content vary vertically, making it thereby difficult to accurately measure the apparent density, conductivity, and thermal capacity that influence thermal diffusivity. A simple way to do so involves considering the surface temperature as a sinusoidal function of time, reflecting a harmonic change of temperature (Chap. 3 and Annex 1) insofar that the characterization of the thermal wave propagation is (Arya 1988)

$$T_{s.t} = \overline{T} + A_s \sin((2\pi/P)(t - t_m)) \tag{6.9}$$

where $\overline{T}$ is the surface temperature or mean subsurface temperature, A_s and P the amplitude and the period (e.g., daily or yearly) of the surface thermal wave, respectively, and t_m the instant when $T_{sT} = \overline{T}$.

The solution to Eq. (6.9) with the boundary conditions such that when z = 0, $T = T_{s(t)}$ and z → ∞, $T \to \overline{T}$ is given by

$$T(z, T) = \overline{T} + A_s exp(-z/D)\sin[(2\pi/P)(t - t_m) - z/D] \tag{6.10}$$

where D known as the damping temperature, is defined as

$$D = (P\alpha/\pi)^{1/2} \tag{6.11}$$

Figure 6.3 is representative of the vertical temperature variation and the first and second derivatives of the soil temperature, with a clear vertical variation of temperature in the top layer with a thickness of an order of a few mm.

The duration of the thermal wave in soil remains constant, while its amplitude decreases exponentially with depth ($A = A_s \exp(-z/D)$); when z = D, temperature amplitude of soil is reduced to about 37% of its surface value, and when z = 3D temperature amplitude is reduced to about 5% of its surface value. The soil temperature lag increases in direct proportion to the depth and when z = πD the phase angle is π and there is a reversal of the wave phase. That is, when the surface soil temperature reaches a maximum, the temperature at z = πD is at a minimum, and vice versa (Monteith and Unsworth 1991). The time lag between the maximum

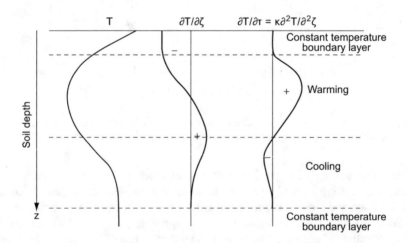

Fig. 6.3 Representative diagram of vertical soil temperature variation and the respective first and second derivatives (after Monteith and Unsworth 1991)

and minimum temperature is proportional to the depth z, of the order of $(zP/2\pi D)$. Equation (6.10) indicates that the soil maximum temperature corresponds to the phase angle $\pi/2$ and the minimum to $-\pi/2$. These results are valid for the propagation of maximum and minimum temperature waves through a homogeneous medium, only if the thermal diffusivity of the substrate layer remains constant throughout the entire period and that the surface temperature variation is sinusoidal. The thickness of the damping layer for an annual thermal wave is of the order of the product of $365^{1/2}$ with the corresponding D. For example, dry sandy soil with depth D is about 0.082 m for daytime and 1.57 m for the annual wave.

The theoretical heat flux is given e.g. by Arya (1988)

$$H_G = -k\left(\frac{\partial T}{\partial z}\right)_{z=0} = \left(2\pi\frac{\rho c k}{P}\right)^{1/2} A_s \sin\left[\frac{2\pi}{P}(t - t_m) + \frac{\pi}{4}\right] \tag{6.12}$$

showing that the amplitude of the heat flow is proportional to the square root of the product of the heat capacity and thermal conductivity and inversely proportional to the square root of the period. From Eq. (6.12), it can be deduced that the time instant corresponding to the maximum surface temperature, lagged the corresponding to maximum energy flow by P/8, or by about 3 h during the daytime, and 1.5 months for the year. The real conditions in soil layers differ from theoretical ones in heat transfer equations above the soil. The daytime variation of soil surface temperature may deviate from the theoretical wave profile due to factors such as soil moisture, plant root systems, or factors relating to the water regime such as irrigation, precipitation, and evaporation.

6.1.3 Thermal Properties of Soils

The thermal properties of the soil relevant to heat transfer through a particular medium, and their effect on the temperature distribution are the density, specific heat, heat capacity and thermal conductivity. Most soils consist of particles of varying dimensions and materials with a high degree of porosity. This can be filled by air or water so that the thermal properties vary depending on these factors. Table 6.2 shows some thermal properties of constituent materials of the soil. The determination of soil heat flux according to the Fourier Law (Eq. 6.2) is not practical due to the high atmospheric temperature vertical gradient in the top thin layer of the soil. The ground heat flux can, however, be estimated at the surface based on methodologies relying on measurements of soil heat flux with flux plates (Foken 2017).

The apparent soil density ρ' is given by

$$\rho' = \rho_s x_s + \rho_l x_g + \rho_g x_g \tag{6.13}$$

Table 6.2 Thermal properties of some soil materials (adapt. Campbell and Norman 1998)

Material	Density (kgm^{-3})	Specific heat ($Jg^{-1}K^{-1}$)	Thermal conductivity ($Wm^{-1}K^{-1}$)	Heat capacity ($MJm^{-3}K^{-1}$)
Soil minerals (including clay)	2.85	0.87	2.92	2.31
Quartz	2.66	0.80	8.80	2.13
Organic material	1.30	1.92	0.25	2.50
Water	1.00	4.18	0.56+0.0018 T	4.18
Ice	0.92	2.1+00073 T	2.22–0.0011 T	1.93+0.0067 T
Air (atmos. pres.)	(1.29– 0.0041 T) $\times 10^{-3}$	1.01	0.024+0.00007 T	(1.3–0.041 T) 10^{-3}

where x is the volume fraction occupied by each component and s, l, and g indexes refer to the solid, liquid, and gaseous soil components. The value of where x for sandy and clay soils ranges from 0.3 to 0.4, increases with organic matter content, and is about 0.8 in organic soils. Since the air density is low, the third term on the right side of Eq. (6.13) can be neglected. The equation for obtaining the apparent bulk density is no longer strictly linear for moisture saturated swelling soils. If ρ_s and x_s are constant, the bulk density increases linearly with the liquid fraction (Monteith and Unsworth 1991).

The specific heat of a material c is defined as the heat absorbed or released per unit mass during a 1 °C change in temperature. The product of mass density and specific heat is called the specific heat per unit volume ($Jm^{-3} K^{-1}$) or thermal capacity. For a given soil, without swelling, its value is the sum

$$\rho'c' = \rho_s c_s x_s + \rho_l c_l x_l + \rho_g c_g x_g \qquad (6.14)$$

Thermal properties of air and water are temperature dependent as shown in Table 6.2. Air has a lower heat capacity and thermal conductivity of all-natural materials, including water which has a heat capacity (4.18 $MJm^{-3}K^{-1}$). The thermal diffusivity of the air is high due to its low density. The thermal capacity of soil increases almost linearly with moisture content (Oke 1992).

The main components of the soil, quartz, and clay have similar densities and specific heats. However, as quartz has higher thermal conductivity than clay, sandy soils have higher thermal diffusivities relative to clay.

The specific heat of organic matter is about twice that for quartz, and the density of organic matter is about half that for quartz density (Table 6.2). As the thermal conductivity of organic matter is low, soils with high organic content have low thermal diffusivity.

The volumetric specific heat of soils varies between 0.5 $MJm^{-3}K^{-1}$ and 3.5 $MJm^{-3}K^{-1}$, and both the heat capacity and thermal conductivity increase with moisture content. The heat capacity increases linearly with moisture content (Campbell and Norman 1988) and depends on soil type, being higher in organic soils, than for mineral soils. In mineral soils, heat capacity is higher in sandy soils compared to clay soils. The thermal diffusivity increases with up to 20% moisture and thereafter decreases (Oke 1992; Arya 1988).

In micrometeorology, soil and ground heat fluxes variations are not taken into account insofar that the large differences in soil physical properties in scales of 10^{-3}–10^{-2} m are often not considered (Foken 2017). Indeed, e.g., soil thermal conductivity varies over time and space, being dependent on the overall conductivity of the constituent particles, porosity, and moisture content which are determinant for short term conductivity variability (Oke 1992). Soil moisture increases the conductivity by coating the individual particles, thereby facilitating contact among the particles, and by replacing the air with water in porous soils. Water in the soil surface sublayers attenuates the daytime variation in the temperature regime due to increased surface evaporation, as well as the increased heat capacity and thermal conductivity, as previously mentioned. In moist soil without canopy cover, a large fraction of the radiation balance is used up in the initial evaporation, after which the thermal capacity of the remaining water also contributes to the reduction in soil heating due to incident radiation (Oke 1992; Arya 1988).

The on-site measurement of soil surface temperature is complicated by the high vertical and horizontal air temperature gradients close to the ground which are difficult to detect due to factors such as finite sensor size. Remote sensing is another methodology for measuring surface temperature when the surface emissivity is known. An example is a radiometer facing the ground, which records long-wavelength radiation flux emitted from the surface, using the Stefan–Boltzmann equation defined in Eq. (6.60) (Arya 1988).

Biological processes are dependent on soil temperature, examples of which are seed germination, and the development of root systems influencing plant growth. In practice, it is difficult to study fauna and flora in undisturbed soil or monitor the development of an undisturbed root system and then quantify its physiological responses to temperature gradients (Monteith and Unsworth 1991). There are, however, empirical ways to improve soil thermal regime, including the use of straw, peat or biochar acting as insulation and reducing heat losses in winter, or covering soils with dark polyethylene to increase surface heat.

During the day, the ground surfaces including plants and urban areas are heated by the incoming solar radiation. Vegetation canopy also attenuates the daytime amplitude of surface temperatures. Some of the incident solar radiation is intercepted by the canopy, reducing the surface radiation intensity, thereby dampening the soil temperature regime. At night, surface radiative cooling is also reduced by downward long-wavelength radiation emitted by vegetation and the surface is also cooler than the air and the deeper soil layers (Arya 1998; Foken 2017).

The biosphere thermal capacity above the soil is relatively low so that thermal storage of solar radiation occurs primarily in soils, rocks, and aqueous substrates. For example, the thermal storage rate in a 1 m layer of humid soil is about 1.7 times higher than for forest canopy 20 m in height and with 500 m^3 ha^{-1} biomass content (Lee 1978).

6.2   Convection

6.2.1   General Principles

Convection involves heat transfer caused by a moving fluid in contact with a warm surface, which then warms further, becomes less dense, and rises by buoyancy. Such a process involving the movement of different density fluids is referred to as free or natural convection. Heat transfer by forced convection occurs when the fluid movement is mechanical, e.g., under atmospheric wind or mechanical ventilation. This happens, for example, when a heated metal plate cools more rapidly when placed in front of a fan than when exposed to still air. In both cases, heat transmission between the solid surface and the fluid takes place by instantaneous contact between the surface molecules and those in the adjacent fluid. If convective heat through a fluid causes a change in temperature, sensible heat is transferred. Sensible heat is transported, for example, from a hot to a cooler surface, by turbulent eddies and released into the surrounding atmosphere after mixing. Latent heat is transferred if, in the absence of a temperature change, convection causes a change in the state of a substance (e.g., liquid to vapor), Latent heat can be released into the air from humid air mass which condenses with the formation of clouds.

Heat convection in environmental systems occurs with the displacement of the air in the active surface close to the ground. This can be soil or another solid surface, e.g., a layer of vegetation where solar radiation is absorbed. The active surface is the atmospheric layer where the main radiation exchanges occur. It is usually warmer than the ambient air during the daytime with positive radiation balance and colder than the air during the night-time with negative radiation balance.

It is well-known that fluids provide resistance to flow over a flat surface due to drag resistance to movement between fluid layers. This resistance is stronger in lubricating oils than in water. Viscosity (Chap. 2 and Annex II) is used to quantify flow resistance between the various layers and between the fluid and contact surface. Although less apparent in gases, this property is common to both liquids and gases. Viscosity is the motion quantity that corresponds to flow velocity (Holman 1983). In laminar flow, fluid molecules move from one layer to another, transporting motion corresponding to the velocity flow. Fluid motions are transported from high to lower velocity zones, generating shear stress in the flow direction.

The dynamic or absolute viscosity μ, in Nsm^{-2} units, is defined as the proportionality constant between the vertical gradient of horizontal velocity and the induced shear stress. The kinematic viscosity, v, is defined as

$$v = \mu/\rho \qquad\qquad (6.15)$$

with m^2/s units, where ρ is the fluid density.

When a fluid flows over on a flat surface at temperature T_f, and velocity v_f, the particles in contact with the surface remain fixed due to viscosity and retard the movement of adjacent fluid layers located at higher levels (Chap. 2 and Annex II).

The region where the viscosity is felt is called the flow boundary layer (Figs. A2-7 and A2-8, Annex II). The thickness of the boundary layer is defined by the vertical ordinate which corresponds to particle velocity of about 0.99 of the flow velocities. Under the described conditions, the flow becomes irregular and disordered at a distance, D_t, from the point of contact with the flat surface. The flow evolves from an ordered laminar to a turbulent state in which fluid particles move randomly. A common dimensionless parameter that quantifies the laminar or turbulent of flow is called the Reynolds number, Re, given by the following expression:

$$Re = \frac{\rho U_e D_{hr}}{\mu} \tag{6.16}$$

where U_e is the flow velocity outside the boundary layer fluid and h_r the length of a flat plate exposed to flow, e.g., in boundary layers. Re expresses a ratio between inertial and viscous forces allowing to quantify the laminar or turbulent status of the flow.

The Reynolds number is the main parameter used to evaluate the transition between laminar and turbulent flow regimes. This transition occurs for Reynolds numbers between 10^3 and 10^6, depending on the surface roughness and homogeneity of the original flow, with a mean value of 5×10^5 (Holmes 1983).

Fluid film in contact with the surface is an integral part of the laminar sublayer where heat transfer takes place mainly via conduction. If the temperatures of the plate T_p and the fluid T_f are different, for example, if $T_p > T_f$, heat will be transmitted through conduction by molecular diffusion to fluid particles in contact with the plate. There will also be heat convective transfer among fluid particles moving above the laminar sublayer. Convection occurs through turbulent diffusion and mass exchange (Chap. 2), according to flow-gradient principles, resulting in the formation of a turbulent thermal boundary layer. The temperature of the particles in this boundary layer lies between that of the surface and the external fluid. Under turbulent flow, increasing the vertical component of the fluid velocity will increase the heat transfer rate in this layer. The thickness of the thermal boundary layer is defined as $(T_p - T)/(T_l - T_f) = 0.99$ where T is the air temperature at the top of the thermal boundary layer.

As mentioned in Chaps. 2 and 3, the turbulent flow region is not made up of distinct fluid layers because turbulence concerns macroscopic volumes of fluid masses, transporting energy and momentum, as opposed to microscopic transport based on individual molecules. This heat turbulent flow is dependent on factors such as the difference in temperatures, fluid velocity, and surface roughness.

It can be inferred that in a flat plate, the thickness of the laminar boundary layer, δ_l, is related to the distance to the initial flow point, D_l (Holman 1983; Mimoso 1987)

$$\delta_l \approx 5 D_l Re^{-1/2} \tag{6.17}$$

On the other hand, the thickness of the thermal boundary layer in laminar flow δ_{tl}, is related to the distance to the initial flow point, D_{tl}, according to the expression (e.g. Özisik 1990)

$$\delta_{tl} \approx 5.5 D_{tl} Pr^{-1/3} Re^{-1/2} \tag{6.18}$$

where Pr is the Prandtl number defined as

$$Pr = \frac{v}{\alpha} = \frac{\mu/\rho}{k/\rho\, c_p} = \frac{c_p \mu}{k} \tag{6.19}$$

which is the ratio between the molecular diffusivity of linear momentum and heat molecular diffusivity. The Prandtl number is the ratio between the viscous forces that retard flow, giving rise to the flow boundary layer, and the thermal conductivity, representing thermal diffusivity that gives rise to the thermal boundary layer. Thus, the Prandtl number can be regarded as the ratio between the thicknesses of the two boundary layers. Equations (6.17) and (6.18) show that in laminar flow, both layers grow proportionally to $D^{1/2}$ and, for a given fluid at a given point, the thickness of the thermal boundary layer will be greater (or smaller) than the thickness of the boundary layer, depending on whether the Prandtl number is smaller (or larger) than the unity.

On a flat plate, assuming turbulent flow (Re between 10^4 and 10^7) shortly after contact between the fluid and the flat surface, the thickness of the turbulent boundary layer $\delta_{t,}$ in analogy with the thickness of the thermal boundary layer is given by (Holman 1983; Mimoso 1987)

$$\delta_t \approx \delta = 0.38\, D_t Re^{-1/5} \tag{6.20}$$

showing that boundary layer thickness increases in proportion to $x^{4/5}$, which is faster than the boundary laminar layers (Eqs. 6.17 and 6.18). Heat transfer in the thermal laminar boundary layer is slower than in the turbulent boundary layer, because as aforementioned, turbulent eddies promote greater homogeneity of the air between the warm and cooler zones.

The transfer rate of sensible heat from a flat surface to a fluid dq/dt, is given by (Lee 1978; Gates 1980)

$$\frac{dq}{dt} = h_c A \Delta T \tag{6.21}$$

where A is the surface area, ΔT the temperature difference between the surface and the fluid, and h_c the surface heat transfer coefficient ($Wm^{-2}K^{-1}$). In the fluid layer adjacent to the surface (e.g., a vegetable leaf) under laminar flow heat is transferred by conduction

$$\frac{dq}{dt} = -kA\frac{dT}{dy}\bigg|_{y=0} \tag{6.22}$$

where k is t fluid thermal conductivity. An estimate of h is based on assuming that the heat flux emerging from the laminar boundary sublayer from thermal conduction is equal to the convective flow of the thermal boundary layer to or from the exterior

$$\frac{dq}{dt} = -kA\frac{dT}{dy}\bigg|_{y=0} = h_c A \Delta T \tag{6.23}$$

Applying Eq. (6.21) to the thermal boundary layer with depth, δ_T:

$$H = \frac{1}{A}\frac{dq}{dt} \approx k\frac{T_s - T_a}{\delta_T} = h_c(T_s - T_a) \tag{6.24}$$

where H is the sensible heat transfer per unit area. Thus

$$h_c = \frac{k}{\delta_T} \tag{6.25}$$

At 20 °C, the thermal conductivity for air, k, is approximately 25.7×10^{-3} W m^{-1}K^{-1}, and considering a thermal boundary layer 1 cm thick, the h_c value is 2.51 Wm^{-2}K^{-1}. In the biosphere, convective heat coefficient for common objects is of the order of 4.19 Wm^{-2}K^{-1}, for a thermal boundary layer thickness of about 6 mm (Gates 1980).

The convective heat transfer rate varies between adjacent points, so that the mean convective heat transfer coefficient $\bar{h}_c$, is

$$\bar{h}_c = \iint_A h_c \, dA \tag{6.26}$$

The heat transfer coefficients for convection can be obtained by dimensional analysis, or directly from laboratory measurements. For example, plant leaves can be considered as flat surfaces so that the heat transfer coefficients can be obtained using the same principles. Tree trunks and branches can be considered cylindrical for heat transfer analysis and many animal species can be considered spherical. However, surfaces of other objects in the biosphere may take on more complex forms and require extended analysis that goes beyond heat transfer basics.

For any object in a fluid, the heat transfer coefficient is a function of many different variables, such as size, shape, object orientation, viscosity, specific heat, laminar or turbulent nature of the flow, etc., all of which have a bearing on fluid properties. Many of the variables involved in the heat transfer process can be combined into dimensionless groups with functional relationships among them.

Heat transfer principles between different fluids are similar, even though different heat transfer rates vary between each other. There are fundamental relationships between various dimensionless groups that make possible generalizations of the heat transfer rate between an object and a fluid (Gates 1980). Dimensional analysis is the method used for deducing dimensionless groups. Equations relating to convective transfer coefficients, hx, can be as follows (e.g., Mimoso 1987; Fox and McDonald 1985):

$$h_x = f(j_1, j_2, j_3, \ldots, j_n) \qquad (6.27)$$

where the coefficients, j_i, are representative of independent factors.

The previous equation can be written as

$$\frac{h_x}{f(j_1, j_2, \ldots, j_n)} = 1 \qquad (6.28)$$

or:

$$F(J_1, J_2, \ldots, J_k) = 0 \qquad (6.29)$$

where $J_1, J_2, \ldots, J_k$ $(k < n)$ are sets of independent variables that form groups or dimensionless numbers. The dimensionless formulations for forced and natural convection are different, giving compact expressions, valid for any of the independent variables, provided that the quantitative similarity relationships between these variables are maintained throughout (e.g., Fox and McDonald 1985; Mimoso 1987).

6.2.2   Forced Convection

6.2.2.1   Dimensionless Parameters

Forced convection relates to heat transfer caused by fluid in motion induced by an external effect. The heat transfer coefficient by forced convection, h_c, is a function of a variable relating to the object, e.g., the characteristic dimension D on a flat plate (abscissa in Fig. 6.4) or cylinder, or dimensionless variables, linking several object variables, for example, the ratio between length and width of a flat surface. A set of five variables related to the fluid are (Gates 1980): flow velocity V_a, thermal conductivity k, density ρ, dynamic viscosity μ, and kinematic v, and the specific heat at constant pressure c_p. The three dimensionless groups derived from these variables are the Nusselt Nu, Reynolds Re, and Prandtl Pr, numbers.

Experimental data can be grouped into dimensionless groups, such as the Nusselt number, defined as

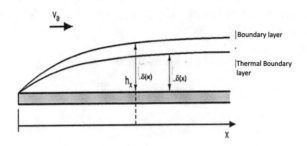

Fig. 6.4 Aerodynamic and thermal boundary layers on a flat plate under laminar flow (after Mimoso 1987)

$$Nu = \frac{h_c D}{k} \tag{6.30}$$

From Eq. (6.25), we can derive

$$Nu = \frac{D}{\delta_t} \tag{6.31}$$

where the Nusselt number becomes the ratio between the characteristic dimension of the object and the thickness of the thermal boundary layer. Equation (6.30) allows interpreting N_u as the ratio between convective heat transfer from a surface and the conductive flow through a fluid plane, per unit of the characteristic dimension.

Equation (6.24) can also be written as

$$H = h_c(T_s - T_a) = \frac{kNu}{D}(T_s - T_a) \tag{6.32}$$

From the knowledge of the Nusselt number (Eq. 6.30) in conjunction with Eq. (6.32), it is possible to determine the convection coefficient and the sensitive heat transfer, under conditions of forced convection.

Based on the heat transfer by convection theory, the following general relationship can be determined (Gates 1980):

$$\overline{Nu} = f(Re, Pr) \tag{6.33}$$

Physical properties of water and air, the most important fluids in environmental physics, allow to obtain their dimensionless flow parameters. For air at 20 °C, thermal conductivity k, is 25.7×10^{-3} Wm^{-1} K^{-1}, kinematic viscosity v is 15.3×10^{-5} m^{2}s^{-1}, dynamic viscosity μ, equals 18.2×10^{-6} Nsm^{-2}, the specific heat at constant pressure c_p, is 1.01×10^3 JKg^{-1}K^{-1}, gravity acceleration is 9.8 ms^{-2}, and the volumetric expansion coefficient, β, is 3.67×10^{-3} K^{-1}. Using these values, the Nusselt, Reynolds, and Prandtl numbers for air are given by the following expressions (Gates 1980):

$$Nu = 38.9\,h_c D \tag{6.34}$$

$$Re = 6.54 \times 10^4 UD \tag{6.35}$$

$$Pr = 0.72 \tag{6.36}$$

where U is the wind velocity ranging from 0.1 to 10 ms^{-1} and D is the characteristic dimension which for many living organisms varies between 10^{-3} to 1 m.

Similarly, for water at 20 °C, k is 59.9 × 10^{-2}Wm^{-1} K^{-1}, v is 10.05 × 10^{-2}m^{2}s^{-1}, μ is 10.05 × 10^{-2} Nsm^{-2}, c_p is 41.8 × 10^2 JKg^{-1}K^{-1}, gravity acceleration is 9.8 ms^{-2}, and the volumetric expansion coefficient β, is 41.9 × 10^{-4}K^{-1}. Nusselt, Reynolds, and Prandtl numbers are defined as follows:

$$Nu = 1.67 h_c D \tag{6.37}$$

$$Re = 99.5 UD \tag{6.38}$$

$$Pr = 701 \tag{6.39}$$

Unlike air, the Prandtl number for water varies considerably with temperature and cannot be considered as constant (Gates 1980).

6.2.2.2 Forced Convection in Laminar and Turbulent Regimes on Flat Plates

Heat transfer coefficients of forced convection for laminar fluid flow, on a flat surface, can be obtained from experimental and theoretical data. On flat surfaces, conditions of constant temperature with variable heat flux differ from those of constant heat flux with varying temperatures. An expression generally used for the average Nusselt number, on a flat surface and forced convection in laminar flow and constant temperature is (Holman 1983)

$$\overline{Nu} = \frac{0.34 Re^{1/2} Pr^{1/3}}{\left[1 + \left(\frac{0.468}{Pr}\right)^{2/3}\right]^{1/4}} \quad \text{for } RePr > 100 \tag{6.40}$$

Using Pr values for air and water at 20 °C, given by Eqs. (6.36) and (6.39), the simplified equations for the heat transfer coefficient to forced convection, in laminar flow at a constant temperature on a flat surface, for air and water at 20 °C are, respectively (Gates 1980)

$$\bar{h}_c = 3.93 \left(\frac{V}{D}\right)^{0.5} \tag{6.41}$$

and

$$\bar{h}_c = 357 \left(\frac{V}{D}\right)^{0.5} \tag{6.42}$$

The general expression for the average Nusselt number on a flat surface and forced convection in laminar flow and constant heat flux conditions, pertinent to biological processes, for $RePr > 100$, is Holman (1983)

$$\overline{Nu} = \frac{0.46Re^{1/2}Pr^{1/3}}{\left[1 + \left(\frac{0.02}{Pr}\right)^{2/3}\right]^{1/4}} \tag{6.43}$$

Constant heat flux predominates in situations in which weak heat conductors, e.g., vegetable leaf surfaces are exposed to a uniform radiation flux. Other strategies for estimating the Nusselt number are given in references such as Monteith and Unsworth (1991). Simplified equations, equivalent to Eqs. (6.41) and (6.42), for heat transfer coefficient under conditions of laminar flux and constant heat flux on a flat surface, for air and water at 20 °C, respectively, are

$$\bar{h}_c = 5.37 \left(\frac{V}{D}\right)^{0.5} \tag{6.44}$$

and

$$h_c = 488 \left(\frac{V}{D}\right)^{0.5} \tag{6.45}$$

In general, the airflow is laminar on a flat plate near the contact end with the flat plate, becoming turbulent downstream in the plate. Incident airflow, under conditions of air mixing, over vegetation bodies with highly irregular geometries such as leaves and branches, will be turbulent. If the mean flow and the obstacles are flat and the air velocity is low, then the flow will remain laminar well after contact with the surfaces (Gates 1980).

With small leaves and calm air, Reynolds number is, as aforementioned, frequently used to measure the transition between laminar and turbulent flows. Experimental data on smooth flat plates, cylinders, spheres, etc., for critical transition, give Re of 5×10^5. Literature values for the Re vary between 10^3 and 10^6.

Using the definition for Reynolds number and a transition Re of 5×10^5, provides a distance d_e for the initial contact zone of the flat plate

$$d_e = \frac{5 \times 10^5 \mu}{\rho V} \qquad (6.46)$$

Equation (6.46) shows that on flat surfaces in natural environments, the turbulent flow is the rule and laminar flow only occurs in a small initial area.

The flow will be laminar over plant leaves at low wind velocities and low Reynolds numbers, whereas turbulent flow predominates at higher wind speeds. In general, the laminar flow will prevail on small plant leaves. On the other hand, the flow will be turbulent on leaves longer than 5 cm with wind velocities >3 ms^{-1}.

The approximate expression for the Nusselt number on a flat plate under turbulent flow conditions is (Mimoso 1987)

$$\overline{Nu} = 0.029 \, Pr^{1/3} Re^{4/5} \qquad (6.47)$$

assuming $0.6 < Pr < 60$, and that the flow becomes turbulent at a distance d_e, about 0.05 of the flat surface lengths. Equation (6.47) can be simplified (Gates 1980)

$$\overline{Nu} = 0.032 \, Re^{4/5} \qquad (6.48)$$

The equations for the heat transfer coefficient in a turbulent regime, for air and water at 20 °C, are respectively

$$\overline{h_c} = 5.85 \, V^{0.8} D^{-0.2} \qquad (6.49)$$

and

$$\overline{h_c} = 76 \, V^{0.08} D^{-0.2} \qquad (6.50)$$

In general, at wind velocities below 1 ms^{-1}, the laminar convective transfer coefficient is greater than the corresponding turbulent one, except for leaves and larger flat plates (Gates 1980).

6.2.2.3 Forced Convection in Cylinders and Spheres

Many objects and organisms studied in environmental physics are either cylindrical or spherical in shape, as is the case with trunks, branches, or animal bodies. Fluid flow in spherical and cylindrical objects is more complex than on flat surfaces, due to wake formation because of boundary layer separation in the area opposite the contact point. Flow on a cylindrical object or spherical object creates a stagnation point (Annex II) at the most exposed flow point, forming a thin laminar boundary layer and an adverse pressure gradient as well. This thin layer contours the circular

Table 6.3 Nusselt numbers for air under forced convection (adapt. from Lee 1978)

Surface/Re range	Nu
Plate	
(a) $Re < 2(10^4)$	$0.6\ Re^{0.5}$
(b) $Re > 2(10^4)$	$0.032\ Re^{0.8}$
Cylinder	
(a) $10^{-1} < Re < 10^3$	$0.32 + 0.51\ Re^{0.52}$
(b) $10^3 < Re < 5(10^4)$	$0.24\ Re^{0.60}$
Sphere	
(a) $Re < 300$	$2 + 0.54\ Re^{0.5}$
(b) $50 < Re < 1.5(10^5)$	$0.34\ Re^{0.36}$

cross-section to about 90° and the adverse pressure gradient is formed outside the boundary layer where viscosity is negligible.

The pressure gradient is established by analogy with the flow in the absence of viscosity, in which the velocity corresponds to pressure decreases and flow separation, leading to a wake. The complexity of convection lies in the fact that wake turbulence promotes recirculation of the fluid and is not very effective in transferring heat.

In these objects, for $Pr = 0.71$ (constant for air), the expression for the Nusselt number is Monteith and Unsworth (1991)

$$Nu = ARe^n \qquad (6.51)$$

where A and n are constants, dependent on the Reynolds number and object geometry. The characteristic dimension for spheres and cylinders is the diameter, although, for irregular animal bodies, it can be more appropriate to apply the volume cubic root (Monteith and Unsworth 1991). The sensible heat transfer rate from a sphere is always higher than for a cylinder with the same diameter, so that a factor of 1.5–1.8 is used (Gates 1980). Table 6.3 gives values of the constants A and n for different ranges of Re with different geometries.

After estimating the heat transfer coefficient by convection h_c and the Nusselt number (Eqs. 6.30 and 6.51), the convective flow of sensible heat can be calculated with Eq. (6.32).

6.2.3 Free Convection

Free convection occurs when an object is warmer or cooler than the surrounding fluid. In free convection, heat transfer depends on the fluid circulation caused by differences in density associated with temperature gradients and the viscosity. In the field areas which are large enough for convection to develop above them, e.g., in horizontal scales >200 m, free convection can be detected through aircraft measurements through the depth of adjacent atmospheric boundary layer (Foken 2017).

On a vertical plate at a temperature T_s, higher than temperature T_a, of the air in contact, heating occurs by conduction of the air layer in contact with the plate, so that it becomes less dense, rises, and heats the upper layers. This process is repeated as long as the plate temperature is higher than that of the air. The velocity profile in this boundary layer differs from the velocity profile in the forced convection boundary layer (Holman 1983). At the wall surface, the velocity is zero, increases to a maximum, and then decreases to zero at the boundary layer frontier. Also, in this case, the boundary layer is initially laminar, evolving into a disordered turbulent regime and at a certain distance from the plate contact surface.

Under free convection, the flow and thermal boundary layers are of equivalent thicknesses, and therefore, the Reynolds number cannot be used as a separation criterion, unlike forced convection. That is, the air surrounding the boundary layer is stationary and so there is no characteristic velocity for comparison (Holman 1983) (Fig. 6.5).

In free convection, a range of dimensionless variables is required representing the ability of a volume of warm or cold air to rise (or descend) in the surrounding environment. The set of variables involves the difference in temperature between the surface and the fluid, the thermal expansion coefficient β, and the gravity acceleration g. Dimensional analysis now provides the Grashof number Gr, necessary for process characterization

$$Gr = \frac{g\beta \rho^2 \Delta T D^3}{\mu^2} \tag{6.52}$$

where D is the characteristic dimension of the surface.

The Grashof number approximates the ratio of ascending or descendant forces due to changes in density and viscous forces. In vertical or horizontal flat plates, if the Grashof number is above the 2×10^7 threshold, forces causing the density differential will predominate and flow may become turbulent (Gates 1980). If the Grashof number is less than the threshold, viscous forces will induce laminar flow.

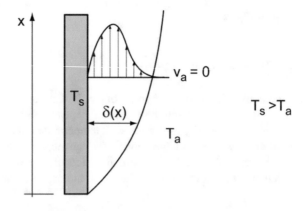

Fig. 6.5 Schematic of flow under natural convection conditions on a vertical flat plate (after Mimoso 1987)

The results of the dimensional analysis for free convection give a general expression of the type

$$\overline{Nu} = f(Gr, Pr) \approx BGr^m \tag{6.53}$$

where B and m are empirical constants (Monteith and Unsworth 1991). Table 6.4 provides values for B and m for air under several ranges of Gr for different geometries. Equations (6.30) and (6.32) that were used to obtain coefficients for convection and transfer of sensible heat can now be applied to free convection.

Considering their physical properties at 20 °C, the Grashof number can be simplified to obtain the following equations for air and water, respectively (Gates 1980):

$$Gr = 15.4 \times 10^7 \Delta TD^3 \tag{6.54}$$

$$Gr = 4.07 \Delta TD^3 \tag{6.55}$$

The separation between free and forced convection can be determined from the ratio G_r/Re^2 (Lee 1978; Gates 1980) between flow buoyancy and inertial forces. Available data indicate that natural convection predominates when G_r exceeds 16 Re^2 and that forced convection predominates when G_r is $<0.1\ Re^2$ and that forced convection predominates when G_r is $<0.1\ Re^2$. As a rule of thumb, air velocity equal to or $>0.1\ \mathrm{ms}^{-1}$, favors forced convection whereas lower air velocity conditions tend to promote free convection (Gates 1980).

Table 6.4 Nusselt numbers for forced convection in air (after Lee 1978)

Surface/Re range	Nu
Laminar flow ($Gr^{0.25} = 3.54d^{0.75}\Delta T^{0.25}$)	
Vertical plate ($Gr < 10^5$)	
Upper surface	$0.50\ Gr^{0.25}$
Lower surface	$0.23\ Gr^{0.25}$
Cylinder ($10^4 < Gr < 10^9$)	
Horizontal	$0.48\ Gr^{0.25}$
Vertical	$0.58\ Gr^{0.25}$
Sphere ($Gr < 10^9$)	$2 + Gr^{0.25}$
Turbulent flow ($Gr^{0.33} = 5.4d^{0.75}\Delta T^{0.33}$)	
Horizontal plate ($Gr > 10^5$)	$0.13\ Gr^{0.33}$
Horizontal cylinder ($Gr > 10^9$)	$0.09\ Gr^{0.33}$
Vertical plate and cylinder ($10^9 < Gr < 10^{12}$)	$0.11\ Gr^{0.33}$

6.3   Radiation

6.3.1   General Principles

All matter emits energy as radiation in the form of moving electromagnetic waves. In contrast with convection and conduction, which require a natural body, radiation may also be transmitted in a vacuum. Radiation energy is transported by photons, with properties like particles and waves (Oke 1992).

Photons move at the speed of light which in a vacuum is 3×10^8 ms^{-1}. The ability to emit and absorb radiation is intrinsic to solids, liquids, and gases, associated with changes in the electromagnetic energy state of atoms and molecules. These changes represent possible changes in the energy state of their electrons, which are converted into radiation emission at specific or ranges of frequencies. Molecular changes of the electromagnetic spectrum corresponding to vibration and rotation of atoms resulting in different energy states, so that radiation can be emitted or absorbed over a wide frequency range forming spectral bands (Monteith and Unsworth 1991).

Photons are discrete quantum quantities proportional to radiation frequency v, or the inverse of the wavelength $1/\lambda$. The proportionality constant h, known as Planck's constant, is $6.626 \times$ s 10^{-34} Js is a universal property of matter. The frequency and wavelength of the radiation vary with the state and properties of the emitting body (solid, liquid, or gaseous), related by the equation

$$c = \lambda v \tag{6.56}$$

where c is the velocity of light (3×10^8 ms^{-1}, in a vacuum). A quantum of energy $Q = h\ v$ (about 4.2×10^{-19} J at $\lambda = 0.5$ μm) is very small, so that the radiative energy is expressed in terms of multiples of Avogadro's number (6×10^{23}), denoted Einstein or quantum mole.

Surfaces emit radiation proportional to the fourth power of their absolute temperature (Eq. 6.59). The amount and type of radiant energy emitted from a surface per unit time and per unit surface area depending on the nature of the surface and its temperature. At low temperatures, surfaces emit radiation in the infrared (IR) range, and a surface at a higher temperature, such as an incandescent filament of a tungsten lamp, emits radiation at lower wavelengths such as in the visible spectrum (0.38–0.76 μm), as well as the near ultraviolet (UV).

Radiative flux emitted by a body is determined by its surface properties. The radiation is emitted over a wide range of frequencies and wavelengths and the magnitude of the flow, for a given wavelength is determined by the relative efficiency or ε, the emissivity of the radiant surface (Lee 1978). Photons are emitted or absorbed due to discrete energy transitions in the middle of emission or absorption, and each transition produces photons at a discrete wavelength (Campbell and Norman 1998). If there are an infinite number of transitions over the entire electromagnetic spectrum, then a black body will be a perfect emitting or absorbent of all the radiation at a given temperature. A perfect black body absorbs all incident radiation at all wavelengths. The blackbody concept deals with theoretically ideal radiative characteristics for comparison with the radiative characteristics of real bodies. About 99% of black body emissions at surface temperatures of the earth and sun of 300 and 6000 K, corresponds to the wavelength ranges of 0.2–4 μm and 4–100 μm, denoted short and long-wavelength radiation, respectively.

In an empty space or homogeneous medium, radiation propagates linearly and energy from a given point reaches an area A, so that the flux per unit area E_i, decreases with the inverse square of distance d, from the source (principle of the inverse square). The product $E_i A_i$ is constant and $E_1 d_2^2 = E_2 / d_1^2$. The concept of point radiation is a theoretical abstraction when the source is small relative to the distance traveled as is the case of the sun in relation to the earth's distance.

The radiant energy incident on a surface is either reflected, absorbed, or transmitted. At a given wavelength, the reflectivity $\rho(\lambda)$, absorptivity, $\alpha(\lambda)$, and transmissivity, $\tau(\lambda)$, are the ratios of the radiation reflected, absorbed, and transmitted and the incident radiation, respectively. At a given wavelength, or on the average across the spectrum, $\rho(\lambda) + \alpha(\lambda) + \tau(\lambda) = 1$ and for a blackbody at all wavelengths, $\rho(\lambda) = \tau(\lambda) = 0$ and $\alpha(\lambda) = 1$.

When transmissivity is zero, a body is referred to as opaque. For many natural surfaces such as soil, water, and vegetation, it can be assumed that the emissivity is one for wavelengths between 4 and 100 μm, at typical earth surface temperatures. Snow is a black body as it emits radiation within this range regardless of the range of the reflected solar radiation, due to its white color.

An object in a sealed vacuum container has a thermal equilibrium so that radiation absorbed at a given wavelength $\alpha(\lambda)$, is equal to the energy emitted at this wavelength $\varepsilon(\lambda)$ (e.g., Monteith and Unsworth 1991).

Planck's Law describes the distribution of radiant energy emitted by a black body per wavelength unit E_λ, as a function of the surface temperature and wavelength (Lee 1978)

$$E_\lambda = \frac{3.74(10^8)\lambda^{-5}}{(\exp(1.44/\lambda T)) - 1} \tag{6.57}$$

expressed in KWm^{-2} μm.

Wien's Law defines the maximum emission wavelength by a black body, as a function of its temperature, simplified as Lee (1978)

$$\lambda_{max} T = 2898 \, \mu mK \tag{6.58}$$

Accordingly, Wien's Law establishes the wavelength at which the energy emitted E_λ is at a maximum, λ_{max}, being inversely proportional to temperature. The λ_{max} value is about 0.48 μm, for solar radiation at a surface temperature of about 6000 K, and 9.7 μm for terrestrial radiation (at about 300 K) (Monteith and

Unsworth 1991). The ratio between $E_{\lambda max}$ values for solar and terrestrial radiation is about 3.2×10^6.

The Stefan–Boltzmann's Law indicates that the emissive power of a black body (Wm^{-2}) is proportional to the fourth power of its absolute temperature

$$E_b = \sigma T^4 \tag{6.59}$$

where σ is the Stefan–Boltzman's constant equal to 5.673×10^{-8} Wm^{-2} K^{-4}. A gray body does not emit radiation with total efficiency at all wavelengths, in which case Eq. (6.59) must include the emissivity coefficient ε, ranging between 0 and 1

$$E_g = \varepsilon \sigma T^4 \tag{6.60}$$

Kirchhoff's Law indicates that if a gray body is suspended in space, surrounded by a black body at a constant temperature, thermal equilibrium implies equality between the energy emitted by the gray body and the energy absorbed by the black body (Lee 1978)

$$\alpha E_b - \varepsilon E_b = 0 \tag{6.61}$$

where α (λ) and ε (λ) are the absorptivity and emissivity coefficients of the gray body at a given wavelength λ. It follows that:

$$E_b(\alpha - \varepsilon) = 0 \text{ or } \alpha = \varepsilon \tag{6.62}$$

Kirchhoff's Law means that the absorptivity and emissivity coefficients at a given temperature and wavelength are equal. A good absorber body will also be a good emitter. Kirchhoff's Law is only valid for bodies at similar temperatures and with absorptivity and emissivity coefficients of the same magnitude (Lee 1978).

In real bodies, the absorptivity of the surfaces varies depending on the wavelength of the incident radiation, the angle of incidence, and temperature. A comparison made between the absorptivity of two surfaces painted white and black as a function of the wavelength of the incident radiation (Mimoso 1987) shows that the absorptivity of the black body is always higher by about one. For the white painted surface, it varies from 0.1 in the visible range because of the reflection of radiation to 0.98 in the infrared range, as all the radiation emitted by a body at room temperature is absorbed (wavelength greater than 3 μm). The roughness and surface finish are also elements that influence absorptivity. The absorptivity of a metallic mirrored surface is practically zero and increases with surface roughness. The corresponding absorptivity for this oxidized surface is >0.9.

6.3.2 Spatial Relationships

Radiant energy rays emitted from a point source (Wm^{-2}) passing through a homogeneous medium are parallel over a circular section if the linear characteristic dimension of this section is small relative to the distance to the emitting source. This principle applies to the sun rays relative to the earth–sun length. Irradiance and emittance can be used to evaluate the radiative balance of surfaces. The irradiance J, of a surface, is the total power incident on the surface per unit area. The emittance of a surface is the radiant energy emitted per unit of time and per unit area. The reflectivity of a surface of ρ (λ) is defined as the ratio between the incident and reflected flux at the same wavelength. If reflectivity ρ, has a unit value, the body is a perfect reflector.

The reflected radiation depends on the surface properties of the bodies, which can be specular or diffuse. For specular reflective surfaces a radiative beam with an incidence angle ψ, relative to the reference, is reflected at the same angle ($-\psi$). Often natural surfaces act as diffuse reflectors, scattering fractions of the incident radiation in all directions. According to the Lambert cosine law, the scattered energy is independent of the angle of incidence, but the radiant flux reflected from a given surface is proportional to $\cos \psi$.

The reflection at the surface of a natural body depends on the surface structure and its electrical properties (Monteith and Unsworth 1991). In general, natural surfaces (e.g., water, leaves, or smooth surfaces) are diffuse reflectors, when the zenith angle (defined between the direction of the solar rays and the normal to the location) is <60° to 70°. Specular reflection predominates when the angles are higher, as specular reflectors absorb less radiation than diffuse reflective surfaces.

The interception of direct sunlight causes great seasonal, daily, and spatial variability in radiation regimes. An opaque body in the sun's path casts a shadow, changing the radiative climate over this area. The local topography, canopy density, as well as other natural bodies, introduce light and shade patterns which are surface energy sources and sinks, influencing biological processes (Lee 1978). The shadow area on a horizontal surface, A_h, multiplied by the horizontal flux density of direct radiation, S_b, equals the total flux intercepted by the body, and

$$S_{b1} = \frac{A_h}{A_b} S_b \tag{6.63}$$

where S_{b1} is the radiative density flux mean over the body area, A_b. The A_h/A_b ratio is the shape factor which can be determined geometrically for real forms, which are representative of natural irregular shapes. For example, tree needles, trunks, and many animals can be regarded as vertical or horizontal cylinders. Plant leaves are considered flat, whereas tree canopies can be spherical or conical surfaces.

For simplified calculation of the shape factors, consider two gray bodies with surface infinitesimal elements dA_1 and dA_2, having normal unit vectors $\bar{n}_1$ and $\bar{n}_2$, as shown in Fig. 6.6. The surfaces are diffuse reflecting radiation in all directions.

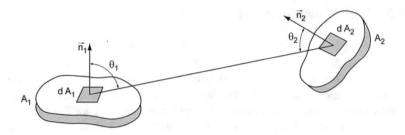

Fig. 6.6 Geometric relationships relating to radiation exchange between two infinitesimal surfaces dA_1 and dA_2

The dA_1 element emits radiation in all directions (emission angle, $|\bar{\theta}_1| < 90°$) and a part of the emitted radiation is intercepted by the dA_2, at an angle $|\bar{\theta}_2|$, with the vector $\bar{n}_2$. The radiation flux, $d_{q1\rightarrow2}$, from dA_1 to dA_2 is proportional to the apparent area of dA_1 seen from dA_2, and inversely proportional to the square of the distance separating the two infinitesimal elements

$$d_{q1\rightarrow2} = \frac{K_1' dA_1 \cos\theta_1 dA_2 \cos\theta_2}{r^2} = K_1' dA_1 \cos\theta_1 d\omega_{12} \qquad (6.64)$$

where K_1' is a proportionality constant and $d\omega_{12}$ the solid angle, expressed in steradians, defined as the ratio between the apparent area dA_2, viewed from dA_1, and the square of the distance separating the two elements. This angle delimits conical or pyramidal spaces, with the base of the vector $\bar{n}_1$ as the apex, and the area of dA_2 viewed from dA_1, as the base. As $dA_1\cos\theta_1$ represents the dA_1 projection in the line direction connecting dA_1 to dA_2, K_1' is defined as the amount of radiation intensity emitted by dA_1 in each direction (in this case $dA_1 \rightarrow dA_2$) per normal unit area, unit solid angle, and unit time (Holman 1983).

Figure 6.7a represents radiation flux, emitted from a point source and oriented according to a solid angle, $d\omega$; Fig. 6.7b is representative of the emitted radiation on a dS plane, oriented at a zenith angle ψ, with the vertical.

Concepts of emissive power, E, emitted radiation intensity K' and radiation flux q, are different but interconnected. Emissive power (expressed in Watts m^{-2}) is a scalar quantity representing the radiant energy emitted in all directions per unit area and per unit time. If the body is black, even at only one wavelength, the emissive power is maximal. Thus, the emissivity of a non-black body can be defined as the ratio of its emissive power and that of a black body, at a given wavelength.

The radiation intensity, as mentioned above, is the radiation that passes through a hypothetical plane, normal to the emission direction, and is delimited by a solid angle expressed in Wm^{-2}sr^{-1} units. It is directional and can be represented by a vector in the direction normal to the hypothetical plane. The radiation intensity is related to radiance, defined as the radiative flux emitted per unit solid angle, divided

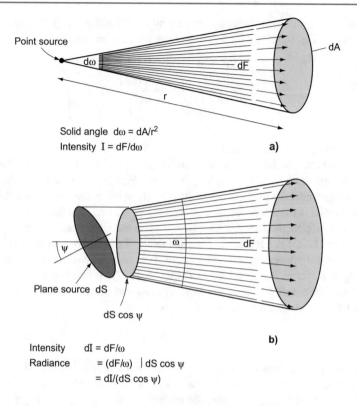

Fig. 6.7 a Representative diagram of radiation geometry emitted by a point source, and **b** representative diagram of radiation geometry emitted by a plane source (adapt. Monteith and Unsworth 2013)

by the projected area $dScos\psi$ (Fig. 6.7b). The radiation flux q, (expressed in W) is the power emitted by a body, intercepted by another per unit time. The radiative flux density is defined as the radiation flux per unit area (Wm^{-2}).

Both E and K' are functions of the characteristics and temperature of the emitting body, while the radiation flux is also dependent on the geometry of the receiving body, as well as the distance from the emitting surface (Mimoso 1987). If this distance increases, the flow decreases, Eq. (6.64), since the radiation is dispersed inversely proportional to the square of the distance, but E and K' remain constant.

For the calculation of the emissive power of a black body, E_n, over a hemisphere surrounding an element with infinitesimal area dA_i, the integration of Eq. (6.64) followed by some manipulation gives (Holman 1983)

$$E_n = \pi K'_n \qquad (6.65)$$

where K'_n is the radiation intensity emitted by dA_1. Taking Eq. (6.64), dividing by dA_1 applying Eq. (6.65) and integrating over the infinitesimal areas

$$\frac{q_{1 \to 2}}{A_1} = E_1 \left[\frac{1}{A_1} \int\limits_{A_1} \int\limits_{A_2} \frac{\cos \theta_1 \cos \theta_2}{\pi r^2} dA_1 dA_1 \right] \tag{6.66}$$

thus, obtaining the total radiation emitted from A_1 to A_2, per unit area of the emitting surface. The left side of Eq. (6.66) represents the radiative flux between A_1 and A_2 per unit area A_1. As the right side term of Eq. (6.66) to the left of the square brackets represents the total energy emitted by A_1 in every direction, the term inside the parenthesis is the shape factor, defined above, which is the fraction of radiative flux from A_1 intercepted by A_2. This term, F_{12}, is the shape factor (Holman 1983).

The shape factor is a function of the geometry of the bodies and can vary from 0 (when A_1 is not viewed by A_2) and 1 when A_1 is only viewed by A_2, for example, when A_1 is a sphere inside a box A_2 (Mimoso 1987). When F_{12} and F_{21}, are the shape factors representing emission from A_1 to A_2 and from A_2 to A_1, the reciprocity theorem can be obtained in its simplified form

$$A_1 F_{12} = A_2 F_{21} \tag{6.67}$$

This relationship was developed for surface blackbodies, but it is also valid for other surfaces, provided that the radiation is diffuse (Holman 1983). Equation (6.67) can also be generalized to systems with more than two gray bodies. Equation (6.66) is a complex way to calculate the various form factors, and in the literature, e.g., Holman (1983), Özisik (1990), Monteith and Unsworth (1991) simplified methodologies are indicated for calculating surface and solid shape factors representative of natural bodies.

Beer's Law, often used in environmental physics, refers to the attenuation of a radiation beam in a system where single-wavelength radiation is absorbed, without being dispersed when transmitted through a homogeneous medium. Beer's Law can be expressed as

$$\Phi(x) = \Phi(0) \exp(-kx) \tag{6.68}$$

where $\Phi(0)$ and $\Phi(x)$ are the incident radiation flow and the flow at distance x from the beginning of a given transmission medium, respectively, and k a proportionality constant or attenuation coefficient. Equation (6.68) can also be applied when k is a constant (homogeneous dispersion of molecules and particles that can absorb radiation), or in systems with a low concentration of scattering centers so that a quantum of energy is likely to be scattered only once.

The multiple scattering of radiation, such as that occurring within plant canopies, is more complex since the proportionality constant k may vary with the direction of the radiative beam so that the direction of the dispersion should be considered. In the simple case where the k constant is independent of direction, the Kubelka–Munk equations can be used. For example, when radiation is uniformly dispersed in all directions (isotropic scattering)

$$\rho' = \left(1 - \alpha^{0.5}\right)/\left(1 + \alpha^{0.5}\right) \tag{6.69}$$

where ρ' is the reflection coefficient and α, the absorption coefficient. Equation (6.69) excludes the situation where $\alpha = 1$, valid for Beer's Law (Monteith and Unsworth 1991).

6.3.3 Radiation Environment

6.3.3.1 Introduction

The ambient radiation applies to the combined solar, atmospheric, and terrestrial radiation. The sun is the primary source of virtually all the energy used and interchanged in the biosphere, and a key objective of environmental physics is to explore the mechanisms by which solar energy is dispersed and stored as thermal, chemical, and mechanical energy. The sun is a sphere of gaseous matter, with temperatures at the core of about 15×10^6 K. These temperatures decrease progressively so that at the surface it is about 5760 K. Solar radiation that reaches the earth's atmosphere is basically radiation emitted by a black body at the sun's surface temperature. The spectra of solar radiation and of radiation emitted by the earth-atmosphere systems fall within the short and long-wavelength ranges between 0.15 and 3 µm and 3 and 100 µm, respectively.

Irradiance on a surface outside the Earth's atmosphere, theoretically perpendicular to solar radiation and at an average Earth–Sun distance of 1.49×10^{11} m, is termed the solar constant, with an average of 1373 W m^{-2}.

The total power emitted by the sun per unit time E is given by the product of the solar constant and the area of a sphere with a radius of about the mean Sun–Earth distance, as follows

$$E = 4\pi r^2 \times 1373 = 3.88 \times 10^{26} \, \text{W} \tag{6.70}$$

For estimation of sun surface temperature, the Stefan–Boltzman's Law is used (Eq. 6.59)

$$\sigma T^4 4\pi r^2 = 3.88 \times 10^{26} \, \text{W} \tag{6.71}$$

where r is the sun's radius (6.69×10^8 m), giving for its surface a temperature of about 5800 K.

The net radiation in the atmospheric layers close to the ecosystems (e.g., constant flux layer) may be seen as the algebraic sum of long and short-wavelength components. High-temperature sources such as the sun, fires, and volcanoes emit radiation below 2.5 μm, as well as long-wavelength radiation. Solar radiation emission per unit area peaks at mid-spectrum, about 0.48 μm. Radiation referred to as thermal or high wavelength is emitted from bodies with surface temperatures below 600 K. At lower temperatures, these bodies emit negligible amounts of short-wavelength energy.

Much of the radiation emitted by the earth's surface is absorbed into specific ranges of wavelengths by atmospheric gases, particularly water vapor and carbon dioxide. According to Kirchhoff's Law, these gases have equivalent emission and absorption spectra. A small fraction of large wavelength radiation emitted by the earth-atmosphere system is lost to outer space, so this energy loss must be compensated by incident solar radiation (Monteith and Unsworth 1991).

The solar radiation spectrum can be divided into several ranges, relating to percentages of the solar constant: 1.2% of 0–300 nm, 7.8% of 300–400 nm (UV), 39.8% of 400–700 nm (vis./PAR), 38.8% of 700–1500 nm (near IR), and 12.2% of 1500 nm to ∞. The visible radiation ranges between 400 and 700 nm, corresponding to blue and red, respectively. Photosynthesis is stimulated by the radiation in the photosynthetically active radiation (PAR) range, corresponding to 21–46% of the total energy of the extraterrestrial solar spectrum. More than 70% of the solar radiation absorbed by the plant canopy is used in transpiration and convective exchanges with the surrounding air, which regulates the temperature of the various plant organs.

About 28% of the total solar energy used in photosynthesis is stored chemically in the form of organic compounds. The remaining, from UV to IR of about 750 nm is used in the regulation and control of growth and development of photomorphogenic processes (Ross 1975). Due to atmospheric attenuation, global or total solar radiation reaching the ground has two components: direct and diffuse radiation. The sun provides direct radiation, including a small fraction of scattered radiation that has not undergone any directional change. The direct radiation incident on the earth's surface is up to about 75% of the solar constant. The remaining 25% is diffuse radiation, including emissions from the sky and clouds via transmission and reflection of attenuated solar radiation. This attenuation is due to absorption and scattering, in similar proportions, by molecules and aerosol (Monteith and Unsworth 1991).

The radiative attenuation decreases direct solar radiation content and changes the spectral composition also. The radiative absorption and dispersion processes vary with wavelength, with absorption causing warming of the atmosphere and dispersion only changing the direction of solar rays. Table 6.5 shows the spectral distribution under temperate conditions, for average fractions of UV, PAR, and near IR radiation in terms of direct, diffuse (clear skies), and total radiation.

Table 6.5 Average levels of UV, PAR, and near IR radiation under clear sky conditions (adapt. from Ross 1975)

Wavelength ranges (μm)	UV (0.29–0.38)	PAR (0.38–0.71)	Near IR (0.71–4)
Direct radiation	0.02	0.42	0.56
Diffuse radiation	0.10	0.65	0.25
Total radiation	0.03	0.50	0.47

6.3.3.2 Solar Radiation Outside the Atmosphere

The incident radiation throughout the day and year, above the Earth's atmosphere, varies with the season, time, and latitude due to the earth's rotation and translation. Three-dimensional geometrical sun–earth relationships can be used to establish the expressions for solar declination, zenith angle, duration of the solar day as a function of time and latitude.

Spherical trigonometry gives expressions relating latitude ϕ, solar declination δ, defined as the angle between the sun rays and the equatorial plane; the zenith angle ψ, defined as the angle between the sun's rays and the line through the center of the Earth, vertical at a given point P; the solar height β, the angle between the local horizon at point P and the sun's rays, complementary to the zenith angle, the azimuth angle α, defined as the angle between the projections of the sun's rays in the horizontal plane and the true north and h, the solar hour angle (Gates 1980)

$$\cos(\psi) = \sin(\phi)\sin(\delta) + \cos(\phi)\cos(\delta)\cos(h) = \sin(\beta) \tag{6.72}$$

$$\sin(\alpha) = -\cos(\delta)\sin(h)/\sin(\psi) \tag{6.73}$$

As an hour is equivalent to a $15°$ rotation of the Earth, the angle h, is given by Oke (1992)

$$h = 15(12 - t) \tag{6.74}$$

where t is the apparent solar time location based on a 24 h period. The value of the apparent solar time location depends on the coordinated universal time, the longitudinal correction, and the time equation. This calculation is developed in Examples 2 and 10 of Chap. 7.

The angle of solar declination δ, between $23.5°$ and $-23.5°$ depends on the Julian day, t_j, given approximately by Campbell and Norman (1998)

$$\sin \delta = 0.39 \sin\left[278.97 + 0.985t_j + 1.92\sin(356.6 + 0.986t_j)\right] \tag{6.75}$$

Instant solar radiation, S_h, at a point outside the atmosphere with zenith angle ψ, is given by the following (Gates 1980)

$$S_h = S_o\left(\frac{\overline{d}}{d}\right)^2 (\sin \phi \sin \delta + \cos \phi \cos \delta \cos h) \tag{6.76}$$

where S_o is the solar constant, d the Sun–Earth distance at a given instant, and $\overline{d}$ the Sun–Earth mean annual distance. Spitters et al. (1986) have suggested the following equation:

$$S_h = S_o\left[1 + 0.033 \cos\left(360\, t_j/365\right)\right] \sin \beta \tag{6.77}$$

Integrating Eq. (6.77) for a period of one day gives the daily irradiance for a surface outside the atmosphere, S_{hd} $(MJm^{-2}d^{-1})$

$$S_{hd} = \left(\frac{84600}{\pi}\right) S_o \left(\frac{\overline{d}}{d}\right)^2 (h_s \sin\phi \sin\delta + \cos\phi \cos\delta \cos h_s) \qquad (6.78)$$

where h_s is expressed in radians.

The solar angle h_s, corresponding to half a day, is given by Gates (1980)

$$\cos h_s = -\text{tg}\,\phi\,\text{tg}\,\delta \qquad (6.79)$$

6.3.3.3 Solar Radiation on the Earth's Surface

The radiation incident on the earth's surface at any point depends on atmospheric conditions and can be analyzed using geometric expressions. When a solar ray beam is transmitted through the earth's atmosphere, it is attenuated and changes in quality and quantity through absorption and dispersion. This is due to photons contacting the medium molecules and particles in suspension. The atmosphere transmittance, or transmissivity τ, is defined as the fraction of the radiation incident at the top of the atmosphere, which reaches the ground vertically at a given point. Transmissivity depends on the concentration of gases, particles, and droplets (ozone, water vapor, dust clouds, smoke, etc.) that can reflect, disperse, or absorb solar radiation. In temperate regions with low sun elevation angles, diffuse radiation makes up to about 50% of the total radiation, whereas for higher solar elevation angles, the corresponding fraction is about 20% (Gates 1980).

Total transmissivity τ_t, is related to the transmissivity to diffuse light, τ_{di}, through the empirical equation (Gates 1980)

$$\tau_{di} = 0.384 - 0.416\,\tau_t \qquad (6.80)$$

Typical values of transmissivity to direct radiation for a dust-free atmosphere range from 0.4 to 0.8, and for diffuse radiation range between 0.037 and 0.153. For a horizontal ground surface with total transmissivity between 0.5 and 0.8, diffuse radiation represents about 6.3–35.2% of the total incident radiation in dust-free clear skies.

A solar beam passing through the earth's atmosphere is modified in relation to the path length and vertical air transmissivity. The magnitude of the absorption and dispersion in the atmosphere depends partly on the solar radiation path length, and also on the content of attenuating components. The path length is usually specified in terms of the air mass m, defined as the ratio of the path length of sun rays in the direction of the sun at zenith angle ψ, and the vertical projection of this zenith direction. For zenith angle values below 60°, the air mass value is given by

$$m = \sec\psi \qquad (6.81)$$

and for zenith angles between 80° and 90°, the m value is lower due to the earth's curvature and atmospheric refraction (Gates 1980). The exact values for m can be found in tables (List 1963). At high altitudes, air mass should be corrected for reduced atmospheric pressure by introducing a multiplying factor, p/p_o, where p is the atmospheric pressure at a given point, and p_o, the atmospheric pressure at sea level.

Beer's Law (Eq. 6.68) is given as

$$S(x) = S(0)\exp(-\tau m) \tag{6.82}$$

enables the characterization of the radiative attenuation process. In Eq. (6.82), $S(0)$ is the incident flux in a given environment, $S(x)$ the corresponding flux at x, after medium attenuation, the attenuation coefficient τ (or turbidity), and the air mass m.

Energy attenuation occurs through absorption by ozone (especially in the UV range), water vapor (in the IR range), carbon dioxide, and oxygen. This process of energy attenuation results in warming of the atmosphere by energy removal of the radiative beam. In the visible spectrum, absorption by atmospheric gases is not as important as dispersion. In the IR spectrum, absorption is, however, more significant than dispersion, and the various atmospheric components absorb radiation between 0.9 and 3 μm. Atmospheric water vapor increases visible radiation in contrast with IR radiation (Monteith and Unsworth 1991). The water vapor content can be described in terms of the precipitable water content, or precipitation column that would be formed if the water vapor were to condense (5–50 mm). Clouds, water droplets, and ice crystals scatter energy in vertical ascending and descending directions. When the cloud depth is high, upward scattering predominates to about 70% of the incident radiation. About 20% can be absorbed, with only 10% remaining for transmission, giving a gray hue to the cloud base (Monteith and Unsworth 1991).

Radiation scatter has two forms. The first relates to Rayleigh scattering in which the diameter of the dispersing agent is lower than the wavelength of the radiation. According to this process, individual quanta incident on any atmospheric gas molecules is homogeneously dispersed in all directions. The efficiency of Rayleigh scattering is inversely proportional to the fourth power of the wavelength. Thus, the dispersion of blue light ($\lambda = 400$ nm) exceeds the scattering of red light ($\lambda = 700$ nm) by a factor of 9. This is the basis for the sky's blue color as seen from the earth's surface. Likewise, under the Rayleigh regime, the PAR radiation fraction incorporated in diffuse light is 1.4 times higher than its fraction of the total radiation (Spitters et al. 1986).

Rayleigh scattering and radiative absorption by ozone cause differences in the spectrum of radiation that reaches the surface, in contrast with the spectrum of a black body at 300 K (Monteith and Unsworth 1991). Thus, spectral irradiance in real conditions provide a constant maximum level in the range of 500–700 nm, whereas black body radiation peaks at 500 nm. When the sun is below 20° from the horizon, the attenuation via scattering increases, and the wavelength at maximum irradiance is in the IR spectrum.

Under clear sky conditions, when the zenith angle is <50°, the ratio of diffuse to total radiation is <0.2. The ratio of diffuse irradiance and total irradiance increases with the increase in cloudiness and maybe one when the sun is fully covered. However, the maximum amount of diffuse light occurs at cloudiness of about 50%. The spectral composition of the scattered radiation is also affected by cloudiness. Under a clear sky, the radiant energy spectrum concentrates in the visible range, and with increasing cloudiness, visible radiation corresponds to about 50% of the total radiation (Monteith and Unsworth 1991).

Rayleigh scattering is not valid for particles such as aerosols (dust, smoke, pollen, water droplets, etc.) where the characteristic dimension d, and the wavelength λ, are the same order of magnitude. In this case prevails the so-called Mie scattering wherein radiative scattering for the various wavelengths is a function of ratio (d/λ). Indeed, for similar (d/λ) ratios, the longer-wavelength radiation is scattered more intensely than the lower-wavelength radiation, unlike what happens with Rayleigh scattering. Radiative dispersion caused by larger particles or aerosols, is more significant in the direction of the incident radiation, without changing direction. Spitters et al. (1986), mention that under a clear sky and total sun height of 45°, about 15% of the diffusive flux consist of radiation without any additional modification.

Under clear sky conditions, Gates (1980) indicates a relationship between the instantaneous transmittance of the atmosphere to direct radiation in the zenith direction, τ, and instant transmittance to diffuse light, τ_{di}:

$$\tau_{di} = 0.271 - 0.294\,\tau^m \tag{6.83}$$

Equation (6.83) shows that the higher the transmissivity to direct radiation, the lower the transmissivity of diffused radiation. The transmissivity coefficient to diffuse light does not vary with the height of the sun or with cloudy conditions (Spitters et al. 1986).

Analysis of the atmospheric absorption spectrum (Oke 1992; Gates 1980) shows wavelength ranges where transmittance is practically one. This atmospheric window in the near IR allows radiation between 8 and 13 μm to escape into space. This window has practical implications in remote satellite sensing involving radiation emitted from the earth-atmosphere surface.

Other radiative ranges are transparent to the atmosphere. These wavelength windows allow solar radiation in the visible range (300 nm to about 720 nm) to reach the earth's surface. Some radiation at wavelengths greater than 5 μm emitted by the earth's surface also escapes into space.

The emission spectrum of the earth's surface-atmosphere system is comparable to that of a black body at 288 K (Campbell and Norman 1998; Gates 1980). This black body shows strong absorption and emission in some radiation bands within the range of 4–24 μm. In the remaining bands, the absorption and emission of radiation are low and so radiation transmission is high.

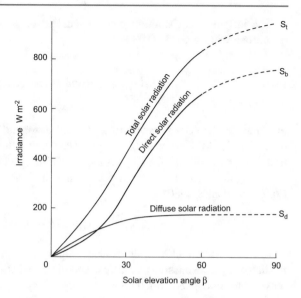

Fig. 6.8 Total solar radiation incident on a surface as a function of solar elevation angle (after Monteith and Unsworth 1991)

It follows that the total or global radiative flux, S_t, incident on soil horizontal surface is given by the following equation (Monteith and Unsworth 1991):

$$S_t = S_p \cos \psi + S_d = S_b + S_d \tag{6.84}$$

where S_b and S_d are the terms for direct and diffuse solar radiation fluxes, respectively, on a surface perpendicular to the solar rays. The S_p term corresponds to direct sunlight irradiance on a surface perpendicular to the sun. Figure 6.8 represents the variation of the direct and diffuse components of total solar radiation on the ground with the angle of solar elevation, in temperate Europe.

On a cloudless day, the S_t curve with time, t (h) is roughly sinusoidal (Monteith and Unsworth 1991)

$$S_t = S_{tm} \sin(\pi t / n) \tag{6.85}$$

where S_{tm} is the maximum irradiance at solar mid-day and the number of sun hours, n.

Under cloudy conditions, S_t variation remains sinusoidal over a month. Integrating Eq. (6.85), the daily irradiance, S_{dt}, becomes

$$S_{dt} = (2n/\pi) S_{tm} \tag{6.86}$$

Equation (6.85) can be adapted to the latitude of a specified site. In temperate climates, the S_{tm} ranges from about 900–1050 Wm^{-2}. Applying Eq. (6.86) for a 14 h day gives a S_{dt} of about 34 MJm^{-2} (Monteith and Unsworth 1991).

The instant direct sunlight content incident on a surface under cloudless sky conditions will be Gates (1980)

$$S_b = S_0 \left(\frac{\overline{d}}{d}\right)^2 (\sin\phi\sin\delta + \cos\phi\cos\delta\cosh)\tau^m \tag{6.87}$$

in which the various parameters are as defined above and τ is the average atmospheric transmissivity to direct radiation, with values between 0.4 and 0.7. The numerical time integration of Eq. (6.87) gives the daily direct incident solar radiation.

The global solar radiation incident on a horizontal surface is obtained from Eq. (6.84) (Gates 1980)

$$S_t = S_0\tau^m\cos\psi + S_0(0.271 - 0.294\tau^m)\cos\psi \tag{6.88}$$

Liu and Jordan (1960) developed empirical diagrams from large data sets, relating daily transmissivity of diffuse and direct radiation with daily solar radiation outside the atmosphere. Such transmissivity values can be used in equations such as (6.87) and (6.88). For practical purposes, such as estimating agricultural productivity, it is enough to use monthly averages of transmissivity $\overline{T}_t$, representing cloudiness indexes. In temperate zones, $\overline{T}_t$ values of about 0.3 correspond to very cloudy areas and values of about 0.7 are associated with clear skies.

Solar radiation received at the surface under a cloudy sky is mostly diffuse. Under these conditions, the average radiance of a totally cloudy sky is about two or three times greater at the solar zenith than at the horizon, due to the greater mass of air on the horizon. An expression relating radiance distribution in overcast skies with the zenith angle is Monteith and Unsworth (1991)

$$N(\psi) = N(0)(1 + b\cos\psi)/(1 + b) \tag{6.89}$$

In this equation, the denominator $(1 + b)$, the ratio between radiance at the zenith and the radiance on the horizon, is about 2.1–2.4 or 3. The reduction in the total solar radiation transmitted because of cloudiness can be calculated by graphs and tables (Gates 1980). Due to cloudiness, the average daily irradiance in Europe is about 15–25 MJm^{-2}, about 50–80% compared to a clear day (Monteith and Unsworth 1991).

An empirical relationship used to establish a relationship between solar radiation incident on the soil surface on cloudy days and the corresponding global solar radiation on clear days, is as follows Gates (1980):

$$S_{tm} = S_t(a + n_i b) \tag{6.90}$$

where S_{tm} is the global solar radiation on cloudy days, n_i the number of monthly hours of clear skies, and a and b are empirical constants, 0.35 and 0.61.

Another empirical expression, for the same purpose, is as follows (Gates 1980):

$$\overline{S}_t = \overline{S}_h\left(0.803 - 0.34\overline{C} - 0.485\overline{C}^2\right) \tag{6.91}$$

where $\overline{S}_t$ is the average monthly solar radiation at the soil surface, S_h the daily average monthly level of solar radiation on a horizontal surface, in the space outside the earth's atmosphere and $\overline{C}$ the monthly average fraction of overcast sky. Equation (6.91) assumes that on a clear day, the maximum fraction of global radiation incident on the ground surface, relative to the incident radiation outside the atmosphere, is 0.803.

Bennet (1965) presents a linear relationship between the monthly averages of daily global solar radiation at the soil surface, the average monthly number of hours of clear skies, $\overline{n}_i$, expressed as a percentage of the total number of hours of clear sky and the daily monthly average of incident solar radiation on a horizontal surface, outside the earth's atmosphere

$$\overline{S}_t = \overline{S}_h\left(203 + 5.13\overline{n}_i\right)10^{-3} \tag{6.92}$$

From Eq. (6.92), if $\overline{n}_i$ is 100% then $\overline{S}_t/\overline{S}_h = 0.716$, and if $\overline{n}_i$ is 0% then $\overline{S}_t/\overline{S}_h = 0.203$ (Gates 1980).

6.3.4 Long Wavelength Radiation

All bodies located in the biosphere or terrestrial atmosphere-surface system emit radiation in proportion to the fourth power of the surface absolute temperature, in accordance with the Stefan–Boltzman's Law. For bioclimatic purposes, thermal or long-wavelength radiation at wavelengths above about 5 μm is that emitted by objects with surface temperature below 600 K (Gates 1980). A body at a temperature of this order of magnitude, or less, emits mainly radiation in the IR, with negligible emission in the visible radiation range. In land surfaces, radiation emission below 2.5 μm occurs in environments such as fires, volcanoes, and other high-temperature sources. Due to higher radiative absorption in the IR, terrestrial bodies have their energy status or temperature tightly coupled to the ambient longer wavelength radiation (Gates 1980). These bodies are loosely coupled to low wavelength radiation due to their low absorbance at wavelengths in the visible radiation range.

Short-wavelength radiation in the natural environment is mainly solar with partial daily duration, whereas long-wavelength radiation is from bodies like clouds, sky, ground, buildings, vegetation, etc., lasting throughout the day, although with varying intensity. The calculation of long-wavelength radiation exchanges in earth-atmosphere systems is complex because of the variability of configuration of surfaces and temperature gradients in natural environments. In clear sky conditions, the radiance of long wavelength is greater on the horizon, decreasing with higher sun elevation angle. This is due to the longer path on the horizon of emitted gases such as CO_2 and H_2O. In general, about half of the radiative flux from the atmosphere comes from gases within 100 m height and more than 90% is from emissions within 1 km. The temperature gradient near the ground contributes largely to determining this energy flux (Monteith and Unsworth 1991).

In temperate zones, the night-time radiative balance is generally negative, with long-wavelength energy losses from the ground to the atmosphere (about 100–140 Wm^{-2}) relating to the heat exchange processes (night-time temperature decreasing with height above ground). The long-wavelength radiation ascending L_u and descending L_d fluxes can be measured or determined based on knowledge of the temperature and emissivity of the bodies. In temperate zones, long-wavelength radiant density flux, L_u, range between 270–430 Wm^{-2} and L_d between 150 and 320 Wm^{-2} (Monteith and Unsworth 1991).

Under clear sky conditions, the apparent atmospheric emissivity of the atmosphere ε_a, is defined from the equation

$$L_d = \varepsilon_a \sigma T_a^4 \tag{6.93}$$

where T_a is the mean environmental air temperature. L_d is usually estimated from empirical expressions as a function of temperature and/or vapor pressure at standard height. A convenient equation for obtaining L_d is

$$L_d = c + d\sigma T_a^4 \tag{6.94}$$

where c and d are empirical constants measured in England, at ambient temperature between −6 and 26 °C, being −119 ± 16 Wm^{-2} and 1.06 ± 0.04 Wm^{-2}, respectively. The estimation error is about ± 30 Wm^{-2} (Monteith and Unsworth 1991).

To quantify ascending and descending long-wavelength radiation, L_u and L_d, in clear skies, a linearization of Eq. (6.94) can be used to obtain the following linear expressions (Monteith and Unsworth 1991):

$$L_d = 213 + 5.5\,T_a \tag{6.95}$$

$$L_u = 320 + 5.2\,T_a \tag{6.96}$$

The loss of long-wavelength radiation is given by the difference between Eqs. (6.95) and (6.96).

Clouds that are sufficiently dense to cause shadows on the ground will emit as blackbodies at the temperature of the constituent water droplets or ice crystals. These increase the thermal radiation flux received at the ground surface, as they contribute to emission between 8 and 13 μm, corresponding to the atmospheric window, at which emission does not occur by atmospheric gases. These emissions complement the radiation emitted by the water vapor and carbon dioxide, in the lower atmospheric layer.

In cloudy sky conditions, the downward thermal radiation can be given by

$$L_d = \varepsilon_a \sigma T_a^4 + (1 - \varepsilon_a)\sigma T_n^4 \tag{6.97}$$

where T_a and T_n are, respectively, temperatures of the ambient air and the cloud base and ε_a is the atmospheric emissivity.

If we consider 283 K and 272 K the mean values for T_a and T_n, representative of temperate regions, then we can see, using Eq. (6.97), that the values for $\varepsilon_a(1)$ and $\varepsilon_a(c)$, for atmospheric emissivity in total cloudiness and partly cloudy conditions, c, can be obtained

$$\varepsilon_a(1) = 0.84 + 0.16\varepsilon_a \tag{6.98}$$

and

$$\varepsilon_a(c) = (1 - 0.84)\varepsilon_a + 0.84c \tag{6.99}$$

The major drawback of this approach is the selection of appropriate values for cloudiness and cloud base temperature, due to variability in cloud type and geometry and their base heights (Monteith and Unsworth 1991).

6.3.5   Radiative Properties of Natural Materials

6.3.5.1   Vegetation and Leaves

In any environment, radiation intercepted by natural bodies needs to be quantified to account for its distribution among the components of the biosphere. The interception of the radiation depends on the spectral and geometric characteristics of natural bodies. Solar radiation intercepted by natural objects is absorbed, reflected, or transmitted at different rates depending on factors such as the predominant wavelengths.

Radiation between 0.35 μm and 3 μm is the predominant means of interaction between the atmosphere and vegetation. The interception of sunlight with three-dimensional objects such as leaves, trees, houses, animals, etc., depends on their shape and geometric relationship with solar rays. Water is a major component of natural materials due to strong absorption bands between 1 and 3 μm. The reflection and transmission by porous materials in the visible spectrum, where absorption by water is negligible, is often closely associated with humidity levels.

At radiation wavelengths longer than 3 μm, most natural bodies behave like black bodies with nearly 100% absorption coefficients, and virtually zero reflectivity. For plant canopies the radiation system depends on:

(i) Direct or diffuse incident radiation scattered in the Earth's atmosphere and incident on canopies in all directions. Fractions of solar radiation reflected and transmitted by a leaf dependent on the zenith angle (or incidence), ψ. For a

Table 6.6 Average values of reflectivity, transmissivity, dispersion, and absorbance for a green leaf (adapt. Ross 1975)		PAR	NIR	Short-wavelength radiation
	Reflectivity	0.09	0.51	0.30
	Transmissivity	0.06	0.34	0.20
	Dispersion	0.15	0.75	0.50
	Absorbance	0.85	0.15	0.50

given wavelength, the reflection coefficient is approximately constant for ψ from 0 to 50°, in relation to the normal position, but as γ increases from 50 to 90°, $\rho(\lambda)$ increases substantially due to specular reflection. Foliar transmission coefficients remain constant between 0 and 50°, decreasing for incidence angles between 50 and 90°. Since reflectivity and transmissivity variations coefficients are complementary, the fraction of radiation absorbed by leaves is practically constant for incidence angles below 80° (Monteith and Unsworth 1991);

(ii) The optical properties of plant canopy in that radiation interacts with plant organs by absorption and dispersion. These processes depend on variables such as leaf structure, leaf age, spectral distribution, etc. Table 6.6 shows the spectral values for a typical green leaf.

In the short-wavelength ranges, radiation exchanges by biological materials are determined by pigments that absorb radiation at wavelengths associated with specific electron transitions (Monteith and Unsworth 1991).

The transmissivity and reflectivity coefficients for a green leaf can be assumed to be equal to 0.1 between 0.4 and 0.7 μm, and 0.7 between 0.7 and 3 μm (Fig. 6.9). As each of these spectral bands contains about half the total radiation in the visible and near IR, it can be assumed that the overall coefficient of leaf transmission and reflection is about 0.25 in this spectral range. Leaves have high absorbances of about 0.9–1, in the range of thermal radiation, typical of the terrestrial surface-atmosphere system. Under these conditions, energy dissipation via convection and transpiration is important for the regulation of leaf temperature.

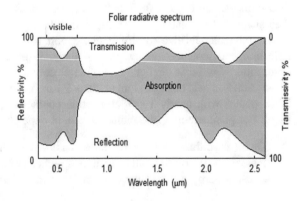

Fig. 6.9 Representative diagram of transmissivity, reflectivity, and absorbance characteristic of a green leaf in visible and near IR spectral bands

The green color of leaves is due to the reflection of green light, at higher angles of radiation of incidence and visual sensitivity, even though the absolute reflectivity is higher in the near IR region. This reflectivity in the near IR region also contributes to the removal of incident radiation (Oke 1992). For much of the foliage, absorption of the green band at 550 nm is about 0.75–0.8, for the blue band between 400 and 460 nm, it is 0.95 and for the red band between 600 and 670 nm, it is 0.85–0.95.

Transmission among thick leaves stems and branches are zero (Ross 1975). Glossy leaves have a very significant reflection component. In general, foliar reflection is higher than transmission except for thin leaves in the green and near IR bands (Fig. 6.9).

Spectral properties of leaves are changed during the growing season. Younger leaves are bright green, have high transmittance and reflection, and low absorption. Mature leaves are dark green, have low reflection and transmission, and high absorption. The older leaves have a higher reflection and lower absorption (Ross 1975).

Radiative properties of the leaf surfaces are not significantly affected by coating with water or waterproofing with compounds such as baseline, nor with coatings used in the calculation of leaf energy budgets (Fig. 6.10) (Rodrigues 1993);

(iii)   The optical properties of soil or surfaces below the plant canopy, for example, by the reflectance coeficient or albedo. In dense vegetation, soil

Fig. 6.10 Instrumentation for calculating leaf transpiration in tomato greenhouse. Left: an instrumented mast for data acquisition. Right: An aspiration hygrometer (adapt. Rodrigues 1993)

reflection can be neglected, as opposed to when the cover is sparse, or the ground covered with snow. The albedo of natural surfaces varies during the day, peaks in the morning, and mid-afternoon and reaches a minimum around noon (Oke 1992). The variation of albedo with the zenith angle is exponential.

This variation is typical of specular reflection, as can be seen from curves of reflection versus solar height (the complementary angle of solar inclination) in clear or cloudy skies over water surfaces (Oke 1992; Monteith and Unsworth 1991);

(iv) The architecture of the canopy, which is the most important factor influencing the dynamics of canopy radiation. The albedo of vegetation is lower than the reflectivity for the individual leaves since the reflection depends not only on the radiative properties of the individual components, but also on the structure and solar elevation angle. The latter two determine the penetration and retention of incident radiation, as well as the shade provided by the canopy components (Oke 1992).

Significant variations of radiation that occur in a clear sky day because of alternating light beams and shaded areas are determined by the canopy structure, as well as plant and leaf distribution. The fraction of radiation intercepted by plant canopy depends on the leaf area index, defined as the projected area of leaves per unit area of soil and the spatial distribution of foliage relative to the radiation direction. The scattered radiation fraction depends on the optical properties of anatomical components of the leaves, such as cuticles, cell walls, and pigments (Monteith and Unsworth 1991).

The highest albedos are recorded on flat surfaces such as lawns. For vegetation layers, with heights ranging between 50 and 100 cm that fully cover the soil, the albedo ranges between 0.18 and 0.25. For forest canopies, the albedo is about 0.1. When the ground is partially covered by the vegetation or when leaves are partially dry, the albedo will depend on foliage reflection and the reflection from the ground. Albedo differences between the different forms of vegetation are due to the complexity of reflection and the scattering of radiation within.

The reflection coefficient for various forms of vegetation also depends on the zenith angle as a key component of specular reflection (Lee 1978). The minimum albedo values of the areas under crops are at noon when the sun is at the zenith and the maximum values occur when the sun approaches the horizon, due to specular reflection of vegetation. Dependence of albedo on solar height also helps to explain the lower albedo values in tropical areas, in relation to the higher latitude areas. On days with cloudiness, or under any other conditions, wherein the diffuse component of the solar radiation is significant, the daily variation of albedo is lower than in clear sky conditions (Lee 1978).

6.3.5.2 Greenhouses, Water, and Soil

Heating inside greenhouses depends on the spectral absorption properties of glass that readily transmits shorter wavelength radiation and is opaque to thermal or long-wavelength radiation. Short-wavelength radiation is readily transmitted into greenhouses and can be absorbed by the soil and plants. However, thermal radiation emitted by soil and plants is retained by total absorption by the glass and emitted back to the interior (Linacre et al. 1964). Glass enhances radiation emitted by the inner surfaces of the greenhouse, thereby increasing heat.

Albedos of greenhouse roofs can be high, especially at angles with low sun height and can reduce transmitted radiation by about 50%. Moreover, factors such as dirt in glass walls or short- wavelength scattered radiation after attenuation can further reduce radiation inside the greenhouse. When recording greenhouse energy balances, it is necessary to also quantify net visible and thermal radiation. It is even possible for the radiative balance in the greenhouse to be lower than that outside (Levit and Gaspar 1987). At night, when only long-wavelength radiation exchanges occur, greenhouse covers serve to minimize radiation losses (Seginer 1984) (Fig. 6.11).

For the aqueous surfaces, radiation incident on a clear and transparent surface at an angle < 45°, the reflection coefficient for the sunlight is approximately equal to 5%. Above 45°, the reflection coefficient increases rapidly with the angle of incidence up to about 100% (Monteith and Unsworth 1991). In the visible spectrum, water is transparent but has a minimum absorption coefficient in the blue-green region between 460 and 490 nm giving natural surfaces with clear water their characteristic color.

In the near IR region, water has several absorption ranges readily identified from the reflection spectra of the soil, leaves, and animal skins. The main absorption bands for water are between 1.45 and 1.95 μm. For wavelengths above 3 μm,

Fig. 6.11 Plant greenhouse (adapt. Rodrigues 1993)

absorbance and emissivity values are about 0.995, decreasing with increasing angle of incidence. For a wavelength of 11 μm and an incidence angle of 80°, emissivity is about 70% (Monteith and Unsworth 1991).

For a given input of solar radiation, the differences of the radiative budget among various crops with adequate water supply are insignificant, due to similar albedo and emissivity, because of the water content and surface temperature homogeneity caused by cooling from evaporation.

Soil reflectivity depends mainly on organic matter content, moisture, and particle size, and incidence angle of solar radiation. The reflection coefficients for the solar spectrum range from about 10% in organic soils to 30% in the sand. Small amounts of organic matter can greatly decrease soil reflectivity. For clay soils, reflectivity is a function of particle size. In the radiative range between 0.4 μm and 2 μm, the reflectivity of kaolinite particles decreases with decreasing particle size. For 1600 and 22 μm particles, reflectivity is 56% and 78%, respectively (Monteith and Unsworth 1991). In general, soil reflectivity is low in the blue region, increasing with the wavelength in the visible and near IR, reaching a maximum value at 1–2 μm. Particles aggregates with irregular shapes retain more radiation by multiple internal reflections as compared to more homogeneous particles, such as fine powder.

The radiation transmission is particularly relevant in snow and ice surfaces, where short-wavelength radiation can reach depths of 10 m in ice and 1 m in snow, as can be verified by Beer's Law. The exponential variation indicates that the attenuation of the radiation intensity is higher at the surface than in-depth (Oke 1992). The albedo of the cover is due to both reflections occurring at the surface and to multiple reflections below the surface. Soil reflectivity decreases with increasing moisture content, mainly because the radiation is retained by internal reflection in the air–water interface formed in the meniscus of the soil's capillary structure (Monteith and Unsworth 1991). Radiation transmission through the soil influences seed germination as well as root development (Hillel 1982).

6.3.5.3   Radiation Over Forest Canopies

Radiation geometry and qualitative principles can be used to estimate radiative energy distribution in the forest canopy, assuming evenly distributed foliage. The canopy leaf content can be determined by the leaf area index. If it is assumed that a thin layer of leaves with leaf area index, dL, is exposed to direct sunlight, then the energy content intercepted by dL is the product of the shadowed area projected by the leaves by horizontal irradiance. By integration, we obtain an equation of the Beer Law type representing the relationship between the global radiation flow in the forest floor and the global radiation above the forest cover (Eq. 6.100)

$$S_{tso} = S_t \exp(-kL) \tag{6.100}$$

in which S_{tso} is the density flux of the total surface radiation, S_t the corresponding flux on top of the outer atmospheric layer of the forest canopy, L the projected area of the canopy of leaves, branches, and stems (receiving area of the radiation beams, with the Sun in the zenith position) per unit area and k the extinction coefficient. Equation (6.100) can be used to calculate the average flux at any level of the canopy, if the k and L, coefficients are defined for the canopy fraction above a specified level. However, Eq. (6.100) has more theoretical than practical value as it requires measurements or precise estimates of the k and L parameters, which are difficult to obtain (Lee 1978).

At the upper limit of the surface layer over the forest canopy, solar radiation is absorbed or reflected, although at lower levels, the dispersion is more complex due to canopy transmissivity. Solar radiation passes through snow, soil, ice, water surfaces, soils, and organic materials presumed to be opaque.

Solar radiation in the visible range can penetrate to depths greater than 100 m in clean water, but this decreases with increasing concentration of impurities in the water. Pure water is opaque to the longer wavelengths in the IR region. The spectral behavior of snow and ice is like water, but the depth of radiation transmission is much lower because about 90% of the visible flux is absorbed in the layer between 10 and 50 cm in-depth. Radiation transmission in forest soil increases with particle size. In coarse sand, a small percentage of the total radiative flux can be transmitted to 1–2 cm, whereas in very fine materials such processes occur within 1–2 mm (Lee 1978).

The transmissivity coefficient of mature leaves of trees ranges between 0 for softwoods and 0.25 for hardwood evergreens. As with other natural bodies, the reflectivity coefficient increases with the zenith angle, simultaneously with the decrease in transmissivity coefficients, so that absorption is not altered. For most evergreens, the average reflectivity for low sun angles ranges from 0.26 to 0.32 and absorbances vary between 0.34 and 0.44. For higher solar angles (low zenith angles), the average reflectivity ranges from 0.2 to 0.26 and the average absorbance varies between 0.48 and 0.56 (Gates 1980).

The foliar transmissivity over forest canopies is variable between wavelengths 0 and 4 μm, in analogy with reflectivity. At wavelengths in the visible range, transmissivity is relatively low, increasing in the green region between 0.5 and 0.6 μm. The total transmissivity is much higher than the average in near-infrared, between 0.7 and 1.1 μm.

Under clear skies, solar radiation flux under the forest canopy is highly variable as the canopy does not form a continuous or uniform shade. When the canopy shade is relatively uniform, the distribution of diffuse radiation of wavelengths will depend on the structure of the forest and canopy density. Under hardwood canopies, the wavelength corresponding to the maximum irradiance is about 0.55 μm (green light) and the wavelength corresponding to the minimum irradiance is about 0.67 μm. Under softwood canopies, there is a relatively uniform decrease in irradiance between the blue and red wavelengths (Lee 1978).

Micro changes in radiation regime have the potential to influence local climates that may have significant socioeconomic implications. Particularly in sloped terrains, the net radiation is non-uniform, as for example, beneficial radiation climates are found in the slopes of river valleys often used for viticulture. Drainage of cold air from locations above the frost sensitive vineyards can be blocked, e.g., through walls to prevent frost deposition in the valleys. Frost-resistant fruit trees should be cultivated in the lowest areas of slopes more prone to cold and frost formation (Foken 2017).

6.4 Transient Heat Balances

6.4.1 The Concept of Time Constant

Heat transmission mechanisms involving conduction, convection, and radiation are key physical processes for heat exchange between constitutive elements in environmental systems. These mechanisms are closely interconnected with the components of the energy balance, and especially with the most relevant, such as latent and sensible heat fluxes, net radiation and heat fluxes, and energy storage in soils or animals or heat production by metabolism in animals.

Heat transfer in natural and modified environments is also a non-equilibrium transient process wherein thermal inertia imposes a time lag, usually termed as the time constant, between energy outputs and inputs and temperature changes of bodies at different energy levels.

The transient energy transfer processes can be simplistically grouped and described as step changes, ramp change, and harmonic oscillations. The latter was discussed above within the process of heat transfer in soil through thermal conduction (Sect. 6.1.3) and thus only step and ramp changes will be shortly evaluated below.

As examples of environmental lagged responses, figure diurnal and seasonal changes of ambient temperatures, following the pattern of solar radiation superimposed with shorter-term fluctuations associated with cloudiness and turbulence. Oscillations in energy budget components occur also in the biosphere, with sinusoidal variations of heat sinking during heating periods (daytime, summer) alternating with similar variations of heat sourcing during the colder periods (night, winter). Surface organic and inorganic materials can thus act as transient thermal reservoirs which attenuate the atmosphere's physical variations near the ground. Also, in the biosphere, while plants and cold-blooded animals with internal energy production can work as passive conductors, warm-blooded animals generate large amounts of internal energy and can, in principle, provide thermal energy to the environment (Lee 1978).

For evaluation of time constant meaning in environmental systems, a simple case of energy budget composed only by net radiation and sensible heat fluxes can be written as follows:

$$R_n = H = \rho c_p (T_0 - T)/r_{HR} \tag{6.101}$$

where R_n and H are the net radiation and sensible heat fluxes per unit area, T_0 is the mean surface temperature, T is air temperature and r_{HR} is a combined resistance for sensible heat loss r_H and longwave radiation r_R as follows:

$$r_{HR} = \left(r_H^{-1} + r_R^{-1}\right)^{-1} \qquad (6.102)$$

The concept of effective temperature, T_f, was created from the need of replacing the value of measured air temperature with another temperature, termed effective temperature, T_f, which incorporated the major components of microclimates involved, net radiation in this case relative simplified energy budget in Eq. (6.1). Its definition, considering net radiation effects only, is as follows:

$$T_f = T + \left(R_n r_{HR}/\rho c_p\right) \qquad (6.103)$$

and from Eq. (6.101), the heat budget equation can be written as

$$T_0 = T_f \qquad (6.104)$$

The ratio $r_{HR}/(\rho c_p)$ corresponds to the slope of a curve relating heat fluxes versus air temperature in micro-environments and boundary layers adjacent to contact surfaces. For a given surface, this curve intercepts the vertical line $x = R_n$ when $y = T_f$ (Monteith and Unsworth 2013).

Considering that the losses of sensible and latent heat occur from the same surface, the respective energy budget can be adapted from the simplest one in Eq. (6.101), with adding of a latent heat term, LE, and a variable including both temperature and vapor pressure effects, termed as apparent equivalent temperature, T_e^* is defined as

$$T_e^* = T + e/\gamma^* \qquad (6.105)$$

were γ^* is the modified psychrometer constant defined as $(\gamma r_{Av}/r_{HR})$ with the psychrometer constant γ already defined in Chap. 4 as equal to $(c_p p/L\varepsilon)$.

The new budget equation will come as follows:

$$R_n = H + LE = \rho c_p (T_{e0}^* - T_e^*)/r_{HR} \qquad (6.106)$$

where T_{e0}^* is the mean value of apparent equivalent temperature at the surface that is conceptualized from air adiabatic cooling considerations in psychometry (Annex A2).

The energy budget in Eq. (6.106) can be then written in a similar way of Eq. (6.104) as follows:

$$T_{e0}^* = T_{eR}^* \qquad (6.107)$$

where from Eq. (6.104), we can deduce

$$T_{eR}^* = T_f + e/\gamma^* = T_0 + e/\gamma^* \qquad (6.108)$$

A relation between Eqs. (6.104) and (6.107), relative to energy budgets in Eqs. (6.101) and (6.106), respectively, with and without latent heat terms is thereby established by Eq. (6.108).

For a general estimation of time constants, it is enough an evaluation of a simple energy budget (Eqs. 6.101, 6.103, and 6.104) with terms for net radiation and sensible heat flux only. In this context, according to Eq. (6.103) and Eq. (6.109), the effective temperature increases from T_f (Eq. 6.103) to T'_f (Eq. 6.110) (Monteith and Unsworth 1991), with increases in air temperature and/or net radiation. An increase in effective temperature will not be although instantaneous due to thermal inertia reflected in a finite heat capacity. The heat capacity per unit area α will allow for thermal storage in the contact surface as follows:

$$R_n = H + \alpha \partial T_0 / \delta t \tag{6.109}$$

Taking Eq. (6.109) and replacing H and Rn from Eqs. (6.101) and (6.103), respectively, we deduce the following relation:

$$(\partial T_0 / \partial t) = (T'_f - T_0)/\tau \tag{6.110}$$

where τ is the time constant, with time units. For example, the time constants in elements of vegetal canopies range between orders of seconds for small leaves, minutes for large leaves and hours for tree trunks.

6.4.2 Transient Responses

As aforementioned, the main types of transient responses in environmental systems to external forcing energy changes in terms of temperature and components of energy budgets can be grouped as step change, ramp change, and harmonic changes. These transient concepts are discussed in Annex 2 and Chap. 3 for instrumental fundaments.

Under boundary conditions, step changes of the effective temperature in environmental systems change instantaneously from T_f to T'_f

$$t = 0 \Rightarrow T_0 = T_f$$
$$t = \infty \Rightarrow T_0 = T'_f$$

From Eq. (6.110), it follows that

$$T_0 = T'_f - \left(T'_f - T_f\right) \exp(-t/\tau) \tag{6.111}$$

In Eq. (6.111) for instant $t = \tau$ results that $\left(T'_f - T_0\right) = \left(T'_f - T_f\right) \exp(-1) = 0.37\left(T'_f - T_f\right)$, so 0.63 is the remaining temperature fractional adjustment after a period of one time constant.

Monteith and Unsworth (2013) reported two interesting case studies on temperature step change transition concerning vine leaves and water streams. The former concerned a study of Linacre (1972) reporting a value of 20 s for τ of small vine leaves, leading to a combined resistance r_{HR} for sensible heat flow and longwave radiation of 33 sm^{-1}. The latter is related to water temperature in streams as irradiation increases with the removal of vegetal canopies shading the watercourse. A step change in the input of net radiation occurs as the water stream enters a clear area after coursing in a forest canopy with environmental implications, e.g., in aquaculture resources.

The estimated time constant in water streams in forest areas was estimated as 40 h for each meter of water depth (Sinokrot and Stefan 1993), clearly indicating an inertial effect in a response to sudden variation in the net radiation inputs. Results from Brown (1969) reported maximum rates of water temperature of 9 and 1 °Ch^{-1} in open and shaded areas, respectively.

The water temperature time variation ($\delta T/\delta t$) within a uniform well-mixed flow in a cross-section and velocity V can be evaluated as follows:

$$\frac{\partial T}{\partial t} = -V\frac{\delta T}{\delta x} + \frac{S_u + S_l}{\rho_w c_p h} \tag{6.112}$$

with h being the stream depth, S_u and stream bed S_l the heat exchanges between water interfaces with the atmosphere and stream bed, respectively, and $\rho_w c_p$ the volumetric heat capacity of water. The heat budgets of upper and lower water interfaces with the atmosphere can be put, respectively, as

$$R_{nu} = H + LE + S_u \tag{6.113}$$

$$R_{nl} = S_l + G_{sb} \tag{6.114}$$

where G_{sb} is the heat flux conducted in the stream bed and the other term S_l is the stream bed. The term G_{sb} is considered significant in bedrocks of shallow clear streams. The main driver of temperature response on watercourse transitions from shadow to clear surfaces will be the step change in net radiation. However, the influence of G_{sb} and S_l in heat transfer in stream systems can be also important and will depend on the thermal conductivity of the streambed material and on the boundary layer resistance of the bedrock water interface.

The ramp change in temperature variations in environmental systems occurs when the rate of change of ambient temperature, φ or effective temperature, under an external heat input is the following:

$$T(t) = T(0) + \varphi t \tag{6.115}$$

so that from solving Eq. (6.110) the change of the mean temperature of the system will be governed by the following equation:

$$T_0(t) = T(0) + \varphi t - \varphi t(1 - \exp(-t/\tau)) \tag{6.116}$$

and thus, the heating rate of the surface will be

$$\frac{\partial T_0}{\partial t} = \varphi(1 - \exp(-t/\tau)) \tag{6.117}$$

so that the rate of heating of the system will be nil in the beginning and will increase to φ in one later instant t much higher than the time constant. When the ratio $(-t/\tau)$ exceeds 3, the exponential term in Eq. (6.116) can be neglected and this equation becomes

$$T_0(t) = T(0) + \varphi(t - \tau) \tag{6.118}$$

so that the system heats at the same rate as the environmental temperature with a time lag τ, corresponding to a temperature lag of φT. Monteith and Unsworth (2013) mentioned for forest canopies that a ramp change regime could command the condensation rate in trunks of trees although imperceptible due to an eventual roughness of the surface. In this context, Unsworth et al. (2004) reported that saturated air in the forest canopies of Douglas fir became saturated in summer mornings shortly after sunrise, defending that condensation on tree trunks could be determinant for bark-dwelling organisms.

6.5   Mass Transfer

6.5.1   Gases and Water Vapor Transfer

In the free atmosphere, turbulent diffusion controls the circulation of gases and water vapor by convective processes like those of heat convection, with less influence on ground surfaces on calm nights. The turbulent transfer of water vapor and carbon dioxide is of utmost relevance for the biosphere and forest canopies. For example, the amount of carbon dioxide absorbed by a green crop canopy is roughly of the same order of magnitude as all the carbon dioxide existing up to a height of 30 m above the canopy.

Although the daily concentration of carbon dioxide changes in the boundary surface layer due to the photosynthesis dynamics, the amount of removed carbon dioxide is of a maximum of 15% of the diurnal mean CO_2 concentration in the surface layer. This is due to an available capacity of turbulent transfer for replenishing CO_2 absorbed by plants from the mixed boundary layer (Chap. 1). Experimental data presented by Monteith and Unsworth (2013) showed that during summer afternoons turbulent transfer enabled the extraction of CO_2 from the top planetary boundary layer of the order of 1–2 km for keeping some balance with carbon dioxide uptake by vegetal canopies.

The mass transfer coefficient is defined in a similar way of the heat transfer coefficient (Eq. 6.21) as follows (Holman 1983):

$$\frac{dm}{dt} = F = h_D A(\chi_2 - \chi_1) \tag{6.119}$$

where F is the mass flux of gas per unit surface area, (g m^{-2} s^{-1}), h_d is the mass transfer coefficient, and χ_2 and χ_1 are the mean concentrations of gas in the surface (gm^{-3}), in the sites among which mass transfer occurs. For example, if the two sites correspond to a given surface χ_s and to free air χ, the term corresponding to the concentration difference will be $(\chi_s - \chi)$.

Mass transfer to or from objects suspended in airflows is an akin process as heat transfer by convection and can be parametrized by the Sherwood number, Sh, a dimensionless parameter representative of the concentration dimensionless gradient at the surface, similar as Nusselt number for heat convection, defined as

$$Sh = \frac{F}{D(\chi_s - \chi)/h} \tag{6.120}$$

where D is the molecular diffusivity of gas in air (m^{2}s^{-1}) and h the thickness of an equivalent still air layer. An expression for the resistance of mass transfer, either for water vapor or for carbon dioxide, can be derived from Eq. (6.120), obeying the analogy with electric resistance (Eq. 2.24), as follows:

$$F = \frac{(\chi_s - \chi)}{r} \tag{6.121}$$

and

$$r = \frac{d}{DSh} = \frac{1}{Sh\,h_D}$$

where h_D is the coefficient of mass transfer in Eq. (6.119). Values for water vapor and carbon dioxide molecular diffusion (D) coefficients range between 20.5×10^{-6} and 28×10^{-6} and 12.4×10^{-6} and 17×10^{-6}, respectively, increasing with air temperatures in the environmental representative range between -5 and 45 °C. The Sherwood number for forced convection is a function of the Reynolds number and of the Schmidt number (Sc), which is defined by the ratio v/D between thermal and mass diffusivities, where v is the coefficient of kinematic viscosity of air $(m^{-2}s^{-1})$ and D the molecular diffusivity of gas in air $(m^{-2}s^{-1})$. The vertical profiles of velocity and mass concentration will be similar when $v = D$ or $Sc = 1$. For mass exchange above a flat plate, Sh is given by the following equation:

$$Sh = 0.66\,Re^{0.55}Sc^{0.33} \qquad (6.122)$$

Equation (6.122) accounts either for turbulence as for the differences in the effective thickness of boundary layers for heat and mass. When mass and heat transfer occur simultaneously, under forced convection, Nusselt and Sherwood numbers are related by the following relations:

$$Sh = Nu(\alpha/D)^{0.33} \qquad (6.123)$$

with the ratio $(\alpha\,/D)$ between thermal and mass diffusivities being the Lewis number, and

$$\frac{h}{h_D} = \rho c_p \left(\frac{Sc}{Pr}\right)^{2/3} = \rho c_p \left(\frac{\alpha}{D}\right)^{2/3} = \rho c_p Le^{2/3} \qquad (6.124)$$

where ρ and c_p refer to the density and specific heat at a constant pressure of the fluid, and Pr is the Prandtl number (Eq. 6.19), representing the ratio between the thicknesses of the two boundary layers. Two ratios of resistances to water vapor and carbon dioxide transfer, D_V and D_C, are the following (Monteith and Unsworth 2013):

$$r_V/r_H = (\alpha/D_V)^{0.67} = 0.93 \qquad (6.125)$$

and

$$r_C/r_H = (\alpha/D_C)^{0.67} = 1.32 \qquad (6.126)$$

The mass transfer of gases and vapors under free convection is determined, e.g., by differences in air density due to temperature gradients and/or by vapor concentration gradients. The Sherwood number will now be related to Grashof (Eq. 6.52) and Schmidt numbers by the following relation:

$$Sh = BGr^m Sc^m = NuLe^m \qquad (6.127)$$

where B is a constant similar to the constants in Eqs. (6.54), (6.55) and Table 6.4 for heat transfer in free convection, and m is a constant of 1/4 and 1 /3 in laminar and turbulent regimes, respectively. Monteith and Unsworth (2013) recommend that for calculation of Grashof number (Eq. 6.52) when buoyancy on lesser dense moist air occur, it is convenient to replace the difference between temperatures of surface and air $(T_0 - T)$ by the differences of virtual temperature $(T_{v0} - T_v)$ (Chap. 1):

$$T_{v0} - T_v = (T_0 - T) + 0.38(e_0 T_0 - eT)/p \tag{6.128}$$

where e and e_0 are the vapor pressures at the surface and air and p is the air pressure.

When the mass transfer is induced by the ventilation of a system such as greenhouses, the flux gradient approach following the electrical resistance analogy (Eq. 2.24) can be used to define a transfer resistance according to the above principles presented in the discussions of heat and mass convection. If the air inside of a greenhouse is ventilated so that the interior carbon dioxide volumetric concentration is homogeneous ψ_i, while the outside air concentration is ψ_o, the rate at which plants in the greenhouse absorb carbon dioxide from the external atmosphere can be put as

$$F = \rho_c vN(\psi_o - \psi_i)gh^{-1} \tag{6.129}$$

where v (m^3) is the volume of air in the greenhouse, h is the height of the house, N is the number of air changes per hour and ρ_c (gm^{-3}) is the density of carbon dioxide.

In terms of resistance analogy, the resistance to CO_2 diffusion r_c can be defined by the following equation:

$$r_c = \frac{\rho_c(\chi_c - \chi_i)}{F} = \frac{A}{vN} = (N\bar{h})^{-1} \tag{6.130}$$

where A is the floor area.

6.5.2 Particle Transfer

Particle mass transfer is relevant in boundary layer dynamics at a wide range of scales from fungal spores or seeds in agroforestry systems to large dust emissions and sandstorms following an outbreak of strong wind, with potentially devastating environmental impacts. Small particles of solids and liquids are transferred in the free atmosphere by turbulent diffusion. Inertial effects in particle mass transfer are relevant close to obstacle surfaces, where particles are thrown against if a fast direction change of airflow carrier occurs. Gravitational forces in particle transport are also more relevant than in molecules transport. Relative motion between air and particles induces drag forces in particles, such as pollen from stamens, allowing to detach them from surfaces where they are attached (Fig. 6.12).

Fig. 6.12 General scheme of
the three main forces acting
on a particle in a fluid flow

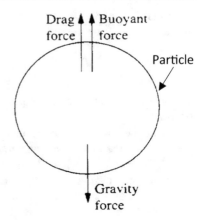

Three special cases of drag forces (Monteith and Unsworth 2013) on particles
are usually considered:

(i) for particles with radius r much smaller than the mean free path λ of gas
     molecules, a case wherein particles behave as larger gas molecules, and the
     drag force is the result of the higher impact of windward surfaces of particles
     than in leeward surfaces. If the mass of the particles is much larger than the
     mass m_g of the gas molecules, considering perfect reflection of collisions, then
     the drag force on a particle moving with a velocity U in a gas is given by

$$F = \frac{4}{3}\pi n m_g \bar{c} r^2 U \qquad (6.131)$$

where n is the number of gas molecules per unit volume and $\bar{c}$ their mean velocity.
The drag force is then proportional to particle velocity and surface area. A typical
example (Monteith and Unsworth 2013) is that at 20 °C there are about 3.10^{25}
molecules of air per cubic meter, a mass of a molecule is about 5.10^{-26} kg and $\bar{c}$ is
about 500 ms$^-$. The mean free path of molecules in the air is about 0.1 μm and thus
this case applies to particles with a radius lower than 0.01 μm;

(ii) for particles with radius r higher than path λ of gas molecules but with a
      particle Reynolds number Re_p is defined as $2rU/v$, with U being the velocity of
      relative motion of the particles relative to air and v the kinematic viscosity of
      air lower than 0.1, drag is derived mainly from the viscous forces resulting
      from the interaction between the particle and the gas molecules. The drag force
      is now governed by the Stokes law

$$F_d = 0.5 c_d \rho_g U^2 A \qquad (6.132)$$

with ρ_g being the gas density, A the cross-sectional and c_d the drag coefficient. In
this case, drag force is proportional to radius and particle density. With spheric

bodies for Re_p higher than 10^3, under the domain of Eq. (6.131), it can be considered that c_d is independent of Reynolds number, being proportional to U^2 and the cross-sectional area. For $1 < Re_p < 400$ another empirical equation was proposed

$$c_d = (24/Re_p)\left(1 + 0.17Re_p^{0.06}\right) \qquad (6.133)$$

For values of Re_p lower than 0.1, under the domain of Stokes Law, the drag coefficient of a spherical body decreases with the increase of Re_p, and a proposed empirical expression for the variation of c_d is

$$c_d = \frac{24}{Re_p} \qquad (6.134)$$

Monteith and Unsworth (2013) mentioned one application example of these particle mass transfer principles about the estimation of the force necessary to detach cylindrical spores of a fungal pathogen, with about 20 μm of diameter, from a stalk with 150 μm length, where they grow. This example showed that 50% of the spores were removed under a steady average wind speed of 10 ms^{-1}. The corresponding drag force obtained with Eq. (6.132) was around 10^{-7} N and it was also mentioned that spores in the field were detached under much lower average wind speeds emphasizing the relevance of brief turbulent gusts in the increase of drag coefficients by tearing down leaf boundary layers and dispersing the pathogen spores in the atmosphere.

The same authors showed that the drag coefficient for a spherical particle explained by Eq. (6.131) was reflected in a linear decrease of drag coefficient between 10^4 and 50 within a range of Re_p between 0.001 and 1 corresponding to a flow governed by Stokes law (Eq. 6.133) was related to Re_p ranging between 400 and 1 corresponding to a drag coefficient decreasing from about 50 to 5. Within the flows with Re_p values higher than 10 up to 10^5 or more, already within a turbulent boundary layer, drag coefficient was almost constant ranging between 2 and 6, and the drag force was shown as proportional to the square of mean velocity U^2 and to cross-sectional area A, according to Eq. (6.132).

At terminal sedimentation velocity, the excessive vertical force on particles, F_g, due to the difference between weight and buoyancy is given by

$$F_g = \frac{4}{3}\pi R^3 g(\rho_p - \rho_f) \qquad (6.135)$$

where ρ_p and ρ_f are the mass densities of the sphere, with a radius R, and g the gravitational acceleration. From Eq. (6.132), the calculation of the drag force of a spherical particle is given by

$$F_d = 0.5 c_d \rho_g U^2 \pi R^2 \qquad (6.136)$$

and a sedimentation velocity, U_s, from the equilibrium of drag, buoyant and gravitational forces in Eqs. (6.135) and (6.136), can be evaluated as

$$U_s^2 \approx 8R\rho g / 3 c_d \rho_f \tag{6.137}$$

In cases of pollutant particles moving in the air, the term ρ_f can be removed because air is much lighter than these particles.

If a spherical particle falls in a viscous fluid under a flow with a Reynolds number lower than 0.1, then a terminal sedimentation velocity is reached when the sum of frictional and buoyant forces on the particle equilibrates the gravitational force. The gravitational effects will increase with heavier particles and will decrease with higher fluid velocity and viscosity. The sedimentation velocity within the Stokes regime is given by

$$U_s = \frac{2}{9} \frac{(\rho_p)}{\mu} g R^2 \tag{6.138}$$

For particles in flows with Re_p higher than 0.1, Eq. (6.135) should be used with the drag coefficient estimated with Eq. (6.133) or (6.134), with an iterative scheme till achievement of a balance between drag and gravitational forces. The initial guess for iterations to calculate the sedimentation velocity can be obtained by the application of Eq. (6.138).

Particles of soil and pollutant materials are not spherical and show variable density. Such particles are characterized by the termed Stokes diameter corresponding to the sphere with the same density and sedimentation velocity. The change of shape and associated physical characteristics are, e.g., noticeable with water drops that flatten with fall increasing drag coefficient if having a radius ranging between 0.4 and 2 mm. For drops with a radius ranging between 2 and 3 mm, increases in weight are compensated by the increase in deformation so that c_d and U_s keep about the same values. Drops with a radius higher than 3 mm break up during fall.

The transient horizontal motion of particles is governed by the following expression:

$$x(t) = \tau\, U_0 \left(1 - \exp\left(-\frac{t}{\tau}\right)\right) \tag{6.139}$$

where τ is the relaxation time of the particle which is the time constant in the exponential decay of the particle velocity due to drag and U_0 is the initial velocity corresponding to dx/dt when $t = 0$. As examples of relaxation time for particle transfer in air figure 3.54 μs for particles with a radius of 0.5 μm or 1.23 ms for particles with a radius of 10 μm. In the case of Stokes flow for Re_p values <0.1, the particle drag coefficient is, as mentioned, inversely proportional to Re_p, so the relaxation time τ can be calculated as follows:

Fig. 6.13 Scheme of the behavior of particles in a fluid (from Li et al. 2018)

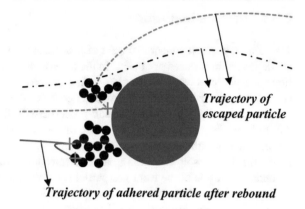

Trajectory of escaped particle

Trajectory of adhered particle after rebound

$$\tau = \frac{\rho_p d_p^2}{18\mu} \tag{6.140}$$

with ρ_p being the particle density, d_p the particle diameter, and μ the gas dynamic viscosity.

Particles suspended in a fluid flow can impact obstacles, due to their significant inertial momentum. Thereby these particles do not follow exactly streamlines which change direction very fast to contour obstacles, avoiding direct contact with them. The transported particles are not as rapid as streamlines to change direction, colliding thereby with the obstacles (Fig. 6.13). In this context, Stokes number (Stk) is a dimensionless parameter for the prediction and quantification of impaction of particles with objects. The Stokes number is defined as the ratio

$$Stk = \frac{\tau u_0}{l_0} = \frac{l}{l_0} \tag{6.141}$$

where τ is the relaxation time of the particle, u_0 is the fluid velocity windward to the obstacle, l is the stopping distance depending on particle size, velocity, and drag forces and l_0 is the characteristic dimension of obstacles, typically their transversal diameter. A particle with a Stokes number <1, follows fluid streamlines under perfect convection but a particle with a large Stokes number has higher inertia continuing along its initial trajectory.

The stopping distance τu_0 is the distance traveled by a suspended particle caused by inertial delay after the horizontal component of the airflow is zero, or after the vertical acceleration of flow is nil due to force equilibrium. In the latter case, vertical sedimentation velocity is as follows:

$$U_s = g\tau \tag{6.142}$$

6.5.3 Particle Deposition

The term deposition concerns the set of processes by which particles in fluid flows are deposited in surfaces or objects. In the case of aerosol, defined as a suspension of fine solid particles or liquid droplets with a diameter lower than 1 μm in the air or another gas, deposition processes, which decreases the concentration of particles in the air, are usually divided into two sub-processes which are dry and wet deposition. Aerosols can be natural such as fog, mist, or forest exudates, or from anthropogenic origins such as particulate air pollutants or smoke. Another classification of aerosols is of primary aerosols containing particles introduced directly into the gas flow or secondary aerosols borne from gas-particle conversion processes.

Dry deposition includes processes such as impaction, abovementioned, gravitational deposition due to the gravity-driven particles fall, the interception related with the unavoidable collision of particles following streamlines strictly with very close obstacles, the collision of particles transferred by turbulent eddies or the electrostatic attraction by particles and obstacles with different electric charges. Electrostatic forces are more effective when the particles are very close to obstacles. Direct interception involves particles whose size is neither large to have inertia nor small to diffuse within the streamlines.

Wet deposition consists of the purging of aerosol by atmospheric hydrometeors such as raindrops or snowflakes. Wet deposition is developed through gravitational, Brownian, or coagulation with water droplets. Types of wet deposition processes are below-cloud scavenging and in cloud scavenging. The former occurs when rain droplets and snow particles hit aerosol particles through Brownian diffusion, interception, impaction, or turbulent diffusion. The latter occurs when cloud droplets or cloud crystals act as nuclei or collide with aerosol particles and capture them. These particles can be brought to the ground surface by transportation with falling rain or snowflakes.

As an example, Monteith and Unsworth (2013) refer that for a short vegetation canopy with a height of around 0.1 m, the dimensional ranges of particles apt for increases from about 0.001 μm to fold scales of dozens of micrometers in the order from Brownian diffusion, interception, impaction, rebound to gravitational settling. Brownian displacement of particles controls the dislocation of sub-micron sized particles with possible coagulation up to characteristic dimensions of about 0.3 μm which largely contribute to the atmospheric amounts of 2.5 fraction of particulate matter. Brownian diffusion will be minimal with particle dimensions higher than 1 μm, especially with more viscous fluids. The Brownian deposition obeys Fick law, relating the diffusive flux to the gradient of concentration (Chap. 2), wherein fluxes are directed from high concentration to low concentration location as follows:

$$J = -D\frac{d\varphi}{dx} \tag{6.143}$$

with J being the diffusion flux defined as the amount of matter per unit area and unit of time, D is the diffusion coefficient or diffusivity defined in units of area per unit time, φ is volumetric mass concentration and x the length position. The particles under Brownian diffusion do not move along streamlines and should decrease with higher fluid velocities because under these conditions particles will have less time to diffuse.

Particle interception involves particles in the mid-range size ranging between 0.2 μm and 1 μm, a range neither large enough to have inertia for leaving streamlines nor small enough to diffuse with the flow streams. Impaction inertial particles correspond to characteristic dimensions ranging between 2 and 5 μm, and particle rebound upon impact with an obstacle surface or with particles on that surface occurs with particles with characteristic dimensions ranging about between 5 and 10 μm.

Particle rebound or bounce-off is more pronounced with relatively large particles, with higher normal impact velocities above a critical velocity, small obstacles with thin boundary layers, or small loss ok kinetic energy on impact. Mechanisms for higher deposition velocity, within ranges of about 10–100 ms^{-1}, are most effective for very small or very large particles. The latter with a characteristic dimension higher than 10 μm will sediment quickly under gravitational settling, with the deposition rate being the settling velocity due to the gravity-induced drag.

Monteith and Unsworth (2013) refer that in vegetal systems the capture of particles larger than about 10 μm is influenced by foliar surface stickiness of wetness, contrasting with smaller particles whose capture is more influenced by hairs or surface irregularities. In vegetal systems, the ratio of particle flux deposited on a surface to the atmospheric concentration at a level above is termed as the total deposition velocity.

Flux measurements (Chaps. 2 and 3) can be used for the calculation of a corrected surface deposition velocity for comparisons of deposition velocities determined on different surfaces and windspeeds. Particles with diameters lower than 50 μm are not prone to recirculate in airflow, after being deposited in vegetal leaf surfaces.

In the desert, Bagnold (1941) found the same for sand particles, although for larger grains displacement by wind would occur with redepositing of displaced particles. The later can induce displacement of other particles in a possible chain reaction through the termed saltation process, described below, with possible causation of dust storms. The practical impossibility of resuspension spores and pollen by wind forces, can explain why pathogens and fungi progressed in the formation of mechanical mechanisms to release spores, such as stalks above the viscous boundary layer of plant leaves.

The deposition processes of interception, diffusion, impaction, and sedimentation are not effective for particles in the middle range between 0.1 and 0.2 μm, a range size very representative of anthropogenic aerosols distributed in the atmosphere. These aerosols can be wet deposited by interaction with hygroscopic particles. For example, soluble sulphate particles in this range can be transported to large distances and persist in the atmosphere till they find conditions of high humidity where, by condensation, they can increase in size into larger drops that are more prone to be captured by falling rain. The particle diameters D_0 at dry state and D_s at a water vapor saturation ratio, S, defined as the product between the % of relative humidity and $1/100$ are related as follows:

$$D_s/D_0 = (1 - S)^{-\gamma} \qquad (6.144)$$

with γ being a hygroscopic growth parameter depending on particle chemical composition with a value of 0.2 for aged European aerosols and larger for marine aerosols. For γ of around 0.2, the size of particles would double for a relative humidity of 97%. Monteith and Unsworth (2013) mention the example of soluble aerosols of sea-salt particles, including sodium chloride interacting with bubbling water drops from waves in oceans, which grow by condensation at smaller saturation ratio than that of pure water. This is because dissolved salt causes a decrease of equilibrium vapor pressure over a water surface. As the drop derived from a dry salt particle grows, salt concentration decreases tending to pure water, with an increase in the equilibrium vapor pressure. In the development of larger hygroscopic droplets, a specific interaction exists, for a given set of environmental conditions, between the equilibrium drop size and variables such as relative humidity or particle mass.

The growth of soluble aerosol particles in instantaneous equilibrium with adjacent relative humidity increases quickly with environmental humidity. Those changes in particle dimensions influence vertical mass fluxes measured with the eddy covariance method. Indeed, considering an increased humidity close to, e.g., a grass canopy surface, particles moving upwards can be larger than those with the same dry diameter moving downwards, implying a correction for the particle growth due to this hygroscopic effect.

6.5.4 Sand and Dust Transfer

Atmospheric motion of sand and dust particles or high concentration of particulate in airflow, e.g. during the outbreak of a sandstorm, are huge environmental physical processes with relevant negative socio-economic impacts.

Transport of solid particles by wind is due to a combination of gravity with/or the flow of fluid where the sediment particles are entrained. The sediment transport can result in ripples or sand dunes.

The dust/sand particles are lifted by airflow under patterns depending on the diameter, density ratio, trajectories of the fluid and particles, kinetic energies, and inertia of the particles and the fluid. Turbulent vortices sweep along the ground surface, pushing particles which can be entrained and lifted if they are light enough. Dust/sand particles follow trajectories ranging between displacement in the atmosphere to great heights or distances with creeping and saltation cycles consisting of lifting and returning to the ground and lift again (Fig. 6.14).

Particle movement under fluid transport can occur under several modalities (Fig. 6.15). Particle creeping consists of particles rolling downstream, keeping

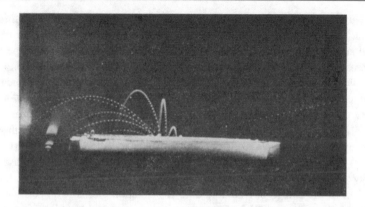

Fig. 6.14 Saltation cycles showing lift-up and trajectories of particles (from Mitha et al. 1986)

contact with the ground surface, under conditions where the forces exerted by the fluid do not allow a lift of particles. Creeping particles along the ground through small jumps might eventually splash and eject much smaller particles.

There have been few studies of creep and little is known about this type of motion, due to the extreme difficulty of direct observation of particle movements close to the ground (e.g., Cheng et al. 2012).

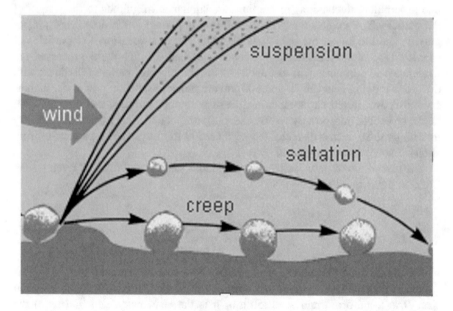

Fig. 6.15 Modalities of suspension, saltation, and creep of sand particle wind transport (from https://en.wikipedia.org/wiki/Saltation_(geology))

Under wind speeds above a critical value, termed as impact or fluid threshold, the drag forces exerted by the airflow are enough to lift some particles from the ground surface generating the termed saltation process. Saltation is a type of particle transport, e.g., by wind or water, consisting of a detachment of loose materials from a floor surface and transport by a fluid before returning to the surface.

Particle saltation is a combination of aerodynamic entrainment, wind field change, particle bed splashing impacts, and variable particle trajectories (Cheng et al. 2012). Aeolian saltation includes three main steps which are: (i) the acceleration by wind and gravity developing an arcing trajectory over the ground surface; (ii) progressive transfer of momentum from airflow with an increase of the quantity of sand/dust involved in saltation, insofar that the airflow becomes dependent of the course of saltation; and (iii) saltation grains, with some incipient creeping, hitting the ground surface with ejection or splashing of additional particles which become engaged in saltation dynamics and propagating thereby the transport process (Mitha et al. 1986).

Saltation particles are accelerated by fluid flow and pulled downward by gravity, causing them to travel in roughly arc ballistic trajectories. These trajectories are modified parabolas with initial angles close to 45° and smaller impact angles of about 10°.

Small particles, with characteristic dimensions smaller than 20 μm are entrained into the air stream, e.g., from the impact of saltation of sandy particles and carried out by large eddies over long-term suspension with potential convective advection up to great heights and long distances. For very small particles, vertical drag forces due to turbulent fluctuations in the fluid are similar in magnitude to the weight of the particle. The smaller the particle, the less important is the downward pull of gravity, and the longer the particle is likely to stay in suspension. For particles of characteristic dimensions ranging between 20 and 70 μm, these processes are moderate and suspension in airflow lasts for shorter-term periods. Particles with diameters ranging between 70 and 500 μm are prone to saltation processes, such as sand-drift over desert surfaces, snowdrift over smooth surfaces, soil blowing over fields, or pebble transport by rivers. Creeping occurs under very low fluid velocity or with particles with a diameter bigger than 500 μm or protected physically from strong influence with the fluids.

The mass flow of particles by saltation under steady conditions is given by the Bagnold equation (1936)

$$q = C \frac{\rho}{g} \sqrt{\frac{d}{D}} u_*^3 \qquad (6.145)$$

where q is the mass transport of sand across a floor strip of the unit ($kgm^{-1}s^{-1}$), C is a dimensionless constant equals to 1.8 for natural graded dunes, d is the mean grain size, D is a uniform grain size 250 μm, ρ is the air density, and u_* (ms^{-1}) the friction velocity considering the shear stress between the wind and the sheet of moving sand and under a condition that this velocity should be higher than the threshold shear velocity u_{*t} for initiation of particle motion. The threshold shear velocity can vary with sediment moisture or the presence of vegetation.

Alternative models for evaluation of particle saltation dynamics is the following (Kawamura 1951):

$$q = c\,\rho/g(u_* - u_{*t})(u_* + u_{*t})^2 \tag{6.146}$$

where c is a constant equals to 2.78 and the threshold shear velocity u_{*t} is given by

$$u_{*t} = A\sqrt{gd(\frac{\rho_s - \rho}{\rho})} \tag{6.147}$$

of Bagnold (1936), where A is a constant of 0.085 for impact threshold, and ρ_s is the sediment density taken as 2.650 kgm^{-3}.

Zingg (1953) proposed another expression

$$q = c\left(\frac{d}{D}\right)^{3/4}\frac{\rho}{g}(u_*)^3 \tag{6.148}$$

with constant c equal to 0.83.

As an example, for experimental application of particle saltation principles, it can be mentioned the data of Granja et al. (2012), regarding the vertical profile of sand saltation mass flow in the non-vegetated sand surface near the top of a smooth parabolic shaped dune at *Esposende* beach, Portugal. The floor site exhibited an unobstructed fetch of 60 m length, an estimated apparent roughness length of 1 mm, and a slope of about 10°.

These authors showed a clear exponential decay of saltation intensity away from the sand surface. Vertical flux profiles were evaluated using a geometrically weighed trap centring method. The sand transport and mass-flux profile measurements were collected during 13 data runs with arrays of seven vertically stacked hose style traps. Over the runs, mean sand grain sizes varied from 0.27 to 0.35 mm, averaging 0.32 mm and friction velocities ranged between 0.35 and 0.49 ms^{-1}. The measured vertical sand mass fluxes ranged between 0.004 and 0.015 kgs^{-1}m^{-1}, averaging 0.005 kgs^{-1}m^{-1}. The estimated fluxes through models of Bagnold (1936), Kawamura (1951) and Zingg (1953) averaged 0.011 kgm^{-1}s^{-1}, 0.020 kgm^{-1}s^{-1} and 0.007 kgm^{-1}s^{-1}, respectively.

The general nonlinear least-squares fitted curves for describing the exponential decay of sand mass-flux gradient, over heights above floor surface ranging between 0 and 600 mm, were as follows:

$$Q_n = \alpha\exp(\beta h) \tag{6.149}$$

where α and β are the regression coefficients, Qn is the normalized flux in percent per 10 mm height, and h is the elevation (in mm) of any point above the surface. The fitted coefficients α and β showed ranges of 13.52–24.78 and -0.010 to -0.024, respectively, with an R^2 of 0.992. These results were very similar to the ones obtained from a study in Guadalupe beach in California, USA, with sand with grain size of about 0.39 mm, giving average values for α and β of 13.65 and -0.01. The two sites in Western Europe and North America, delivered very similar patterns of sand mass normalized fluxes over heights above floor surfaces ranging from 0 to 60 cm, with a very good similarity in the range between 20 and 70 cm, revealing a common decay pattern of exponential vertical mass flux. The methodology applied was supposedly valid for long-term analysis, under available representative databases for wind velocity, sediment accumulation, weather, and micro-topographic data.

Cheng et al. (2012), followed a Lagrangian modeling approach for analyzing the dynamics of entrainment in the atmosphere of dust particles after the outbreak of strong winds storms. Lagrangian turbulent flow modeling is a common pathway for the treatment of dispersion problems consisting of the description of a given flow of molecules and particles overtime periods, instead of Eulerian analysis in the fixed control volume. Given the specificity and uniqueness of turbulent flows and patterns, which depend on the initial and boundary conditions under a nonlinear way, numerical models are not strictly universal without a need to readjustment of model parameters. During a period of strong winds, soil erosion with sand and dust emissions is due to descending strong average airflow and fluctuations, with dust entrainment due mainly to the coherent structure of wind gusts.

A first theoretical step for the analysis of particle trajectory in airflow is the quantification of the instantaneous fluid velocity of the air surrounding each particle, or the velocity of fluid seen by the particle, along with successive time instants. Initially, the fluid and the particle are at the same position, but because of the inertia of the particles and gravity effects, the two trajectories will be eventually de-correlated with different velocities and the rising of relative velocity. Throughout its trajectory, the solid particle will leave the initial single air parcel slipping instead between parcels and eventually leaving the initial turbulent structure and moving to another one.

For the calculation of sand-dust particles entrainment in a quasi-two-dimensional incompressible flow, Cheng et al. (2012) showed that lagrangean autocorrelation for particle acceleration is much smaller than for velocity, considering large Reynold numbers and time periods Δt higher than the Kolmogorov time scale τ_K

$$\Delta t > > > \tau_K = (v/\bar{\varepsilon})^{1/2} \qquad (6.150)$$

In Eq. (6.150) v is the kinematic viscosity of air and $\bar{\varepsilon}$ the ensemble average rate of TKE dissipation. The particles velocities between two instants t and $t + \Delta t$ can be considered independent, with the latter depending on the velocity and acceleration of the fluid, and not at any other mechanisms acting on the acceleration for time periods smaller than Δt.

These authors used measured values of air velocity in surface boundary layers related to the airflow velocity field and turbulent energy for parametrization of initial and boundary conditions for the simulations of sand–dust particles entrainment.

The numerical simulations assumed particle densities of 1500, 1700, and 2650 kgm^{-3} for dust, sand and dust mixtures, and sand, respectively, with a stationary horizontal velocity of 13 ms^{-1}. Particles were injected into the flow randomly over the entire ground surface, with the injection angle and particle vertical velocity being considered as Gaussian randomly distributed. The trajectories of particles were computed over 5000 m with three mean diameters of 5, 20, and 40 μm, considering suspension as the only transportation mechanism without rebound or ejection of new particles after returning to the ground surface.

The first simulation of sand–dust entrainment in a purely horizontal flow with three suspensions with 50 hypothetical particles each regarding dust, sand, and sand and dust mixtures, showed that under the boundary turbulent velocity field assumed very few particles could reach a 150 m height. This height was about 100 times greater than the thickness of the estimated initial sand/dust 1.5 m two-phase boundary layer of fluid flow. The calculations estimated that about 34, 48, and 88% remained in the 1.5 m layer. The greater height that dust or mixtures of dust and sand could reach was about 250 m. For sand, the estimated maximum height achieved was 180 m.

The second simulation concerned entrainment of particles injected in a descending airflow over the ground. A vertical component was introduced in the wind velocity profile, and it was shown that in three suspensions with 50 particles each, about 92, 98, and 72%, of particles of dust, sand, and sand and dust mixtures remained in the 1.5 m double phase layer. The larger height achieved by dust was 20 m, with only one sand particle reaching 50 m and the remaining lift by no more than 1 m. For dust and sand mixture, only one particle reached 50 m height and the remaining reached 30 m at maximum.

The third type of simulation concerned wind gust, as the only element of the airflow velocity field with a downward velocity of 1 ms^{-1}. A major change in the boundary and initial conditions was imposed, with new variables concerning vertical and horizontal amplitudes of wind gust changes with height above the ground surface. The results of these simulations with three hypothetical particle suspensions showed that, with the coherent structure of wind gust, particles could indeed reach the top of the atmospheric boundary layer.

As expected, lower quantities of heavier particles should reach the top height by comparison with lighter particles. Sand/dust particles with a diameter ranging between 5 and 40 μm can overcome the descending air motion, penetrate the middle and upper levels of the atmospheric boundary layer, and then propagate further and diffuse into the troposphere where ascending air motion prevails. This is a typical scenario for soil erosion, sand/dust emission, and entrainment, e.g., in eastern Asia during spring (Cheng et al. 2012).

6.5.5 Sediment Transportation in Rivers

Another typical environmental management issue where mass transfer principles are fundamental is the quantification of sediment fluxes in rivers. The typical dimension of transported particle sediments is around 1 mm or smaller because air is a light fluid with low viscosity, and therefore, not able to deliver much shear on its bed. Fluvial transport with flowing water in natural systems such as rivers or streams with higher density and viscosity is larger than air.

Water transport of sediments in rivers can result in the formation of structures like ripples, dunes, fractal shapes, or floodplains. This problem is strictly related to increases in, e.g., channel erosion and deposition potential or in flood risks or in the degradation of the benthic zone. The movement of bed material begins as soon as the critical values of shear stress ($\tau = \rho g h S$ on average terms with ρ being the water density, g the gravity acceleration, h the mean flow depth, and S the channel slope) or specific stream power ($\omega = \tau U$ with the U representing the mean water velocity) are surpassed. Above this threshold, the transport rate increases nonlinearly with excess stress or power, meaning that estimates of sediment transport flow duration or total bed material flux are highly sensitive to the values of critical shear stress or specific stream power.

In this context, effective management of fluvial resources requires a quantitative catchment-scale approach which must be based on variables such as discharge rates, slopes, channel widths, or grain sizes. Stream bedload transport is highly intermittent in streams and rivers with beds consisting mainly of gravel or boulders. These are examples of easily estimable and/or collectible variables, e.g., by remote sensing, existing databases, or direct measurement. Many sediment transport models, following empirical or theoretical approaches, exist with each one developed under a specific context and objective, and thereby a major goal to achieve should be the evaluation of model sediment transport with limited data available (Lammers and Bledsoe 2018).

Fluvial sediment fluxes depend on variables in theoretical transport equations, related to shear stress, velocity fields, or flow depths. However, due to their complex temporal and spatial variability, these variables are difficult to measure or model turning thus impracticable their application in quantification, e.g., of load transport in equations at the catchment-scale level. A main attempt to overcome these handicaps was the introduction of an alternative variable termed as specific stream power, ω, that can be defined as the available power supply in a stream as follows:

$$\omega = \frac{\Omega}{w} = \frac{\rho g Q S}{w} \qquad (6.151)$$

where Ω is the unit-length power stream by unit-length ($\mathrm{Wm^{-1}}$ or in $\mathrm{Nm^{-1}s^{-1}}$), w is the channel width (m), Q is the discharge rate ($\mathrm{m^3s^{-1}}$) and ω the specific stream power ($\mathrm{Nm^{-2}}$) and g is the gravitational acceleration ($\mathrm{ms^{-2}}$). Specific stream power in rivers or water stream is an expediting estimation as mentioned above using, e.g., streamflow gauges or remote sensing which allow quantifying channel slope.

A traditional pathway for characterizing the potential for sediment transportation in river basins is based on the abovementioned assumption that the amount of flowing water energy needed for the onset of an incipient particle motion and sediment transport should be higher than a critical threshold of shear stress. The rationale is that solid particles remain without acceleration if the acting forces such as gravity, buoyancy, drag, or friction are in equilibrium. Shear stresses produced by vertical velocity gradient deliver a drag force against solid particles in bedrock that, if higher than the balance of forces, induce the beginning of the acceleration of these particles. The magnitude of shear stresses is dependent on the surface slope, channel geometry, and flow dynamics. This critical shear stress can be expressed by a dimensionless ratio which is the Shields parameter with its near-constant critical value θ_c, expressed as

$$\theta_c = \frac{\tau_c}{(\rho_s - \rho)D_s} = \frac{hS}{(s-1)D_s} \tag{6.152}$$

where τ_c is the critical shear stress (Pa), ρ_s and ρ are sediment and fluid density (kg m^{-3}), g is the gravitational acceleration (ms^{-2}), h, is the flow depth, D_s is the representative grain size (m) usually considered as the median grain size D_{50}, and s is the sediment specific gravity.

The application of this methodology is limited by the reliance on flow depth, which can be calculated from discharge and flow resistance, under assumptions on channel geometry not always applicable. Another limitation is that Eq. (6.152) is based on mean values, averaged over the width of the channel, instead of the real forces on individual solid particles due to practical difficulties in accurately estimating the real force acting on individual particles. This reduces the accuracy of this shear stress-based methodology for evaluation of sediment transport contradicting the assumption that real shear stress should be more representative of sediment transportation mechanisms than specific stream power. Another limitation of the average formulation of shear stress in Eq. (6.153) may be the fact, reported by meta-analysis works, that specific stream power was more correlated with bedload transport rates than shear stress or velocity (e.g., Parker 2010; Parker et al. 2011; Martin and Church 2000).

Another handicap with the Shields parameter can be a bias with channel slope which could deliver an over-prediction of sediment transport rates. Stream power-based transport equations can be more accurate with higher channel slopes due to the absence of correlations between stream power and channel slope (Lammers and Bledsoe 2018).

The following empirical bedload equation presented by Bagnold (1980) was pivotal for the discussion and estimation of the critical stream power, ω_c, expressed in kg m^{-1}s^{-1}, for delivering incipient motion of sediments

$$\omega_c = 5.75 \left(\frac{g}{\rho}\right)^{0.5} [(\rho_s - \rho)D_{50}0.04]^{3/2} log\left(\frac{12h}{D_{50}}\right) \qquad (6.153)$$

with the dependent variables defined above. This equation still relies on the flow depth and requires additional data not easily quantifiable at the catchment scale.

An attempt to improve this equation for obtaining a non-dimensional version of critical stream power ω^*, by the elimination of flow depth, was proposed by Parker et al. (2011)

$$\omega^* = \frac{\omega}{\rho(gRD_{50})^{3/2}} \qquad (6.154)$$

where $R = (\rho_s-\rho)/\rho$ is the termed submerged specific gravity of the sediment grains. Equation (6.154) was found by these authors as useful for comparing empirical estimates of critical power with different values of D_{50}. Parker et al. (2011), considered that dimensionless critical stream power could be considered weakly correlated with channel slope and with a mean value of around 0.1. Despite its simplicity, and without relying on the depth of the flow or some measure of roughness, these approach for critical specific stream power has shown similar accuracy to more complicate and data demanding equations. The assumption of a constant value for ω^*, or distribution of values for ω^* centered on that value allows ω_c to be calculated using only grain size represented, e.g., by D_{50}.

Lammers and Bedsoe (2018) based on a broad meta-database analysis work corroborated these findings delivering average values for ω_c^* of 0.085 ± 0.03 and 0.1 ± 0.065, similar to the 0.1 average of Parker et al. (2011) from flume and field data, respectively. These results are not totally incompatible with those of Camenen (2012), wherein ω_c^* varied by a factor of only about 1.5 for a range of channel slope magnitudes between 0.02 and 30%.

Alternative robust empirical approaches for quantifying ω_c were proposed. The empirical equation presented in Bagnold (1980) for sediment transport was the following

$$\frac{q_b}{q_{b,ref}} = \frac{s}{s-1}\left[\frac{\omega - \omega_c}{(\omega - \omega_c)_{ref}}\right]^{3/2}\left(\frac{h}{h_{ref}}\right)^{-2/3}\left(\frac{D_{50}}{D_{50,ref}}\right)^{-1/2} \qquad (6.155)$$

wherein q_b is the unit bedload transport rate (kg m^{-1}s^{-1}, dry mass), ω, and ω_c are the specific and critical specific stream power, respectively (kgm^{-1}s^{-1}). The subscript *ref* means "*reference values*" which Bagnold used for making its empirical equation dimensionless, and values $q_{b,ref}$ = 0.1 kgm^{-1}s^{-1}; $(\omega - \omega_c)_{ref}$= 0.5 kgm^{-1}s^{-1}; h_{ref} = 0.1 m, and $D_{50,ref}$= 1.1 E^{-3} m. Equation (6.156) shows that the sediment transport rate is inversely related to flow depth and to the relative roughness (h/D_{50}).

Lammers and Bledsoe (2018) based on Bagnold sediment relationship, introduced the unit discharge q defined as

$$q = Q/w \qquad (6.156)$$

which is directly related to flow depth

$$\frac{Q}{wU} = \frac{q}{U} = h \qquad (6.157)$$

where q is the unit discharge (m^2s^{-1}), U is the average water flow velocity (ms^{-1}), w is the channel width (m) and Q is the discharge rate (m^3s^{-1}). Thus, flow depth and unit discharge are only equal at a constant velocity. Flow depth is also influenced by surface roughness and this influence is neglected by replacing depth with unit discharge.

In all this context, Lammers and Bledsoe (2018) based on extensive meta-analysis and database treatment, proposed two empirical equations for bedload q_b ($kgm^{-1}s^{-1}$) and total sediment load transport Q_t (ppm) as follows:

$$q_b = a_1[\omega - \omega_c]^{3/2} D_s^{-1/2} q^{1/2} \qquad (6.158)$$

and

$$Q_t = a_2[\omega - \omega_c]^{3/2} D_s^{-1} q^{-5/6} \qquad (6.159)$$

where the coefficient a_1 is 0.0044 if ω and ω_c are expressed in kg $m^{-1}s^{-1}$ units. The value of ω_c, indicative of a threshold of a motion, can be obtained from Eq. (6.155) by assuming the value of 0.1 for ω_c^*. In Eq. (6.160), expressing ω and ω_c in kg $m^{-1}s^{-1}$ units, the coefficient a_2 is 0.66. These two equations for values of bedload and total sediments transport rates ranged between 10^{-4} and 10^2 $kgm^{-1}s^{-1}$ and 10^{-1} and 10^5 ppm, respectively, showed good agreements, under residual metrics, with data from direct measurements on available databases and from four alternative models.

Sensitivity analysis showed that discharge and channel slope had the largest influence on transport rates either with bedload or with total load sediments. On transport rates of both loads, channel width, grain size, and ω_c^* showed having a small effect only.

Among theoretical approaches for analyzing sediment tQransport in rivers it is worth mention those developed by Lamb et al. (2008) and Ferguson (2005, 2012). Lamb et al. (2008) established equations for quantifying the influence of critical shear stress in fluvial sediment transport on balance equilibrium in the coordinate system parallel to the streambed between the system forces composed by buoyancy,

F_B, lift, F_l, drag, F_D, and gravity forces F_G on grain particles in streamflow as follows:

$$F_D + (F_G - F_B) \sin \beta = [(F_G - F_B) \cos \beta - F_l] \tan \phi_0 \qquad (6.160)$$

where ϕ_0 is a friction angle between grains and β is the bed slope angle. These authors through an extensive meta-analysis evaluation, of flume and field data, established that for particle Reynolds number $R_{ep} > 10^2$, the critical shear stress τ_c^*, corresponding to an onset of incipient motion can be given by the following expression:

$$\tau_c^* = 0.15 \, S^{0.25} \qquad (6.161)$$

thereby showing a trend of increasing critical shear stress with channel slope.

Ferguson (2012) conceptualized a model for sediment transport under a theoretical approach whose main assumption is based on the hiding and protrusion of solid grains in streambeds with distinct grain sizes. It was considered that some of the total shear stress in a river with an almost absent sorted coarse bed is unavailable for sediment transportation because it is expended on the form drag on protruding standing grains in bedforms or obstacle clasts with less energy remaining for acting on individual grains which are potentially mobile. The predominant grain sizes correspondent flow resistance and flow entrainment are not the same, with coarser grains dominating in the former and finer grains prevailing in the latter, e.g., in the context of huge amounts of transport under big floods. The protruding grains which generate additional resistance to flow belong to the coarser half of the size distribution.

The traditional critical value of the Shields parameter, θ_c, regards mainly beds with smooth topography such as the generated in flume experiments at low shear stresses. The minimum degree of bed irregularity, termed as "grain roughness", generates the base resistance conditions without grain hiding or protrusion. The total stress at the onset of motion will be higher than the base resistance created by immobilized coarse grains, as with macroscale bedforms such as barriers or dunes, to an extent that depends on the ratio on the ratio of total resistance to the base level of flow resistance. In this context, the total shear stress in a channel with a slope S and a range of distribution of grain sizes is given by

$$\tau = \rho g h S \qquad (6.162)$$

is partitioned into the components related with the base and additional resistance, with the mean velocity U of a stream depends on a balance between total flow resistance and shear driving force which can be represented by the shear or friction velocity

$$u_* = (gdS)^{1/2} = \left(\frac{\tau}{g}\right)^{1/2} \tag{6.163}$$

and the total flow resistance can be expressed as

$$\frac{U}{u_*} = \frac{U}{(ghS)^{1/2}} = F\left(\frac{h}{k}\right) \tag{6.164}$$

with F being a general monotonically increasing resistance function and k, a roughness height generally scaled representative diameters such as D_{84} or D_{90}. Different velocities at the same flow depth, caused by different additional resistances of protruding grains in bed material, correspond to different depths for the same velocity due to the exponential vertical velocity profiles of the boundary layer. Under base resistance with lower roughness height and probably smaller k', a given mean fluid velocity, U, would occur at lower mean depth h', and unit discharge $q = h'U$, with the ratio (h/h') increasing with decreasing of relative submergence defined by the ratio (h/k). Steep channels tend to typify a pattern of coarser bed material and shallow flow, giving higher values for (h_c/h'_c) and θ_c. The ratio of critical shear stress to critical stress at base resistance conditions without particle hiding and protrusion can be related to the ratio of critical flow depths as follows:

$$\frac{\theta_c}{\theta'_c} = \frac{h_c}{h'_c} \tag{6.165}$$

with θ'_c and h'_c being the critical Shields parameter and flow depth at the base resistance condition.

Each resistance equation expressed in the general form of Eq. (6.165) is derived from databases with, e.g., distinct velocities and resistances under the same flow depth. The key criterion for choosing the total resistance function is the good fit of roughness and velocity data mainly under intermediate to low submergence ratios (h/k) typical of thresholds conditions in coarse bed streams in accordance with simulations with the expectable patterns of shallow flows.

The base level of the flow resistance supposedly follows a vertical logarithmic profile of the type

$$\frac{U}{u_*} = \frac{1}{\kappa} \ln\left(12.2\frac{h}{D_{50}}\right) \tag{6.166}$$

with κ being the von Karman constant of about 0.41 which, under the assumption that the vertical velocity profile is logarithmic, can describe the grain resistance of a plane bed. The logarithmic equation can be written as a power law equation as follows:

$$\frac{U}{u_*} = a_0 \left(\frac{h}{D_{50}}\right)^{1/6} \tag{6.167}$$

One possible formulation for the flow resistance function is the variable power flow resistance (VPE) equation as follows:

$$\frac{U}{u_*} = \frac{a_1\left(\frac{h}{D_{84}}\right)}{\left|\left(\frac{h}{D_{84}}\right)^{5/3} + \left(\frac{a_1}{a_2}\right)^2\right|^{1/2}} \tag{6.168}$$

with a_1 and a_2 constants of about 6.5 and 2.5, respectively. The VPE equation is limited by two asymptotes which are

$$\frac{U}{u_*} = \frac{a_1}{\left(\frac{h}{D_{84}}\right)^{1/6}} \tag{6.169}$$

in deep flows and

$$\frac{U}{u_*} = \frac{a_2 h}{D_{84}} \tag{6.170}$$

in shallow flows.

The deep flow asymptote is equivalent to the Gauckler–Manning–Strickler equation for calculation of average velocity of water in turbulent flows in open channels written as follows:

$$U = S^{1/2} D_{84}^{-1/6} h^{2/3} \tag{6.171}$$

The shallow flow asymptote is consistent with the linear velocity vertical profile within a roughness layer. The VPE equation allows for differences in eddy viscosity and velocity profile between deep and shallow flows. It was shown that this equation delivers accurate and unbiased predictions of flow velocity throughout the natural ranges of channel slope between less than 0.001 and more than 0.2, and of relative submergences between 0.1 and 80.

A given velocity can be achieved with lower depth and unit discharge under lower resistance to flow. The total resistance function above (Eq. 6.169) can be used for obtaining the average velocity corresponding to a given flow depth. This velocity can be thereafter used in the power approximation to vertical flow

(Eq. 6.168) to obtain the lower flow depth corresponding to the base level of flow resistance.

The ratio (h_c/h_c') after some manipulation can be written as a function of the relative roughness (D_{84}/h), or the inverse of relative submergence, and grain sorting (D_{84}/D_{50})

$$\frac{h_c}{h_c'} = \left(\frac{a_0}{a_1}\right)^{3/2} \left(\frac{D_{84}}{D_{50}}\right)^{1/4} \left[1 + \left(\frac{a_1}{a_2}\right)^2 \left(\frac{D_{84}}{h}\right)^{5/3}\right]^{3/4} \tag{6.172}$$

This equation predicts that as relative submergence increase, (h/h') decline towards asymptotic values between 1.4 and 2 depending on bed sorting, for deep flows. From a set of values of (h_c/D_{84}), the correspondent sets of values of (h_c/h_c') and (θ_c/θ_c') can be obtained from Eq. (6.172) and Eq. (6.165). The corresponding set of channel slopes can be obtained using the definition given above for the Shields parameter (Eq. 6.152)

$$S = \frac{(s-1)\theta_c D_{50}}{h_c} = \frac{(s-1)(\theta_c/\theta_c')\theta_c'}{(h_c/D_{84})(D_{84}/D_{50})} \tag{6.173}$$

Equation (6.173) allows obtaining critical Shields parameter with trial values of (h_c/D_{84}).

Ferguson (2012) follows the dimensionless form of Parker et al. (2011) in Eq. (6.155), for derivation and comparison of relationships between dimensionless critical specific stream power ω_*, and e.g., channel slope or median grain size D_{50}.

A dependency between ω_c^* and (U/u_*) was proposed, e.g., by Eaton and Church (2011) as follows:

$$\omega_c^* = \theta_C^{3/2} \frac{U}{u_*} \tag{6.174}$$

with ratio (U/u_*) obtained from Eq. (6.168). Equation (6.172) was also explored by Ferguson (2012) for prediction of curves for critical shear stress variation, θ_c obtained from Eq. (6.174) or (6.153), with channel slope, different grain sizes, and with trends from mechanistic models of Lamb et al. (2008) relating critical Shield shear stress as a function of channel flow.

Ferguson (2005) proposed a theoretical based approach for the estimation of critical stream power, ω_c, as follows:

$$\omega_c = \frac{2.3\rho}{\kappa}[\theta_c(s-1)gD_S]^{3/2}\log\left[\frac{30h_c}{\exp(1)\kappa_s}\right] \tag{6.175}$$

with κ being the von Karman constant of 0.4, s the relative sediment density of about 2.65, d the grain size, h_c the critical water depth at the onset of motion, and κ_s the roughness height of a $2D_s$ order of magnitude.

Camenen (2012) developed the model concept for Eq. (6.175) by introducing:

(i) a critical Shields parameter, $\theta_{c,0}$, without channel slope effects and depending on the grain size

$$\theta_{c,0} = \frac{0.3}{1 + 1.2} + 0.055[1 - \exp(-0.02D_*)] \tag{6.176}$$

with

$$D_* = \left[\frac{g(s-1)}{v^2}\right]^{1/3} D_s \tag{6.177}$$

(ii) optimizing the nonlinear relationship between the critical relative flow depth defined as R_h/D_s, corresponding to the flow depth at the onset of motion, with R_h being the hydraulic ratio as follows:

$$\left(\frac{R_h}{D_s}\right)_c = \frac{(s-1)\theta_{c,0}}{s}(0.5 + 6S^{0.75}) \tag{6.178}$$

and

(iii) introducing an effect of channel slope, through an angle of repose φ_S, accounting for the friction between the sediment and slope and by definition of critical value for the angle of repose $\varphi_{cr,S}$ without sediment flow, allowing for a development as follows:

$$\frac{\theta_{C,S}}{\theta_{c,0}} = \cos(\arctan S)\left[1 - \frac{S}{\tan(\varphi_S)}\right] \tag{6.179}$$

Hydraulic radius, R_h, is defined as the ratio between the cross-sectional area of the flow and the perimeter of the wetted perimeter of the cross-section, wherein the wetted perimeter includes all the surfaces directly influenced by the shear stress of the fluid. This parameter is used in turbulent flows with the advantage of being a single unidimensional variable useful in dimensionless fluid flow variables such as Reynolds number.

Equation (6.179) represents a correction of θ_c' in Eq. (6.166), relative to critical Shields stress parameter, under the base resistance conditions concept, θ_c' (Ferguson 2012), to the downslope component of grain particle weight which aids particle dislocation and to a certain extent further reduces θ_c'. In this context $\varphi_{cr,S}$ is considered as equal to 52°, which is a value relative to irregularly packed beds.

In the context of Eqs. (6.175) to (6.178), the theoretical model for critical stream power, ω_c, of Camenen (2012) was

$$\omega_c = \frac{2.3\rho}{\kappa} \left[\left(\frac{R_h}{D_S}\right)_c \left(\frac{\theta_{c,S}}{\theta_{c,0}}\right) sgD_S \right]^{3/2} \log\left[\frac{15}{\exp(1)} \left(\frac{R_h}{D_S}\right)_c\right] \qquad (6.180)$$

The ω_c values estimated by Eq. (6.180) can be used, e.g., in empirical Eqs. (6.158) and (6.159) of Lammers and Bledsoe (2018) for quantification of bedload and total sediment load transport.

Camenen (2012) extended the calculations for specific critical stream power as follows:

$$\omega_c^* = \frac{\theta_c F_r}{\sqrt{s-1}} \sqrt{\frac{R_h}{D_s}} \qquad (6.181)$$

where $F_r = U/\sqrt{gR_h}$ in (Eq. 5.6) is the Froude number that represents the ratio of inertial to gravity forces (Chap. 5).

The introduction of the Froude number into this analysis, allows the implementation and definition of a critical flow ($F_r = 1$) curve in a log graphical analysis of h/D_{84} in ordinates vs. channel slope in abscissas, as derived by Ferguson (2012). These authors established that the curve of $F_r = 1$ is circa an upper limit of likely combinations of slope and relative submergence, defined in this case as h/D_{84}, in alluvial channels.

Below the curve of $F_r = 1$, it was shown an increase with channel slope, of the flow depth at which sediment transport is suppressed by typical flow resistance, above the threshold base resistance. This tendency is illustrated by the differences along increasing channel slopes, between the thresholds of smooth beds with base resistance and constant value of θ_c and of the predicted threshold for typical total flow resistance.

For example, at slopes as low as 0.001, the predicted flow depth required to transport bedload in a river with typical resistance is about twice the depth required to do the same in a river with base resistance only. For channel slopes of around 0.05, the equivalent ratio was predicted as of about 5. In steeper fluvial channels bed load movement is restricted to a narrow range of hydraulically subcritical flows close to $Fr = 1$ critical flow curve above supercritical flow area. The complexity of the process is higher for a poorly sorted bed particle with a ratio (D_{84}/D_{50}) of 3.

A more homogeneous bed sorted particles, e.g., with (D_{84}/D_{50}) at around 2, can be reflected in higher ratios of flow depths corresponding to fluvial typical total and base resistances. This is simultaneous with a crossing of the typical flow resistance inside the supercritical flow area (Ferguson 2012).

Empirical and theoretical approaches to fluvial sediment transport can be considered complementary under the above context. Empirical equations such as those of Lammers and Bledsoe (2018) concerning sediment transport equations based on stream power have the advantage that allows obtaining reliable information for catchment-scale analysis despite being parsimonious insofar that can be expeditiously evaluated, e.g., by remote sensing, measurements or available databases such as channel slope, discharge, width, or bed grain size, with higher influence on the first two. An important detail is that these models are not dependent on particle size distribution as much as with topographic data of discharge.

Empirical formulations are, however, biased by uncertainties related to the multitude of variables and interactions involved, which justify the development of alternative theoretical approaches for deepening the knowledge of processes involved as for the betterment of empirical models.

For example, Eq. (6.159) relative to the transport of bed sediments was shown as more accurate to fine-grained sizes with coarse-grained streams likely to be near-threshold conditions where the possible incertitude in critical stream power may deliver greater variability in transport rates. Another example, presented by Parker et al. (2011), of the source of complexity is the interpretation of the positive relationship between slope and critical mean bed shear stress. According to these authors, the possible location in steep headwater streams of prominent stabilizing bed structures, hiding effects, or form roughness or of flow aeration at high slopes are reflected in an increase of flow resistance with slope and in a decrease of local flow velocity around bed particles. All these factors add sources of uncertainty in the complex problem of flow analysis that must be subjected to continuous analysis.

References

Arya, S. P. (1988). *Introduction to micrometeorology* (Vol. 42, 307 pp.). International geophysics series. Academic Press.

Bagnold, R. A. (1936). The movement of desert sand. *Proceedings of the Royal Society of London. Series A, 157*, 594–620.

Bagnold, R. A. (1941). *The physics of blown sand and desert dunes* (p. 265). London: Methuen.

Bagnold, R. A. (1980). An empirical correlation of bedload transport rates in flumes in natural rivers. *Proceedings of Royal Society of London A, 372*(1751), 453–473.

Bennet, I. (1965). Monthly maps of mean daily insolation for the United States. *Solar Energy, 9*, 145–152.

Brown, G. W. (1969). Prediction temperatures of small streams. *Water Resources Research, 5*, 67–75.

Camenen, B. (2012). Discussion of understanding the influence of slope on the threshold of coarse grain motion: Revising critical stream power by C. Parker, N. J. Clifford and C.R. Throne. *Geomorphology, 126*, 51–65.

Campbell, G. S., & Norman, J. M. (1998). *An introduction to environmental biophysics* (293 pp.). Springer.

Cheng, X. L., Zeng, Q. C., & Hu, F. (2012). Stochastics modeling the effect of wind gust on dust entrainment during sand-storm. *Chinese Science Bulletin, 57*, 3595–3602.

Eaton, B. C., & Church, M. (2011). A rational sediment transport scaling relation based on dimensionless stream power. *Earth Surface Processes and Landforms, 36*, 901–910.

Ferguson, R. I. (2005). Estimating critical stream power for bedload transport calculations in Gravel-Bed rivers. *Geomorphology, 70*, 33–41.

Ferguson, R. I. (2012). River chanel slope, flow resistance, and gravel entrainment thresholds. *Water Resources Research, 48*, 1–13.

Foken, T. (2017). *Micrometeorolog* (2nd ed., 362 pp.). Berlin: Springer.

Fox, R. W., & McDonald A. T. (1985). *Introduction to fluid mechanics* (742 pp.). Wiley.

Gates, D. M. (1980). *Biophysical ecology* (611 p.). Springer.

Granja, H. M., Farrel, E. J., Ellis, J. T., & Sherman, D. J. (2012).Eolian Saltation at Esposende Beach, Portugal. *Journal of Coastal Research, 56,* 327–331.

Hillel, D. (1982). *Introduction to soil physics* (364 pp.). Academic Press.

Holman, J. P. (1983). *Transferência de calor* (639 pp.). McGraw-Hill (in portuguese)

Kawamura, R. (1951). *Study of sand movement by wind* (Vol. 5, No. 34). Institute of Science and Technology, University of Tokyo, Technical Reports. Translated as University of California Hydraulics Engineering Laboratory Report HEL 2–8, 1964.

Lamb, M. P., Dietrich, W. E., & Venditti, J. G. (2008). Is the critical shields stress for incipient sediment motion dependent on channel-bed slope? *Journal of Geophysical Research: Earth Surface, 113,* FO2008. https://doi.org/10.1029/2007JF00083.

Lammers, R. W., & Bledsoe, B. (2018). Parsimonious sediment transport equations based on Bagnold's stream power approach. *Earth Surface Processes and Landforms, 43,* 242–258.

Lee, R. (1978). *Forest micrometeorology* (276 pp.). Columbia University Press.

Levit, H. J., & e Gaspar, R. (1987). Energy budget for greenhouses in humid temperate climate. *Agricultural and Forest Meteorology, 42,* 241–254.

Li, W., Li, H., Shen, S., Cui, F., Shen, B., & Huang, Y. (2018). Study of particle rebound and deposition on fibre surface. *Environmental Technology.* https://doi.org/10.1080/09593330. 2018.1509137.

Linacre, E. T. (1972). Leaf temperature, diffusion resistance and transpiration. *Agriculture Meteorology, 10,* 365–382.

Linacre, E. T., Palmer, J. H., & Trickett, G. S. (1964). Heat and moisture transfer from trimmed glasshouse crops. *Agricultural Meteorology, 1,* 66–72.

List, R.J. (1963). *Smithsonian meteorological tables* (6th ed.). Smithsonian Inst., EUA.

Liu, B. Y., & Jordan, R. C. (1960). The interrelationship and characteristic distribution of direct, diffuse and total solar radiation. *Solar Energy, 4,* 1–9.

Martin, Y., & Church, M. (2000). Re-examination of bagnold's empirical bedload formulae. *Earth Surface Processes and Landforms, 25,* 1011–1024.

Mimoso, J. M. (1987). *Transmissão de calor, bases teóricas para aplicação à térmica de edifícios* (157 pp.). Lisboa: LNEC (in portuguese).

Mitha, S., Tran, M. Q., Werner, B. T., & Half, P. K. (1986). The grain-bed impact process in Aeolian saltation. *Acta Mechanica, 63,* 267–278.

Monteith, J. L., & Unsworth, M. H. (1991). *Principles of environmental physics* (2nd ed., 291 pp.). Edward Arnold.

Monteith, J. L., & Unsworth, M. H. (2013). *Principles of environmental physics* (4th ed., p. 403). Oxford: Academic Press.

Oke, T. R. (1992). *Boundary layer climates* (2nd ed., 435 pp.). Routledge.

Özisic, N. M. (1990). *Transferência de Calor. Um Texto Básico* (661 pp.). Editora Guanabara Koogan S.A. (in portuguese).

Parker, C. (2010). *Quantifying catchment-scale course sediment dynamics in British rivers.* Thesis (Ph.D. dissertation). University of Nottingham, Nottingham

Parker, C., Clifford, N. J., & Thorn, C. R. (2011). Understanding the influence of slope on the threshold of coarse grain motion: Revising critical stream power. *Geomorphology, 126,* 51–65.

Rodrigues, A. M. (1993). *Balanço energético foliar em estufas.* M. Sc. thesis (Environment, Energy profile). Instituto Superior Técnico, U.T.L., Lisbon (in portuguese).

Ross, J. (1975). Radiative transfer in plant communities, pp 13–52. In: J. L. Monteith (Ed.), *Vegetation and atmosphere* (Vol. I, 277 pp.). Academic Press.

Seginer, I. (1984). On the night transpiration of the greenhouse roses under glass or plastic cover. *Agricultural Meteorology, 30,* 257–268.

Sinokrot, B. A., & Stefan, H. G. (1993). Stream temperature dynamics measurements and modelling. *Water Resources Research, 29,* 2299–2321.

Spitters, C. J. T., Toussaint, H. A. J. M., & e Goudriaan, J. (1986). Separating the diffuse and direct
    component of global radiation and its implications for modeling canopy photosynthesis, Part I,
    components of incoming radiation. *Agricultural and Forest Meteorology, 38,* 217–219.
Unsworth, M. H., Philips, N., Link, T. E., Bond, B. J., Falk, M., Harmon, M., et al. (2004).
    Components and controls of water flux in an old-growth douglas-Fir Western Hemlock
    ecosystem. *Ecosystems, 7,* 468–481.
Zingg, A. W. (1953). Wind tunnel study of the movement of sedimentary material. In *Proceedings
    of the 5th Hydraulics Conference* (Iowa City) (vol. 34,pp. 111–135).

Examples of Applications

7

Abstract

This chapter presents 13 exercises solved on the topics of Chaps. 1–6, aiming at consolidating the physical principles to solve applied problems. Exercises 1 and 2 provide the calculation of the energy of one mole of photons and the modeling of continuous solar radiation in the eventual absence of measured data. Exercise 3 shows the calculation of transmissivity using the radiation equation. Exercises 4 and 5 estimate the daily radiation flows in locations with measured data, also considering the emissivity and reflectivity of the surface. Exercise 6 calculates the sensitive heat flow on very small scales based on the wind profile and heat transfer. Exercises 7, 8, and 9 estimate heat and water vapor flows using the aerodynamic, eddy, and Bowen covariance methodologies, considering speed, atmospheric stability, soil heat flows, and turbulent fluctuations. Exercise 10 offers a complete assessment of the radiative thermal load in a building. Exercise 11 parameterizes the components of the turbulent kinetic energy balance. Exercise 12 shows the calculation of the particle sedimentation speed, following the principles of mass transfer in Stokes and inertial domains. Finally, Exercise 13 offers an application of the Bernoulli and mass conservation principles for estimating air velocity and pressure variation in open plains followed by downward restrictive valleys.

7.1 Concepts of Energy Budget

The energy budget can be done on a surface or on a control volume. A large agricultural or forest area may be considered a surface, but an individual tree can also be considered a volume. Determining the energy budget involves identifying and quantifying the types of energy exchanges between the system and the exterior environment.

© The Author(s), under exclusive license to Springer Nature Switzerland AG 2021   237
A. Rodrigues et al., *Fundamental Principles of Environmental Physics*,
https://doi.org/10.1007/978-3-030-69025-0_7

Most objects on the Earth's surface are exposed to solar radiation and in addition, receive thermal radiation emitted by the atmosphere and other bodies. These objects also emit thermal radiation. The balance of these radiative exchanges between the object and the external environment is the radiative budget, or net radiation, Rn. This net radiation is a major cause of the Earth's warming and cooling.

An ideal surface has no thickness and so cannot store energy. The net radiative exchange is distributed through other forms of energy including conduction/energy storage (G), energy used for evaporation or gained by condensation (λE, latent heat), energy gained to heat the air from adjacent layers or gained by air cooling (H, convection, or sensible heat), and energy associated with biological processes such as photosynthesis and respiration (M). For plants, M is usually small compared to the other forms of energy and can be neglected.

The energy budget is

$$R_n + M = H + \lambda E + G \tag{7.1}$$

Each term is the mean value of the respective heat flux per unit area during a given time interval, typically from 30 to 60 min. The terms of this general energy budget equation, follow the principle that the sum of all energy exchanges in the steady-state system is zero. The signal convention assumed is that received fluxes are assigned with positive values, whereas fluxes exiting the system have negative values. So, on the left side of Eq. (7.1), R_n and M are positive when there are gains, and negative when losses are represented. On the right side of the equation, H, λE, and G are positive when they represent heat loss and negative when they represent gains. The concept of transient heat balance characterized by non-equilibrium transient heat transfer processes with time lags between outputs and inputs of energy was assessed in Sect. 6.4.

7.2 Example 1: Calculation of Energy in a Light Photon

As discussed in Sect. 6.3, solar radiation can be characterized by evaluating the different wavelength bands. It is useful to divide the spectrum of solar radiation into six wavelength ranges as shown in Table 7.1.

Table. 7.1 Energy distribution in the radiation spectrum emitted by the sun (Monteith and Unsworth 1991)

Wavelength (nm)	Energy (%)
0–300	1.2
300–400 (UV)	7.8
400–700 (vis /PAR)	39.8
700–1500 (near IR)	38.8
1500–∞	12.4

The UV spectrum can be divided into three bands (Table 7.2, below):

Photosynthesis only takes place with visible light in the range of 400–700 nm, commonly referred to as radiation photosynthetically active radiation (PAR). PAR radiation may be expressed as energy flux units (Wm^{-2}) or photosynthetic photon flux density (PPFD) given by the number of photons (400–700 nm) incident per unit area and per unit time ($mol\ m^{-2}\ s^{-1}$).

The relationship between these two units can be obtained if Planck's law is used (Eq. 6.57), expressed as

$$E_\lambda = hv = hc/\lambda_{va} \tag{7.2}$$

Equation (7.2) indicates that a photon's luminous energy, E_λ, is directly proportional to its frequency, v, where h is the Planck constant, 6.626×10^{-34} J s.

In this context, this exercise envisages the following: (i) to calculate the energy associated with, for example, a blue light photon $\lambda = 460$ nm, corresponding to a frequency of $6.52 \times 10^{14}\ s^{-1}$; (ii) the conversion of radiation units Wm^{-2} into $\mu mol\ m^{-2}\ s^{-1}$, commonly used in biology, for a PAR radiation flux of 420 Wm^{-2}, considering PAR radiation with an average wavelength of 570 nm (yellow), corresponding to a photon energy of 210 kJ mol^{-1}.

Solution:

(i) $E_\lambda = (6.626 \times 10^{-34}\ J\ s) \times (6.52 \times 10^{14}\ s^{-1}) = 43.2 \times 10^{-20}\ J$

The energy per mole of blue light photons is given by.

$E_\lambda = (6.022 \times 10^{23}\ mol^{-1}) \times (43.2 \times 10^{-20}\ J) = 260\ kJ\ mol^{-1}$.

where $6.022 \times 10^{23}\ mol^{-1}$ is the Avogadro number, representing the number of photons per mole and the corresponding radiant energy will be 260 kJ mol^{-1};

(ii) $(420\ Wm^{-2})/(210\ kmol^{-1}) = 2.0 \times 10^{-3}\ mol\ m^{-2}\ s^{-1} = 2000\ \mu mol\ m^{-2}\ s^{-1}$

delivering a radiant energy 2000 $\mu mol\ m^{-2}\ s^{-1}$ under the assumptions assumed.

Table. 7.2   Ranges of UV solar radiation (Monteith and Unsworth 1991)

UVA	400–320 nm	Produces tanning in the skin
UVB	320–290 nm	Skin cancer and the synthesis of vitamin D
UVC	290–200 nm	Potentially dangerous, but almost totally absorbed by the ozone layer in the stratosphere

7.3 Example 2: Estimation of Solar Hourly Radiation

Often there are records of total daily solar radiation but data for radiation accumulated over successive half-hour periods is lacking. In this example, solar radiation is estimated over successive half-hour periods based on daily solar radiation values, obtained during 4 days in 2000.

Solution: Statistical studies of the temporal distribution of the total radiation on horizontal surfaces throughout the day show that it is expressed by Eq. (7.3), where r_t is the ratio between the total half-time radiation, and the total daily radiation is (Duffie and Beckman 1991)

$$r_T = (a + b \cos h) \frac{\pi}{48} \frac{\cos h - \cos h_S}{\sin h_S - h_S \cos h_S} \tag{7.3}$$

where h and h_S are the hourly solar angles and the solar angle at mid-day expressed in radians, and a and b are variables:

$$\begin{aligned} a &= 0.409 + 0.5016 \sin(h_S - 1.05) \\ b &= 0.6609 - 0.4767 \sin(h_S - 1.05) \end{aligned} \tag{7.4}$$

The h_s angle is given by Eq. (6.79):

$$\cos h_s = -tg\phi tg\delta \tag{7.5}$$

and the solar angle h is given by Eq. (6.74):

$$h = 15(12 - t)$$

where t represents the apparent local solar time:

$$t = t_{uc} + CL + ET \tag{7.6}$$

where t_{uc} is the Coordinated Universal Time, CL is the longitude correction given by the product of $-1/15(60)$ min for each degree of longitude east of Greenwich, and ET is the time equation:

$$ET = 9.87 \sin(2B) - 7.53 \cos(B) - 1.5 \sin(B) \tag{7.7}$$

where B is given by

$$B = 360 \frac{t_j - 81}{364}$$

The total daily radiation S_{dt} (MJm^{-2} day^{-1}) is used to calculate the half-hour radiation S_{SH}, (MJm^{-2}30 min^{-1}). From the definition of the r_t ratio, Eq. (7.3) is given as

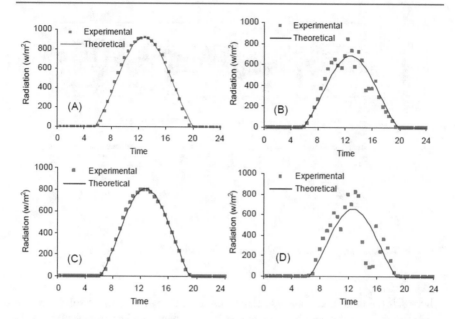

Fig. 7.1 Experimental and theoretical radiation calculated using Eq. (7.9) for October 8 (**a**) August 17 (**b**), September 7 (**c**), and September 17, 2000 (**d**)

$$S_{SH} = r_t \times S_{dt} \tag{7.8}$$

Equations (7.3) and (7.8) are applied to radiation data collected over a period of 4 days. The data was measured over daily periods as well as half-hour periods. The results are shown in Fig. 7.1 and give a comparison between the radiation values measured over successive half-hour periods with the corresponding values, calculated using Eqs. (7.3) and (7.8) applied to daily measured values. On days that were slightly cloudy or with clear skies (August 8 and September 7), the values obtained from the equations show good agreement with the measured data. However, on cloudier days (August 17 and September 17) the modeled data does not accurately reflect the experimental data.

7.4   Example 3: Estimation of Atmospheric Transmissivity from Solar Radiation Values Measured at the Surface

From known values of measured total solar radiation (direct + diffuse) incident on the Earth's surface, the transmissivity of the atmosphere to solar radiation τ can be calculated using the inverse of Eq. (6.88) from Sect. 6.3.3.3, as

$$S_t = S_0 \tau^m \cos \psi + S_0 (0.271 - 0.294 \tau^m) \cos \psi \tag{6.88}$$

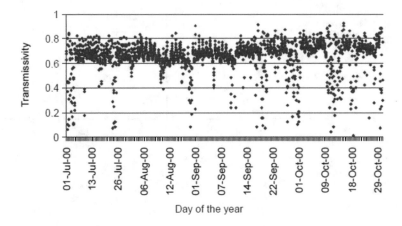

Fig. 7.2 Transmissivity of the atmosphere calculated using Eq. (6.88)

where S_t is the total instantaneous solar radiation to the surface, S_O the solar constant (1373 W m^{-2}), ψ the zenith angle, and m the air mass number, given by sec ψ in Eq. (6.81). As previously mentioned, transmissivity is influenced by the quality of air in terms of suspended particles and gases (including water vapor) as well as by solar and height location coordinates. This needs to be considered when calculating atmospheric transmissivity, in this case for a period of 4 months (July to October 2000).

Solution: The calculated transmissivity is shown in Fig. (7.2) considering that Eq. (6.88) is mainly for zenith angles below 80°. This figure shows only the τ values calculated for the periods between 1 h after sunrise and 1 h before sunset.

The values of atmospheric transmissivity can be considered reasonable about those indicated in Sect. 6.3. From September 14 onwards, there was a slight increase in transmissivity from mid-day onwards, which coincided with the start of the rainy season. Water precipitation cleans out the atmosphere removing some of the particles, causing the atmospheric transmissivity to increase (Fig. 7.3).

7.5 Example 4: Calculation of Short- and Long-Wavelength Incident Radiation on a Given Surface

Estimate the value of the various terms of incident radiation on a horizontal surface, Lisbon, (lat.38.75 °N, lon.9.2 °W) for January 28, 2009, at 10 h (local time), with clear sky conditions at a temperature of 10 °C and relative humidity of 60%. The considered transmissivity for solar radiation was 0.7.

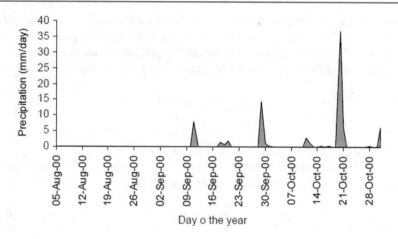

Fig. 7.3 Daily precipitation between August 5 and October 31

Solution: Firstly, solar declination, δ, between 23.5° and −23.5° needs to be calculated using Eq. (6.75), as per Sect. 6.3.3.2:

$$\sin \delta = 0.39 \sin\left[278.97 + 0.985t_j + 1.92 \sin(356.6 + 0.986t_j)\right] \qquad (6.75)$$

where t_j is the 28th Julian day. The solar declination angle is calculated at −20.73°. At 10 h, the solar hourly angle h can be calculated using Eqs. (6.74), (7.5), (7.6), and (7.7). The calculation for h gives −41.77°.

The zenithal angle is calculated using Eq. (6.72):

$$\cos \psi = \sin \phi \sin \delta + \cos \phi \cos \delta \cosh \qquad (6.72)$$

where ϕ is the location latitude. Substituting ϕ, h, and δ with their values gives the zenithal angle ψ at 10 h as 71.19°.

Solar radiation at the surface is then calculated using Eq. (6.88):

$$S_t = S_0 \tau^m \cos \psi + S_d(0.271 - 0.294t^m) \cos \psi \qquad (6.88)$$

Substituting the values of the variables, the direct solar radiation components (first on the right side) and diffuse (second on the right side) are given, respectively, by 151.06 and 79.33 $\mathrm{Wm^{-2}}$. The total solar radiation is 230.39 $\mathrm{Wm^{-2}}$.

For the calculation of the radiation incident at long wavelengths, firstly calculate the atmospheric emissivity of long-wavelength radiation, using Eqs. (7.9) and (7.10). The empirical Eq. (7.9) gives the saturation vapor pressure (kPa):

$$e_s(T) = aexp\left(\frac{bT}{T+c}\right) \qquad (7.9)$$

where T is the air temperature (°C) and a, b and c (°C units) are given by 0.611, 17.502 and 240.97, respectively. The saturation vapor pressure is 1.23 kPa. This value multiplied by the relative humidity of 60% gives the actual vapor pressure, e_a, 0.74 kPa. The atmospheric emissivity ε_{atm}, is then given by Eq. (7.10):

$$\varepsilon_{atm} = 1.72\left(\frac{e}{T}\right)^{1/7} \tag{7.10}$$

where T is the air temperature (K). The value of ε_{atm} is 0.73.

The downward long-wavelength radiation, E_b, emitted by the atmosphere and incident on the surface is given by Eq. (6.60), the Stefan–Boltzmann equation.

$$E_b = \varepsilon_a \sigma T^4 \tag{6.60}$$

where σ is the Stefan–Boltzmann constant, 5.673×10^{-8} Wm^{-2} K^{-4}, and E_b becomes 267.3 W m^{-2}.

This methodology was generalized for the entire day January 31, 2009, when meteorological measurements were made, as shown in Table 7.3 (temperature and relative humidity, solar radiation incident on the surface, wind speed and direction, and atmospheric pressure).

Figure 7.4 gives the calculated solar radiation and long-wavelength radiation incident on the surface. For comparison, the measured global solar radiation is given in Table 7.3.

7.6 Example 5: Calculation of Radiation Budgets in a Eucalyptus Forest

Meteorological data for the site "Herdade da Espirra" (lat. 38.63° N, long. 8.6° W), measured throughout the day on January 5, 2010, in a eucalyptus forest are given in Table 7.4. The aim was to compare the estimated radiative budget with field values, measured using a net radiometer. It is also intended to estimate the total incident solar radiation and net radiation in the sameday. Values assumed for several variables were 0.9 for soil emissivity, 0.8 for atmospheric transmissivity to solar radiation, and 0.15 for the albedo of forest canopy.

Solution: The incident solar radiation flux components are calculated as for the above problem. Firstly, calculate the angle of solar declination (Eq. 6.75) and zenith angle (Eq. 6.72). For this, the solar hour angle h is calculated using Eqs. (6.74), (7.6), (7.7), and (7.8). Incident solar radiation at the surface is given by Eq. (6.88). The flux of long-wavelength incident radiation throughout the day was calculated by the Stefan–Boltzmann Eq. (6.60) using the air temperature values. Equation (6.60) is also used to calculate long-wavelength radiation emitted by the soil, for each temperature (Table 7.4) and emissivity.

The variation in the radiative budget terms throughout the day is shown in Fig. 7.5.

Table. 7.3 Meteorological variable measured on 31st January 2009 in an automatic weather station in Lisbon

Year	Day	Minutes	Temperature °C	Relative humidity %	Solar radiation W/m²	Wind velocity m/s	Wind direction °	Atmospheric pressure mbar
2009	31	0	4.3	76.8	0.0	5.6	297.0	993
2009	31	100	4.0	73.7	0.0	5.5	307.9	993
2009	31	200	4.0	74.6	0.0	5.8	312.9	993
2009	31	300	4.3	72.7	0.0	5.5	307.1	994
2009	31	400	3.8	71.0	0.0	5.9	302.0	995
2009	31	500	3.9	67.9	0.0	5.6	299.8	995
2009	31	600	3.9	62.8	0.0	6.2	316.5	995
2009	31	700	3.6	67.8	0.0	5.2	304.0	996
2009	31	800	4.0	62.5	6.5	5.3	313.0	997
2009	31	900	4.4	60.4	89.7	3.9	306.3	997
2009	31	1000	4.8	60.0	163.9	3.6	289.3	997
2009	31	1100	6.1	56.5	239.0	3.6	270.5	997
2009	31	1200	6.8	52.3	359.9	4.6	258.7	996
2009	31	1300	7.1	52.5	478.9	6.4	246.2	995
2009	31	1400	7.4	53.7	460.4	7.0	229.3	994
2009	31	1500	7.3	54.8	271.0	8.1	232.5	993
2009	31	1600	7.1	58.7	154.2	8.5	221.3	991
2009	31	1700	7.1	61.1	49.5	9.6	223.3	991
2009	31	1800	6.7	61.9	6.8	10.9	219.3	990
2009	31	1900	5.4	79.3	0.0	9.6	215.2	988
2009	31	2000	5.6	82.2	0.0	9.4	200.5	986
2009	31	2100	5.8	84.4	0.0	10.8	202.6	984
2009	31	2200	6.0	84.3	0.0	11.8	201.1	982
2009	31	2300	5.9	85.2	0.0	13.7	194.4	980
2009	32	0	6.0	86.3	0.0	14.0	193.1	977

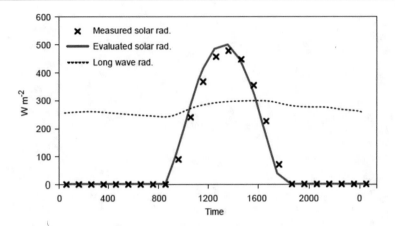

Fig. 7.4 Measured and calculated solar radiation and long-wavelength radiation on January 31, 2009, in Lisbon

Given an albedo of 0.15 for the forest, the radiative balance can be calculated (R_n in Eq. 7.1, the algebraic sum of the various radiation components) and shown to compare with the values measured using the net radiometer throughout the day. The estimated and measured radiative budget is given in Fig. 7.6.

The values of the total estimated daily incident solar radiation and net radiation were calculated by a simple trapezoidal integration over successive half-hourly periods $(t_{i+1} - t_i)$ with values from the weather station which was added to obtain the daily net radiation. Considering general variables $BRad_{(daily)}$ and Rad, the numerical integration proceeded as follows (Eq. 7.11):

$$BRad_{(daily)} = \sum_{i=0}^{47} \left(\frac{Rad_{i+1} - Rad_i}{2} \right) (t_{i+1} - t_i) \tag{7.11}$$

which provided that the incident solar energy flux over the forest canopy was 10 MJ m^{-2} and that the net radiation was 4.3 MJ m^{-2}.

7.7 Example 6: Calculation of Sensible Heat Transfer from the Low Canopy to the Adjacent Atmosphere

In a region covered with low vegetation, a 20 m high instrumented mast was set up and on a given day, the following average wind velocities were determined at four levels above the undergrowth (Table 7.5).

The dry and humid air temperature measured at 2 m were 23 °C and 20 °C. The soil surface temperature was 24.5 °C. Calculate the loss of sensible heat from the soil to the adjacent atmosphere at 2 m.

Table. 7.4 Meteorological data for 5th January 2010 from a weather station in Espirra site

Year	Day	Hour	Air temperature °C	Relative humidity %	Solar global radiation W/m²	Soil temperature °C	Wind direction °	Measured net radiation W/m²
2010	5	0	7.1	68.5	0	6.6	81.6	−50.0
2010	5	1	6.8	79.9	0	6.9	74.8	−50.4
2010	5	2	6.6	80.5	0	6.8	84.7	−51.8
2010	5	3	6.7	79.2	0	6.7	82.5	−52.9
2010	5	4	5.9	80.9	0	6.4	76.2	−52.0
2010	5	5	4.9	82.6	0	6.2	62.3	−48.2
2010	5	6	4.1	84.3	0	5.9	57.7	−39.9
2010	5	7	3.8	84.8	0	5.8	72.0	−36.2
2010	5	8	2.9	85.9	0.61	5.4	67.1	−25.7
2010	5	9	4.3	96.6	90.4	6.2	47.6	43.6
2010	5	10	8.7	78.9	238.3	7.4	49.6	153.8
2010	5	11	11.0	71.1	367.1	7.6	61.8	314.8
2010	5	12	13.0	63.8	454.7	8.1	52.9	398.5
2010	5	13	14.5	55.8	477.8	8.8	58.1	411.5
2010	5	14	15.2	53.0	445.5	9.1	65.6	367.2
2010	5	15	15.1	52.0	355.8	9.1	71.9	269.4
2010	5	16	14.9	51.9	225.6	9.1	98.2	141.9
2010	5	17	14.2	53.5	70.9	8.7	90.5	−1.7
2010	5	18	11.6	59.4	0	7.9	64.9	−56.3
2010	5	19	10.3	68.5	0	7.9	58.0	−50.8
2010	5	20	10.2	70.5	0	7.9	41.9	−49.7
2010	5	21	9.8	70.7	0	7.7	44.6	−52.7
2010	5	22	8.2	73.6	0	7.2	41.1	−54.6
2010	5	23	7.7	75.0	0	7.1	44.1	−53.5
2010	6	0	7.0	62.5	0	5.4	45.7	−53.5

Solution: As the soil surface is at a higher temperature than the atmosphere, there is heat transfer between the surface of the canopy and the surrounding atmosphere. The amount of heat exchanged mainly depends on this temperature difference and the wind speed along the surface.

Firstly data of u (ms⁻¹) and Ln(z) in Table 7.5 are graphed in Fig. 7.7 delivering a linear equation:

$$y = 0.69x - 0.85 \tag{7.12}$$

Assuming that the atmosphere is neutral, the loss of sensible heat from the soil to the atmosphere can be calculated from the difference between soil and air temperatures by

$$H = \frac{\rho c_p}{r_{aM}} (T_{soil} - T_{atm}) \tag{7.13}$$

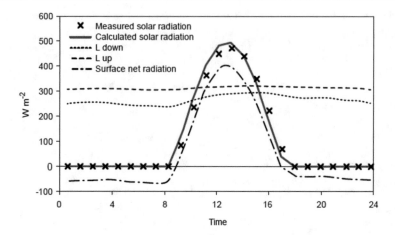

Fig. 7.5 Measured and calculated values for components of the radiative budget. Solar rad, L up. and L down. Refer to incident solar and long-wavelength radiation measured and calculated for January 5, 2010

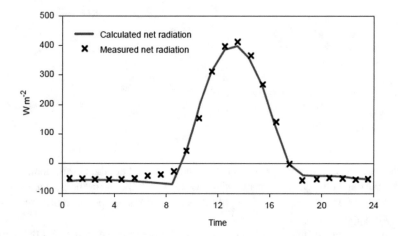

Fig. 7.6 Estimated and measured radiative balance on January 5, 2010, at Espirra site

Table. 7.5 Average wind velocity at four levels above the ground surface	Z(m)	5	8	10	20
	U(m/s)	3.48	4.43	4.66	5.50

Fig. 7.7 Velocity profile over low canopies

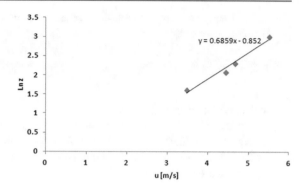

with $T_{soil} = 24.5\ °C$ and $T_{atm} = 23\ °C$.

The velocity profile can be used to calculate u* with Eq. (2.16) from Chap. 2:

$$u(z) = \frac{u_*}{k} ln \frac{z - d}{z_{oM}} = \frac{u_*}{k} ln(z) - \frac{u_*}{k} ln(z_0) \qquad (2.16)$$

The slope of Eq. (7.12) is 0.6859 which is equal to u*/k, where k is the von Karman constant of 0.41. So, u* will be 0.69*0.41 = 0.28 ms⁻¹.

On the other hand, the intercept of Eq. (7.12) is −0.85 which is equal to (u*/k)* ln(z₀) and therefore z₀ will be nil. Now the aerodynamic resistance at 2 m can be calculated from Eq. (2.33):

$$r_{aM} = \frac{u(z)}{u_z^2} \qquad (2.33)$$

giving a r_{aM} of 5.83 sm⁻¹. Transferred sensible heat at 2 m height by Eq. (7.13) will then be 308.7 Wm⁻².

7.8 Example 7: Application of the Aerodynamic Method

A mast with two sets of devices was placed over a pasture (low vegetation) canopy on a clear day, at 11 am. Each set of sensors was made up of anemometer cups, and a thermocouple and a hygrometer were set at z = 1 m and z = 4 m, above the ground. A pyranometer for the measurement of solar global radiation and a thermocouple were used at the canopy surface.

The measured hourly means (between 11 am and 12 pm) are given in Table 7.6.

Given that the specific heat at constant pressure and air density are 1000 JKg⁻¹ K⁻¹ and 1.2 Kgm⁻³, respectively, and the latent heat of vaporization is 2400 Jg⁻¹, it is necessary, considering the zero-plane displacement, d = 0, to:

i. classify the atmospheric stability;
ii. calculate the sensible heat flux over the vegetation canopy, given tangential stress of 0.35 m;

Table. 7.6 Hourly means of meteorological variables at 1 and 4 m of height

$z(m)$	T (°C)	u (m/s)	q (specific humidity) Kg vapour/kg air
1	23.31	2.5	0.0110
4	22.27	3.9	0.0120

iii. calculate latent heat flux; and
iv. using the latent heat flux, calculate the albedo of the cover ρ, given that the overall solar radiation flux S_T is 790 Wm^{-2}, the ground surface temperature T_s is 28 °C, the soil heat flux is about 10% of the surface radiative balance, and the surface emissivity is 0.95.

Solution: (i) Atmospheric stability is given by the Richardson gradient number Ri_g, according to Eq. (2.45) in Chap. 2:

$$Ri_g = \frac{g}{T} \frac{(T_2 - T_1)(z_2 - z_1)}{(u_2 - u_1)^2} \tag{2.45}$$

Substituting the variables gives Ri_g equals to -0.0526, indicating that the atmosphere is unstable.

(ii) The sensible heat flux is calculated using Eq. (2.54) in Chap. 2:

$$H = -\rho c_p k^2 \left(\frac{\Delta u \Delta T}{(\ln((z_2 - d)/(z_1 - d)))^2} \right) (\phi_M \phi_H)^{-1} \tag{2.54}$$

where, for $-0.1 < Ri < 1$, the parameter $F_{est} = (\phi_M \phi_H)^{-1}$ is given by Eq. (2.57):

$$F_{est} = (1 - 5Rig)^2 \tag{2.57}$$

giving a sensible heat flux of 157.42 Wm^{-2}.

(iii) For the latent heat, according to Eq. (2.55) in Chap. 2:

$$LE = -Lk^2 \left(\frac{\Delta u \Delta \rho_v}{(\ln((z_2 - d)/(z_1 - d)))^2} \right) (\phi_M \phi_v)^{-1} \tag{2.55}$$

The estimation of the vapour concentration, ρ_v, or absolute air humidity, for calculation of latent heat flux with Eq. (2.55), is carried out with Eq. (4.13). The latter requires the calculation of vapour partial pressure through Eq. (4.16) considering atmospheric pressure as 100 kPa. The estimation of ρ_v requires Eq. (4.14) for calculation of the density of moist air, at the two temperatures in vertical profile in Table 7.6. The estimation of the density of moist air, is based on the weighted sum of partial pressures of dry air and water vapour. Assuming $-0.1 < Ri < 1$, the stability function $F_{est} = (\phi_M \phi_H)^{-1}$ is given by Eq. (2.57), so that the latent heat flux is 376.6 Wm^{-2}.

(iv) By the energy balance of the surface, given that the flux to the soil is 10% of the radiative balance, we have

$$0.9\left[S_T - \rho S_T + \left(\varepsilon_{atm}T_{atm}^4 - \varepsilon_s T_s^4\right)\sigma\right] - H - LE = 0 \qquad (7.14)$$

where the T_{atm} and T_s referred to average air temperature and ground surface temperature, respectively. The variables ε_{atm} and ε_s refere to air and surface emissivity, respectively. The emissivity of the atmosphere is obtained by Eq. (7.10) which involves calculating the vapor pressure using Eq. (4.16) from the measured specific humidity, as mentioned. Assuming ε (ratio between the molecular mass of water vapor and air) is 0.622 and the atmospheric pressure is 100 kPa at vapor pressure e, using Eq. (4.16) gives 1734.73 Pa:

$$q = \frac{\varepsilon e}{p} \qquad (4.16)$$

Substituting this value into Eq. (7.10), we get

$$\varepsilon_{atm} = 1.72\left(\frac{e}{T}\right)^{1/7} \qquad (7.10)$$

We obtain an air emissivity value of 0.83.
Replacing several variables in Eq. (7.14), we obtain a surface albedo, ρ, of 0.27.

7.9 Example 8: Application of the Eddy Covariance Method

A sonic anemometer and an analyzer were used to sample at a frequency of 0.1 Hz, to measure temperature, velocity, and absolute humidity in a vegetated area. Table 7.7 shows four sets of instantaneous measurements.
    The aim is to calculate:

(i) Sensible and latent heat fluxes over vegetation

    and,
(ii) Canopy aerodynamic resistance.

    Consider that the air specific heat at constant pressure and the density are 1000 $Jkg^{-1} K^{-1}$ and 1.2 kgm^{-3}, respectively, and the value for the latent heat of water is 2500 Jg^{-1}.

Solution:

(i) The sensible heat and latent fluxes are given by Eqs. (3.22) and (3.23) in Chap. 3 as

Table. 7.7 Instantaneous values of the variables listed in the left column

u (ms^{-1})	2.5	2.3	1.9	2.0	2.3	2.1	2.8	2.9	2.4	2.9	2.3	2.1	2.2	2.1
w (ms^{-1})	0.4	1.8	1.5	−1.8	0	−1.6	−1.4	−1.0	−1.9	1.0	1.6	1.1	0.5	−0.2
t (°C)	20.0	20.2	20.1	19.8	19.0	19.5	20.1	19.7	19.1	19.8	20.0	20.4	20.2	19.8
q (gkg^{-1})	10.3	10.1	10.2	9.6	9.8	9.7	10.1	10.4	10.1	9.8	9.9	10.2	10.2	10.4

$$H = \rho c_p \overline{w'T'} \tag{3.22}$$

$$LE = \rho L \overline{w'q'} \tag{3.23}$$

The average measurements of data in Table 7.7 are 2.34 ms^{-1} for the horizontal component of the wind velocity speed u, 0 ms^{-1} for the vertical component of the wind speed w, 19.85 °C for the air temperature, and 10.05 gkg^{-1} for the air specific humidity.

Instant fluctuations and fluctuation products are given by the differences between the instantaneous values of the variables and the average values, and by the respective products (Tables 7.8 and 7.9, respectively).

Replacing and then applying Eqs. (3.22) and (3.23), we obtain for the sensible and latent heat fluxes, 336 Wm^{-2} and 203.6 Wm^{-2}, respectively.

(ii) The aerodynamic resistance of the canopy is given by the application of Eqs. (2.33) and (3.25):

$$r_{aM} = \frac{u(z)}{u_*^2} \tag{2.33}$$

$$u_*^2 = -\frac{}{\overline{u'_w'}} \tag{3.25}$$

where $u(z)$ is the average of the horizontal velocity 2.34 ms^{-1}. After calculation from instantaneous fluctuations of u in Table 7.9, the value of the aerodynamic resistance will be equal to 42 sm^{-1}.

7.10 Example 9: Calculation of Vertical Fluxes Using the Bowen Method

Over a low canopy, the average values measured for air temperature, radiative balance R_n, and heat flux in soil, G, were 300 K, 600 Wm^{-2}, and 54 Wm^{-2}, respectively. The height difference between the two levels (2 and 4 m), specific humidity, and the temperature were 0.818 g/kg and 1.58 K, respectively. It is intended to calculate the sensible and latent heat fluxes by the Bowen method. The values of specific heat at constant pressure, air density, atmospheric pressure, and psychrometric constant are 1000 JKg^{-1} K^{-1}, 1.2 Kgm^{-3}, 100 kPa, and 66.2 PaK^{-1}, respectively.

Solution: Firstly, we calculate the Bowen ratio β, using Eq. (4.15) as

$$\beta = \frac{c_p \Delta T}{L \Delta q} = \frac{\rho c_p \Delta T}{L \varepsilon \Delta e} = \gamma \frac{\Delta T}{\Delta e} \tag{4.17}$$

Table. 7.8 Instantaneous fluctuations of the variables listed in the left column

$u'\,(ms^{-1})$	0.16	-0.04	-0.44	-0.34	-0.04	-0.24	0.46	0.56	0.06	0.56	-0.04	-0.24	-0.14	-0.24
$w'\,(ms^{-1})$	0.40	1.80	1.50	-1.80	0.00	-1.60	-1.40	-1.00	-1.90	1.00	1.60	1.10	0.50	-0.20
$T'\,(^{\circ}C)$	0.16	0.36	0.26	-0.04	-0.84	-0.34	0.26	-0.14	-0.74	-0.04	0.16	0.56	0.36	-0.04
$q'\,(gkg^{-1})$	0.24	0.04	0.14	-0.46	-0.26	-0.36	0.04	0.34	0.04	-0.26	-0.16	0.14	0.14	0.34

Table. 7.9 Products of the instantaneous fluctuations of the variables listed in the left column

$u'w'$	0.06	−0.08	−0.66	0.62	0.00	0.39	−0.64	−0.56	−0.11	0.56	−0.07	−0.27	−0.07	0.05
$w'T'$	0.07	0.66	0.40	0.06	0.00	0.54	−0.37	0.14	1.40	−0.04	0.26	0.62	0.18	0.01
$w'q'$	0.10	0.08	0.21	0.82	0.00	0.57	−0.06	−0.34	−0.08	−0.26	0.25	0.16	0.07	−0.07

where γ is the psychrometric constant (equal to $c_p \, p/L\varepsilon$) 66.2 PaK^{-1} and the ratio $\partial T/\partial e$ is obtained from temperature and vapor pressure data at two levels. The vapor pressure is obtained from the specific humidity using Eq. (4.16), previously used in Example 7:

$$q = \frac{\varepsilon e}{p} \qquad (4.16)$$

where ε (ratio between the molecular masses of water vapor and air) is 0.622. Replacing the variables, we have

$$\beta = 66.2 \frac{1.58}{(0.818 * 100)/0.622} = 0.8$$

The LE and H are calculated with the Bowen method using Eqs. (4.18) and (4.19), respectively:

$$LE = \frac{(R_n - G)}{1 + \beta} \qquad (4.18)$$

$$H = \beta \frac{R_n - G}{1 + \beta} \qquad (4.19)$$

Hence, replacing the values of the variables will give a LE value equal to 300 Wm^{-2} and 240 Wm^{-2} for H.

7.11 Example 10: Calculation of the Solar Radiation Intensity Components and Long Wavelength Incident on a Building with a Known Geometry

Calculate the intensity of direct and diffuse solar radiations, which covers the building depicted in Fig. 7.8 on June 21 at 12 h (legal time), given the coordinates (latitude 38.75° N, longitude 9.2° W, Northern Hemisphere), dimensions, 15 m long, 10 m wide and 8 m high, 10 m span, and the roof slope angle of 21.8° of a building whose width runs parallel to the north. Assume atmospheric transmissivity to direct radiation τ, of 0.7, and a surface albedo ρ, of 0.25.

Solution: Use the same approach as outlined in Examples 2, 3, and 4. Begin by calculating the solar declination angle δ, Julian day function t_j (172 in this case), using Eq. (6.75), from Campbell and Norman (1998):

$$\sin \delta = 0.39 \sin \left[278.97 + 0.985 t_j + 1.92 \sin(356.6 + 0.986 t_j) \right] \qquad (6.75)$$

giving a δ angle of 23.44°.

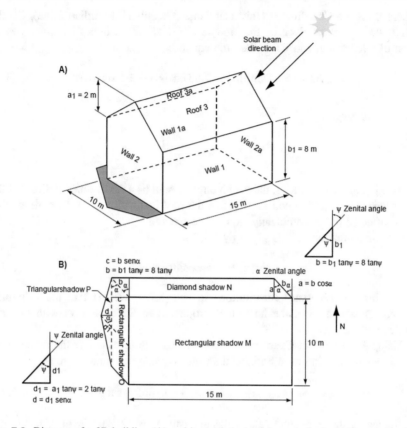

Fig. 7.8 Diagram of a 3D building (**A**) and its total shadowed area in horizontal projection (**B**)

The zenith angle ψ is given by Eq. (6.72) and depends on the latitude of the location, the solar declination and solar angle, h, as

$$\cos\psi = \sin\phi\sin\delta + \cos\phi\cos\delta\cos h = \sin\beta \qquad (6.72)$$

where β is the solar height angle, complementary to the zenith angle.

At 12 h, the solar hour angle h is calculated with Eqs. (6.74), (7.6), and (7.7). Repeating the above, we get

$$h = 15(12 - t) \qquad (6.74)$$

where t, the apparent solar hour, is given by Eq. (7.6):

$$t = t_{uc} + CL + ET \qquad (7.6)$$

where t_{uc} is the Coordinated Universal Time (Greenwich Meridian Time), CL the longitude correction given by the product of $-1/15$ (60) min per longitude degree west of Greenwich, and ET is the time equation:

$$ET = 9.87 \sin(2B) - 7.53 \cos(B) - 1.5 \sin(B) \tag{7.7}$$

where B is given by

$$B = 360 \frac{t_j - 81}{364} \tag{7.8}$$

Making the substitutions, the zenith angle ψ will be 17.34°. This implies that the solar height is 72.66° or considering that the roof inclination is 21.8°, it means that the roof at 12 pm is completely exposed to the sun.

The azimuthal angle α is given by Eq. (6.73):

$$\sin(\alpha) = -\cos\delta \sin h / \sin\psi \tag{6.73}$$

Making the necessary substitutions in Eq. (6.73), we get that the azimuthal angle, representative of the horizontal projection of the solar rays with the true North, will be equal to 30.79°.

Using Eq. (6.76) in Chap. 6, it becomes possible to calculate the intensity of instant solar radiation on a horizontal surface outside the atmosphere as

$$S_h = S_o \left[1 + 0.033 \cos\left(360 t_j / 365\right)\right] \sin\beta \tag{6.77}$$

Replacing the respective values, the intensity S_h is 1268.04 W/m^2.

Next, the calculation of the global direct and diffuse solar radiation on a plane horizontal to the Earth's surface (Eq. 6.88) is as follows:

$$S_t = S_0 \tau^m \cos\psi + S_d(0.271 - 0.294 t^m) \cos\psi \tag{6.88}$$

where τ is the atmospheric transmittance to direct radiation with an arbitrary value of 0.7, S_o the solar constant 1373 Wm^{-2}, and m the mass of air given by Eq. (6.82):

$$m = \sec\psi \tag{6.81}$$

which in this case gives 1.05.

The global solar radiation (Eq. 6.88) per unit area (flux density) is 959.75 Wm^{-2}. This corresponds to 872.67 Wm^{-2} for the direct solar radiation (first term on the right side) and 87.08 Wm^{-2} for diffuse solar radiation (second term on the right side). The total direct incident radiation on the building (Fig. 7.8A) can now be calculated. Its value is the product $S_b A_h$ obtained from Eq. (6.63) between the direct solar radiation flux per unit area S_b, 872.67 Wm^{-2}, previously calculated and the shadowed area on a horizontal surface A_h:

$$S_{b1} = \frac{A_h}{A_b} S_b \qquad (6.63)$$

First calculate the shadowed areas of the walls exposed to the sun, considering that for any object located in the Northern Hemisphere, the sun is located south regardless of the solar declination angle which changes throughout the year. In Fig. 7.8A, the walls are rectangular: Wall 1 (facing south) with the length as the largest dimension, Sidewall 2a with a rectangular component, and a triangular wall (located opposite to the orientation of Fig. 7.8A facing east with width as the largest dimension and the total roof area with two inclined parts (roofs 3 and 3a)). The total shadowed area is horizontally projected in Fig. 7.8B. This rectangular area M is equal to the product of the length and width of the building, corresponding to the roof projected area.

The shaded area in the form of a parallelogram (area N), with the largest dimension oriented along the building length, is due to lateral wall components located opposite to the orientation of Fig. 7.8A and the roof. The shadowed area O likewise a parallelogram, with the largest dimension oriented across the width of the building, is from rectangular wall components facing south and from the roof. The triangular shaded area P is due to the roofs and the lateral wall facing outward. Going back to the calculations of the several shadowed areas gives a shaded rectangular M area of 150 m², equal to the product of the length and width of the building.

The shaded area N, in the form of a parallelogram, with the largest dimension oriented along the length of the building, is given by

$$N = 15 \times \underbrace{\underset{b_1}{\overbrace{8}^{b} \times tan(\psi) \times \cos(\alpha)}}_{a} = 32.19m^2 \qquad (7.15)$$

The shaded area O, shaped like a parallelogram, with the larger dimension oriented along the width of the building is given by

$$O = 10 \times \underbrace{\underset{b_1}{\overbrace{8}^{b} \times tan(\psi) \times sin(\alpha)}}_{c} = 12.79m^2 \qquad (7.16)$$

The triangular shaded area P is given by

$$P = 10 \times \underbrace{\underset{d_1}{\overbrace{2}^{d_1} \times tan(\psi) \times sin(\alpha)/2}}_{d} = 1.6m^2 \qquad (7.17)$$

The total shadow area given by the sum of four components is 196.6 m².

The term $S_b A_h$, representing direct solar radiation incident on the building, is then the product of 872.67 Wm^{-2} and 196.6 m^2, giving 171.550 kW. This gives the total direct solar radiation incident on the building.

For diffuse solar radiation, Eq. (7.18) is used with $\overline{S}_{dif}$ representing the total diffuse radiation incident on the surface facing the atmospheric semi-hemisphere:

$$\overline{S}_{dif} = \cos^2(\alpha/2)S_d + \cos^2(\alpha/2)\rho S_T \qquad (7.18)$$

where the albedo ρ, of the surface under consideration, is 0.25.

In the case of the building, this expression applies to two groups of surfaces with different inclinations relative to the atmospheric half-hemisphere: the two surfaces of the sloping roofs with 21.8° and the four walls with 90° inclination. In each case, the overall diffuse light is the product of $\overline{S}_{dif}$ and the areas of buildings exposed to the atmospheric semi-hemisphere.

Applying Eq. (7.18) to the roofs gives

$$\overline{S}_{dif} = \cos^2(0.38/2) \times 87.08 \times \underbrace{((5/\cos(0.38)) \times 15 \times 2)}_{\text{rooftotal area}}$$
$$+ \sin^2(0.38/2) \times 959.75 \times 0.25 \times ((5/\cos(0.38)) \times 15 \times 2) = 14950\,\text{W}$$

Similarly, applying Eq. (7.18) to the walls gives

$$\overline{S}_{dif} = \cos^2(90/2) \times 87.08 \times \underbrace{(15 \times 8 \times 2 + 10 \times 8 \times 2 + (10 \times 2/2) \times 2)}_{\text{total wallarea}}$$
$$+ \sin^2(90/2) \times 87.08 \times 0.25 \times (15 \times 8 \times 2 + 10 \times 8 \times 2 + (10 \times 2/2) \times 2) = 68646\text{W}.$$

The total diffuse radiation (on the walls and roofs) will thus be 83.596 kW.

7.12 Example 11: Calculation of Kinetic Energy Budget Components for a Flat Surface

From a height of 4 m from a flat surface, it was observed that the vertical variation in the horizontal speed du/dz was 0.02 s^{-1}, the average potential temperature was 20 °C, the average of the product of the instantaneous fluctuations over a period of half an hour $\overline{w'T'}$ was 0.25 K m/s, and the mean vertical moment of fluctuations $\overline{u'w'}$ was -0.04 m^{2}s^{-2}.

Neglecting the terms for turbulent transport and by pressure correlation (Eq. 3.89 in Chap. 3), calculate the TKE dissipation rate required to obtain a steady state and the dimensionless TKE equation.

Solution: The starting expression is obtained from Eq. (3.89) in Chap. 3 as

$$\underset{I}{\frac{\partial \overline{e}}{\partial t}} = \underset{III}{\frac{g}{\theta_v} \left(\overline{u_3' \theta_v'} \right)} - \underset{IV}{\frac{\partial \left(\overline{u_3' e} \right)}{\partial x_3}} - \underset{V}{\left(\frac{1}{\overline{\rho}} \right) \frac{\partial \left(\overline{u_3' p'} \right)}{\partial x_3}} - \underset{VI}{\overline{u_1' u_3'} \frac{\partial \overline{u_1}}{\partial x_3}} - \epsilon \tag{3.89}$$

where term I represents the storage rate of the kinetic energy, III the term for production or consumption by buoyancy. The latter is either production or loss depending on whether the heat flux is positive (daytime) or negative (nighttime); term IV is the turbulent kinetic energy transport caused by u_j' fluctuations and V is transport or correlation term indicative of pressure as the TKE redistribution by pressure fluctuations. This term is associated with the circulation of large eddies. Term VI refers to the superficial boundary layer and normally has a sign opposite to the mean velocity vector and the term VII corresponds to the viscous dissipation and thermal conversion of kinetic energy.

For the conditions described, terms IV and V for turbulent transport and pressure perturbations are omitted. Thus, the dissipation rate needed to maintain stationarity conditions $\frac{\partial \overline{e}}{\partial t} = 0$ is

$$\varepsilon = \underset{III}{\left(9.8 \text{ms}^{-2}/293.15\text{K}\right) \times 0.25 \text{Kms}_.^{-1}} - \underset{VI}{\left(-0.04 \text{m}^2 \text{s}^{-2} \times 0.02 \text{s}^{-1}\right)} = 7.56 \times 10^{-3} \text{m}^2 \text{s}^{-3}$$

Equation (3.89) can be written in a dimensionless form by multiplying its members by $\left(k(z - d)/u_*^3\right)$, as discussed in Chap. 3, obtaining then, under steady-state conditions, Eq. (3.92) as

$$0 = \underset{II}{-\frac{z - d}{L}} - \underset{III}{\phi_t} + \underset{IV}{\phi_p} + \underset{V}{\phi_M} - \underset{VI}{\phi_\varepsilon} \tag{3.92}$$

where terms II, III, IV, V, and VI represent buoyancy, transport, pressure correlation, shear stresses, and dissipation. In the example given, terms III and IV cancel out. The remaining terms can be calculated using the following equations:

$$(II)\frac{z - d}{L} = \xi = -\left(\frac{g}{\overline{\theta}} \right) \frac{\left(\overline{w'\theta'} \right)}{u_*^3/k \, (z - d)} \tag{3.93}$$

$$(V)\phi_M = \begin{vmatrix} (1 + 16|(z - d)/L|)^{(-1/4)} & \xi \leq 0 \\ (1 + 5(z - d/L)) & \xi > 0 \end{vmatrix} \tag{3.97}$$

$$(VI)\phi_\varepsilon = \begin{cases} \left(1+0.5|\xi|^{(2/3)}\right)^{(3/2)} & \xi \le 0 \\ 1+0.5|\xi| & \xi > 0 \end{cases} \tag{3.99}$$

Friction velocity can be obtained from Eq. (3.25) as

$$u_*^2 = -\overline{u'w'} \tag{3.25}$$

implying that u_* is 0.2 m.

For the calculation of the term II for buoyancy, we use Eq. (3.93):

$$-\left(9.8 \text{ms}^{-2}/293.15\text{K}\right) \times 0.25 \text{Kms}^{-1}/\left(\left(0.2\text{ms}^{-1}\right)^{-3}/(0.41 \times 4)\right) = -1.713.$$

The ϕ_M term is calculated by Eq. (3.97):

$$1 + 16|(z-d)/L|^{(-1/4)} = \left(1 + 16x1.713^{(-1/4)}\right) = 0.43$$

The ϕ_ε term is given by Eq. (3.99):

$$\left(1+0.5|\xi|^{(2/3)}\right)^{(3/2)} = \left(1 + 0.5 \text{ x } 1.713^{2/3}\right)^{3/2} = 2.24.$$

The results show higher TKE production via buoyancy (Eq. 3.92), compared to mechanical production by shear stresses, characteristic of an unstable atmosphere. They also indicate a balance between the sum of the product terms and the dissipative term, which according to Eq. (3.92) is negative.

7.13 Example 12: Calculation of Sedimentation Velocity of a Particle

Calculate the sedimentation velocity in two typical environmental spheric objects which are:

(1) An element with a diameter of 10 μm and a density of 1.28 gcm^{-3};
(2) An element with a diameter of 10 mm and a density of 910 kgm^{-3}.

This exercise aims to apply a direct approach in item (1) and an interactive approach for the calculation of sediment velocity on conditions defined in item (2).

Solution:

(1) Assuming that the particle sediments under Stokes regime, Eq. (6.138), are valid,

$$U_s = \frac{2}{9} \frac{(\rho_p)}{v\rho_a} gr^2 \qquad (6.138)$$

where the kinematic viscosity v of air is assumed as $16*10^{-6}$ m^{2}s^{-1}, ρ_a is the air density assumed as 1.16 kgm^{-3}, g the gravitational acceleration of 9.8 ms^{-2}, the particle radius is $5*10^{-6}$ m, and the density of particle σ_p of 1.28 gcm^{-3} is given by $1.28 * 10^{-3} * 10^6 = 1.28 * 10$Kgm^{-3}.

Inserting these values in Eq. (6.138), we have

$$U_s = \frac{2}{9} \frac{(1.28 * 10^3) * 9.8 * (5 * 10^{-6})^2}{1.164*(16 * 10^{-6})} \sim 0.003 \text{ms}^{-1} = 3 \text{mms}^{-1}$$

The corresponding Reynolds number of particle Re_p of $2rU/v$ is

$$Re_p = \frac{2 * 0.003 * 5 * 10^{-6}}{(16 * 10^{-6})} \sim 0.002$$

A value within the Stokes regime (Sect. 6.5.1) justifying the application of Eq. (6.138).

(2) Applying Stokes formulation, Eq. (6.138), for the estimation of sedimentation velocity, we have

$$U_s = \frac{2}{9} \frac{910 * 9.8 * (0.005)^2}{1.16 * (16 * 10^{-6})} \sim 2669 \text{ ms}^{-1}$$

corresponding to a Reynolds number of particles of

$$Re_p = \frac{2 * 0.005 * 2669}{(16 * 10^{-6})} \sim 26.7$$

This value fails the transitional regime between Stokes and inertial domain, so that the Stokes formulation, Eq. (6.138), is not valid, for the estimation of the sedimentation velocity requiring a trial-and-error iterative approach until achieving almost equilibrium between drag and gravitational forces. The main equations now relevant are the following:

$$F_d = 0.5c_d\rho_g V_s^2 \pi R^2 \qquad (6.132)$$

$$F_g = \frac{4}{3}\pi R^3 g(\rho_{p-}\rho_f) \qquad (6.135)$$

$$U_s^2 \simeq 8R\rho_p g/(3c_d\rho_f) \qquad (6.137)$$

$$c_d = \left(24/Re_p\right)\left(1 + 0.17Re_p^{0.06}\right) \qquad (6.133)$$

$$c_d = \frac{24}{Re_p} \qquad (6.134)$$

representing the drag and gravitational forces, the sedimentation velocity of particles under inertial domain, with Re_p higher than 1, and the drag coefficients under airflows with Re_p higher than 1 and lower than 0.1.

So, after obtaining an initial guess for Re_p of 26.7, the first estimation of drag coefficient, with Eq. (6.133), is as follows:

$$c_d = \left(24/26.7\right)\left(1 + 0.17 \times 26.7^{0.06}\right) = 9.9$$

allowing to use Eq. (6.137) to obtain the first estimation of sedimentation velocity as follows:

$$U_s^2 = 8 \times 0.005 \times 9.9 \times 910/(3 \times 9.9 \times 1.16) = 10.35 m^2 s^{-2} \text{ or } V_s = 3.21 ms^{-1}$$

delivering a value of 0.03 for Re_p:

$$Re_p = \frac{2 * 3.21 * 0.005}{(16 * 10^{-6})} \sim 0.03$$

corresponding again to a flow in the Stokes domain. The drag coefficient will be thus calculated with Eq. (6.134):

$$c_d = \frac{24}{0.03} = 800$$

and drag and gravitational forces will be calculated applying Eq. (6.132) with a c_d of 800 and Eq. (6.135), respectively, as follows:

$$F_d = 0.5 \times 800 \times 1.16 \times 10.35 \times 3.14 \times 0.005^2 = 0.38\,N$$

$$F_g = \frac{4}{3} \times 3.14 \times (0.005)^3 \times 9.8(910 - 1,16) = 0.0046\,N$$

The drag and gravitational forces firstly estimated are thus in a non-equilibrium state with distinct order of magnitudes. So, we need to carry on successive estimations of drag forces with tentative values of sedimentation velocity lower than $3.21\ ms^{-1}$.

A trial with V_s of 1 ms^{-1} gives a low Re_p, within the Stokes domain, of

$$Re_p = \frac{2 * 1 * 0.005}{(16 * 10^{-6})} = 0.0099$$

and a correspondent drag coefficient c_d, of

$$c_d = \frac{24}{0.009} \sim 2666$$

Calculating now the drag force F_d, we have

$$F_d = 0.5 \times 2667 \times 1.16 \times 1 \times 3.14 \times 0.005^2 = 0.12\,N \qquad (7.37)$$

still with an order of magnitude for F_d of 0.0046 N higher than the gravitational force.

A further calculation with a U_s of 0.0038 ms^{-1} delivers a low Reynolds number of

$$Re_p = \frac{2 * 0.0038 * 0.005}{(16 * 10^{-6})} = 0.00038$$

a drag coefficient of

$$c_d = \frac{24}{0.00038} \sim 63158$$

and a drag force of

$$F_d = 0.5 \times 63158 \times 1.16 \times 0.00038^2 \times 3.14 \times 0.005^2 \sim 0.0044\,N$$

very close to the gravitational force. Thus, a theoretical very low sedimentation velocity of 0.0038 ms^{-1}, within the Stokes domain, is achieved for a 1 cm diameter falling typical element with a density of 910 kgm^{-3}.

7.14   Example 13: Calculation of Variation of Velocity and Pressure in Airflow in a Plain and a Valley

For an airflow with a velocity of 10 ms^{-1}, and freely circulating in a 30 km plain, calculate the variation of the wind velocity and of pressure, Δp, at a 3 km width downstream valley contracting the flow. Assume that air density is 1.16 kgm^{-3}.

Solution:

The resolution is based on the application of the Bernoulli equation Eq. (A2.45, Annex 2):

$$P + \frac{1}{2}\rho v^2 + \rho g h = constant \tag{A2.45}$$

and of the principle of mass conservation:

$$A_1 v_1 = A_2 v_2 \tag{A2.34}$$

Equation (A2.45) basically stipulates that if air velocity is high, its pressure is low and vice versa with the major pressure variation concerning the second term which is the dynamic pressure term. Equation (A2.34) stipulates that there is no loss at the airflow in two sections A_1 and A_2 and thus the respective velocities must be v_1 and v_2, respectively.

So, v_1 is 10 ms^{-1}, the width of 30 km of the plain is proportional to A_1, and the width of 3 km of the valley is proportional to A_2. From Eq. (A2.34), we get

$$v_2 = (A_1/A_2)*v_1 \Rightarrow v_2 = (30/3)*10\,\text{ms}^{-1} = 100\,\text{ms}^{-1}$$

and

$$\Delta p = \frac{\rho}{2}(v_2 - v_1)^2 = (1.16/2)*(10^2 - 100^2) = -5.7\,kPa$$

So, the airflow in the valley corresponds to a velocity of 100 ms^{-1} and a pressure drop of 5.7 kPa.

References

Campbell, G. S., & Norman, J. M. (1998). *An introduction to environmental biophysics* (pp. 293). Springer.

Monteith, J. L., & Unsworth, M.H. (1991). *Principles of environmental physics* (2nd ed., pp. 291). Edward Arnold.

Fundamentals of Global Carbon Budgets and Climate Change

8

Abstract

Global climate grid.9983.b change and GHG emissions are essential topics for understanding physical processes ongoing in the atmospheric boundary layer, within a control volume extending from the top ground to the troposphere. This chapter aimed therefore to deliver a qualitative and quantitative approach to climate change processes and carbon balance components. This approach is facilitated by the grounding on fundamentals of atmosphere and microclimate dynamics, such as the radiation windowing, spectral turbulence, and heat transfer processes, previously discussed. The first question addressed was the quantification over the last decades of the main components of the global carbon budgets including fossil fuel emissions, land-use change, carbon uptake by land, and oceans or CO_2 atmospheric concentration. The climate change predictive scenarios of IPCC are described, mainly focusing on the interactions among atmospheric carbon, carbon sequestration, environmental temperature, and precipitation. Also discussed is the occurrence of extreme events, such as heatwaves and precipitation episodes. This discussion reports modeling results in the literature characterizing essential parameters of extreme events such as duration, peaking, and returning periods in the context of natural and/or anthropogenic driving. Finally, some elaboration is given to biochar as a promising technology for carbon sequestration and mitigation of carbon emissions.

8.1 Introduction

Emissions of greenhouse gases (GHG), steadily increasing since the Industrial Revolution, have been the main driver of global atmospheric warming and climate change. Climate change is a global reality, and environmental heating is unequivocal since the 1950s and many of the observed changes are unprecedented

© The Author(s), under exclusive license to Springer Nature Switzerland AG 2021   267
A. Rodrigues et al., *Fundamental Principles of Environmental Physics*,
https://doi.org/10.1007/978-3-030-69025-0_8

over decades. The atmosphere and the oceans have warmed which is reflected, for example, in a combined land and ocean surface temperature as calculated by a linear trend of about 0.85 °C [0.65; 1.06 °C] over the period 1880–2012, along with a set of extreme events such as a decreasing of snowfall and ice, a rise in sea level, and rural wildfires and hurricanes that took an unprecedented expression.

In addition to the disruption of ecological systems, it is important to acknowledge the heavy economic and social consequences that have been hitting today in various parts of the globe. To cite an example among many already available today, it is noted that in the US the estimated costs attributable to climatics events in 2017 reached an amount of the order of $ 360 billion (Jackson et al. 2018).

Global climate change takes place when the earth–atmosphere system responds to counteract the radiation flux changes. As early as the beginning of the twentieth century, Svante Arrhenius suggested with precocious wit that the Earth could be warmed due to an anthropogenic increase of carbon dioxide in the atmosphere. Today's knowledge shows that human influence on the climate system is clear, and recent anthropogenic emissions of greenhouse gases are the highest in history.

Awareness, sentiment, and regulatory action related to climate change have accelerated over the last years, and there is a common sentiment that it is time for governments, business leaders, and society to step up and put climate risks at the forefront of their strategy and ensure that they could address the challenges that will be central to all the world for the future. Embodying this feeling, the Paris Agreement (November 2016) built upon the United Nations Framework Convention on Climate Change (UNFCCC) brought all nations in a common cause to undertake efforts to combat climate change and to strengthen the global response to keeping the global temperature rise in this century well below 2 °C above pre-industrial levels and to pursue efforts to limit the temperature increase even further to 1.5 °C.

Being a global threat, and considering regional and socioeconomic differences among countries, the Agreement was not specific on regional thresholds, meaning that regional climate shifts and extreme events will not follow the global mean (Perkins-Kirkpatrick and Gibson 2017).

COP26 (to be held in Glasgow in November 2020) marks the 5-year anniversary of the Paris Agreement and is expected to accelerate change in the way governments think and regulate around climate change. Countries will be assessed on how they are meeting their global warming mitigation targets agreed in Paris. The Paris Agreement was the culmination of a long trajectory that can be traced back to 1988 with the establishment of the Intergovernmental Panel on Climate Change (IPCC) through the cooperation between the World Meteorological Organization (WMO) and the United Nations Environment Programme (UNEP). Its mandate was to assess scientific information related to climate change, to evaluate the environmental and socioeconomic consequences, and to formulate realistic response strategies.

Since its inception, IPCC has played a major role in assisting governments to adopt and implement policies to address climate change and gave authoritative advice of the Conference of the Parties (COP) to the United Nations Framework

Convention on Climate Change (UNFCCC) established in 1992, and to the 1997 Kyoto Protocol. IPCC organized periodic meetings, resulting in successive complete published documents such as the Fourth Assessment Report (AR4 Report) and the Fifth Assessment Report (AR5 Report).

The climate system includes five interacting components which are the atmosphere, hydrosphere, ice and cryosphere with permafrost, biosphere, and lithosphere. Climate change is associated with modifications in the Earth's climate system, resulting in newer climate patterns and long-term averages of climate variables with time scales from decades to thousands of years. These changes can derive from internal variability when natural processes linked to the various components of the climate system modify the global energy budget. Examples of these processes are cyclical ocean patterns such as the El Niño southern oscillation, the Pacific decadal oscillation, and the Atlantic multi-decadal oscillation. There is also external forcing on climate change which includes solar output and volcanism.

Climate change usage refers to a change in the state of the climate that can be statistically assessed, whether due to human activity or because of natural variability. This variability can be identified by changes in the mean and/or fluctuations of its properties, and that persists for an extended period, typically decades or longer. It is worth noticing that this usage differs from that in the United Nations Framework Convention on Climate Change (UNFCCC), where climate change refers to a change in climate that is attributed directly or indirectly to human activity altering the composition of the global atmosphere, and that is in addition to natural sources of climate variability observed over comparable time.

From 1880 to 2019, the global surface average temperature increase was 0.07 °C per decade, but has accelerated since 1981 to an average of 0.18 °C per decade. This tendency is depicted in Fig. 8.1. Considering a 2 °C mean global warming until 2100, compared with pre-industrial conditions, larger increases in temperature extremes are expected relative to that average threshold, but with considerable regional variation. Global climate models project that annual minimum temperatures in the Arctic will reach 5.5 °C, while maximum annual temperatures will be at least 3 °C warmer over much of the Northern Hemisphere, Central America, and South Africa (Perkins-Kirkpatrick and Gibson 2017).

In recent decades, changes in climate have caused cascading global impacts in natural and human systems, indicative of the sensitivity of these systems to climate change. The evidence supporting the importance of anthropogenic factors over climate change has grown from the AR4 IPCC to the IPPC AR5 Reports. IPCC conclusions point to the influence of anthropogenic driving on the average global surface temperature, the melting and retreat of glaciers, Greenland and Arctic ice surface loss, and an increase in the energy content of oceans across the 0–700 m layers and a global mean sea-level rise since the 1970s. Global warming over the twentieth century was widespread and rather uniform. Warming was about 0.4 °C in the tropics, 0.6 °C at middle latitudes, and an average of 0.5 °C globally over 100 years. In the temperate Northern Hemisphere, the warming was about twice as much in the cool season compared to the warm season (Hansen et al. 2001).

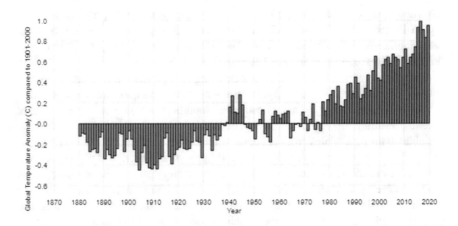

Fig. 8.1 Records of average global surface temperature from 1880 to 2019, the period with reliable information. The zero line is indicative of the long-term global average surface temperature. The blue and red bars show the average differences above or below for each year (adapted from NOAA/CLIMATE.GOV)

According to Mann et al. (1998), the years 1990, 1995, and 1997 were warmer than any other since 1400 AD, within a 99.7% level of certainty. In the past, unforced natural climate variability, relevant in century time scales, was due to natural variations of solar irradiance or explosive volcanism. If a credible empirical description of climate variability could be obtained these authors noted that for the last centuries, it would be possible a more confident estimation about the roles of different external and internal forces of variability on the past and recent climate. Figure 8.2 depicts the estimates of the increases of CO_2 in the atmosphere (red line) along with human emissions (blue line) since the start of the Industrial Revolution in 1750. Emissions rose slowly to about 5 billion tons a year in the mid-twentieth century before a sharp increase to more than 35 Gty^{-1} forescast by the end of the century (Mitra 2020).

Cumulative emissions of CO_2 and other greenhouse gases will largely determine global mean surface warming by the late twenty-first century and beyond. There is also evidence that the negative impacts of climate change on crop yield will override any positive impacts. Available evidence suggests that in higher latitudes there could be positive effects, although it is not yet clear if the overall balance is positive (AR5 Report 2015).

In tropical and temperate regions, the balance of climate change effects is negative, especially in relation to the cultivation of cereal crops such as wheat, rice, and maize. Projections of greenhouse gas emissions vary over a wide range, depending on socioeconomic factors, developmental state, and climate policy. Improvements in climate models, since the AR4 to AR5 Reports, were evident in topics such as the simulations of continental-scale surface temperature, large-scale precipitation, monsoon, Arctic sea ice, ocean heat content, some extreme events, the

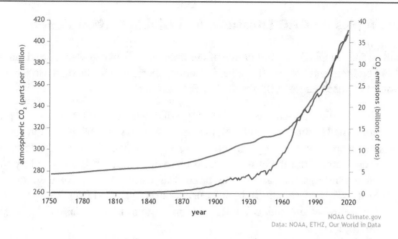

Fig. 8.2 Carbon dioxide annual emissions evolution in the period 1750–2020 (adapted from Howard Diamond—NOAA ARL)

carbon cycle, atmospheric chemistry, and aerosols or the effects of stratospheric ozone or the El Niño–Southern Oscillation.

The atmospheric layer is opaque to infrared radiation but transparent to visible radiation. As mentioned in Chaps. 4 and 6, terrestrial bodies absorb solar radiation with small wavelengths, and radiant energy is then emitted from the surface to the atmosphere as thermal radiation with longer wavelengths absorbed by GHG. GHG allows downward transmission of solar radiation but traps a significant fraction of upward infrared radiation, thus exerting a critical influence on the Earth's global energy budget. Furthermore, accumulations of GHG as carbon dioxide, nitrous oxide (N_2O), methane (CH_4), and chlorofluorocarbons can close the atmospheric window referred to in Chap. 6, by absorbing more infrared radiation emitted from the Earth's surface, enhancing the global warming effect.

Improved experimentation on climate change is possible with numerical weather forecast complex models, designated as global climate models (GCM). Insights into topics such as atmospheric physics and feedback processes were improved with these models, although many rough assumptions in most GCMs turn forecasts of climate change somewhat uncertain (Stull 2000). A thorough discussion about these issues is necessary, and scientists should have a responsibility in this discussion, to avoid distortion of facts by political, industrial, and commercial interests.

8.2   Topics on GHG Emissions and Global Carbon Budget

Global trends in GHGs are indicative of the imbalance between sources and sinks in the gas budgets and are strictly related to atmospheric emissions on a global scale. Among the gases targeted by the Kyoto Protocol are

(i)   Carbon dioxide, emitted by burning solid waste, fossil fuels, woody and non-woody products and biomasses, agricultural residues, and certain chemical reactions (e.g., manufacture of cement). Carbon dioxide is also sequestered from the atmosphere, for example, when it is absorbed by plant photosynthesis as part of the biological carbon cycle. As mentioned in Sect. 4.6, the CO_2 atmospheric concentration increased from 280 ppm at the beginning of the Industrial Revolution around 1750, to about the current 407 ppm.

(ii)   Methane emitted from livestock and other agricultural practices, decomposition of organic municipal waste landfills, or production and transport of coal, natural gas, and oil.

(iii)   Nitrous oxide (N_2O) emitted by combustion of fossil fuels and solid wastes from agricultural and industrial activities.

(iv)   Fluorinated gases such as hydrofluorocarbons, perfluorocarbons, sulfur hexafluoride, and nitrogen trifluoride which are synthetic, powerful greenhouse gases that are emitted from a variety of industrial processes. These gases are typically emitted in smaller quantities but because they are potent greenhouse gases, they can be referred to as high global warming potential gases (e.g., https://whatsyourimpact.org/high-global-warming-potential-gases; global warming potential, Wikipedia).

The Earth has been at a nearly constant temperature during the past 100 million years, with less than a 4% variation over that period. During this long period, the incoming solar radiation was practically compensated for by emitted infrared radiation, maintaining the temperature of the Earth's system in a stable equilibrium. Over the last 50 million years (Stull 2000), the Earth's temperature oscillated by ± 1 °C around the current average of 15 °C. These small variations of temperature can induce significant changes in, for example, the sea level and glaciations. Based on the available evidence, there is a larger consensus that GHG emissions will continue to rise, under the prevalent climate change mitigation policies. For example, AR4 reported that between 2000 and 2030 the increase of GHG emissions would be about 25 and 90%, with fossil fuels maintaining the dominant position by 2020 and beyond.

As GHG concentrations rise, net radiation flux changes, due to an increase in infrared radiation which induces atmosphere warming (e.g., Stull 2000). In this context, a fundamental concept for assessing the degree of climate change is the so-called radiative forcing (RF) (e.g., AR4 Report 2008). This metric has been used for many years by IPCC to evaluate the strength of the various mechanisms affecting the Earth's radiation balance. Aerosols partially offset the effects of greenhouse gases and are the main contributors to the uncertainty associated with the estimates of total radiative forcing.

CO_2 emissions were the largest contributors to the increased anthropogenic radiative forcing or net radiation, in every decade since the 1960s. According to the AR5 Report (2015), GHG emissions, driven by economic and population growth, have increased since the pre-industrial era leading to unparalleled atmospheric concentrations of CO_2, methane, and nitrous oxide.

The long-term increase in CO_2 atmospheric concentration was mainly due to accumulated emissions of about 2040 ± 310 Gt CO_2, from which about 880 ± 35 Gt CO_2 remained in the atmosphere. The last 40 years accounted for about half of the accumulated emissions over the two and a half centuries. It is estimated that global emissions, which may be permissible worldwide, without a high likelihood of dangerous climate change are about 2900 Gt CO_2 (e.g., Quiggin 2019).

Despite increased public awareness and climate mitigation policies, the global yearly anthropogenic GHG emissions increased between 1970 and 2010, reaching an estimated 49 Gt CO_2-eq. (e.g., AR5 IPPC Report). Also, the yearly growths of GHG emissions, in the period 1970–2000 and in the decade 2000–2010, were about 1.3 and 2.2%, respectively. The contribution from fossil fuel combustion and conversion was about 78% in these two periods.

While the contribution of population growth for CO_2 emissions from fossil fuels remained constant in the period 1980–2010, the global economic growth rose drastically over the same period. Therefore, according to the AR5 Report, typical values of total annual anthropogenic GHG emissions were 27 Gt CO_2-eq.y^{-1}, 38 Gt CO_2-eq.y^{-1}, and 49 Gt CO_2-eq.y^{-1} in 1970, 1990, and 2010 respectively, indicating an increasing trend over this period.

Emissions of GHG, mainly CO_2 and CH_4, have continuously increased since the eighteenth century to 405 ppm and 1803 ppb in 2018, both higher by 40 and 150% comparable to 1750 (e.g., AR5; Le Quéré et al. 2018). The stabilization of CO_2 atmospheric concentration to 550 ppm would theoretically increase the atmospheric temperature by about 2–3 °C above the current level. Also, since the onset of the industrial era, ocean uptake of CO_2 has resulted in increased acidification of the oceans with a decrease of the pH of the ocean surface at the order of 0.1, corresponding to a 26% rise in acidity, measured as hydrogen ion concentration (e.g., AR5 Report 2015).

In general, the global carbon budget is considered to have six main components: emissions by fossil fuel applications (EF), emissions from land-use due to land human activities, including those leading to land-use changes (ELU), variations in the atmospheric CO_2 concentration (ΔCAt), and uptake of carbon dioxide by the oceans (UCO) and by land (UCL). A sixth closure budget component (CBC) is due to imbalances resulting from divergences in the estimates for the other components.

The available information about these carbon budget components (Le Quéré et al. 2018) relies on data sets from experimental observations or measurements and from modeling projections that consider several dynamic global vegetation models (Fig. 8.3).

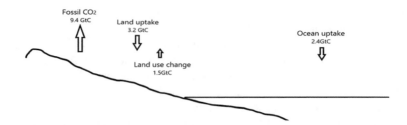

Fig. 8.3 Estimated aggregate components of CO_2 (GtCyr^{-1}) budget for the period 2008–2017 (adapted from Le Quéré et al. 2018)

Over the period 1870–2017, estimates of average accumulated EF and ELU were about 1550 Gt CO_2 and 690 Gt CO_2, respectively. For this period, the average carbon sinking in the atmosphere (ΔCAt), oceans (UCO), and land (UCL) is discriminated in Fig. 8.4.

Between 1959 and 2017, about 82% of total carbon emissions were due to EF components and the remaining 18% to ELU. For the decade 2008–2017, the correspondent CO_2 values were 87 and 13%, respectively. Over this period the emitted carbon partitions were 44% to the atmosphere, 22% to the ocean, and 29% to land (Le Quéré et al. 2018).

The correspondent average closure during the two periods was about 92 Gt CO_2, without accounting for an additional loss of around 73 Gt CO_2 from reduced forest cover over the decades.

The contributions of fossil fuels to regional and global increases in carbon emissions are different, because natural gas, the cleanest of fossil fuels, is the main source of the global increase in CO_2 emissions. The demand for natural gas has globally grown by 2% yr^{-1}, a higher growth rate than any other fossil fuel. It has risen over the past 5 years across many countries, including China and the US, and has taken place at the expense of coal, with the resultant containment in CO_2 emissions.

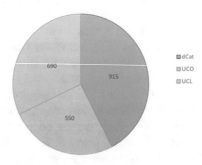

Fig. 8.4 Average carbon sinks relative to the atmosphere (ΔCAt), oceans (UCO), and land (UCL) in GtCO$_2$-eq. for the period 1870–2017

Since 2012, the demand for oil in the transportation sector has grown steadily by 1.4% yr^{-1}, following a tendency of decades, despite fast increases in electric and hybrid vehicles around the world. China and India showed oil consumption growth rates of 4% yr^{-1} and 5% yr^{-1} and were responsible for the largest part of this global demand increase. Even the West saw an increase in oil demand of 1.3% yr^{-1} in USA and 0.4% yr^{-1} in UE, despite the high demand for electrical and hybrid vehicles (e.g., Jackson et al. 2018). Commercial air travel has aggravated the situation with about a 27% increase in fuel demand over the last decade, which has outpaced the higher fuel efficiency of commercial aircraft.

Global coal demand declined steadily over the period 2000–2017 by about 0.9% yr^{-1}, corresponding to about 156 $GJyr^{-1}$ in the latter year. Canada and the USA are the paradigms of this tendency with a combined 40% decrease since 2005. Within the EU, renewable energies are expected to surpass coal as the source of primary energy by 2021. However, this situation may be offset by higher coal consumption in Asia/Pacific and Central/South America, with an increase of 3%. As for India, the current coal consumption is higher than that in the USA and the EU. Overall, the International Energy Agency points out to an inevitable decrease in coal demand in the decades following 2030. Despite a global trend for increased energy efficiency, per capita energy demand in countries such as the US and EU is still much higher than in India by about five- to tenfold (Jackson et al. 2018).

From 1959 to 2017, the emitted carbon from the atmosphere, oceans, and land accounted for average values of 45%, 24%, and 30% for ΔCAt, UCO, and UCL, respectively. Over this period, the average global fossil carbon emissions increased over every decade. For example, the average fossil emissions in the 1960s, 1990, and over the past 10 years (2008–2017) were 11.4 Gt CO_2 yr^{-1}, 23.1 Gt CO_2 yr^{-1}, and 34.4 Gt CO_2 yr^{-1}, respectively. On the other hand, and for the same decades, the average ELU, due to land-use change and forestry, was stationary of 5.5 Gt CO_2 yr^{-1}, 5.1 Gt CO_2 yr^{-1}, and 5.5 Gt CO_2 yr^{-1}. During the second half of the twentieth century, the rate of carbon dioxide emissions decreased every decade, for example, from 4.5% yr^{-1} in the 1960s to 1% yr^{-1} in the 1990s. In the 2000s, the global carbon emissions showed an average growth rate of 3.2% yr^{-1} followed by a decrease of 1.5% yr^{-1} in the period 2008–2017 and with a low yearly growth during the period 2014–2016.

For the 1960s, 1990s, and the period 2008–2017, Fig. 8.5 depicts average atmospheric concentration growth rates of carbon dioxide in the three major ecosystems (atmosphere, ocean, and land).

It shows a tendency toward a strong worsening of the growth rate of the concentration of carbon dioxide in the three ecosystems considered and clearly more pronounced over land.

Average ocean and land carbon sinks increased with the atmospheric carbon augment, and carbon land sinks showed higher variability than ocean land sinks. In 2017, ΔCAt, UCO, and UCL were 16.8 Gt CO_2 yr^{-1}, 9.1 Gt CO_2 yr^{-1}, and 14 Gt CO_2 yr^{-1}, respectively. Over the decade 2008–2017, the budget closure averaged 1.83 Gt CO_2 yr^{-1}, with 1.1 Gt CO_2 in 2017, suggesting an overestimation of the

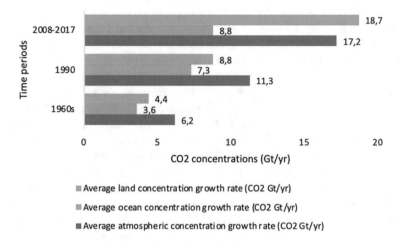

Fig. 8.5 Average concentration growth rates of CO_2-eq. over time periods

emissions and/or underestimations of the sinks. The latter was more likely due to the high variability of land sinks and to an underestimation of the ocean sink.

In the decade 2008–2017, China's and India's average carbon emissions increased by 3 and 5.2% yr^{-1}, whereas emissions from the EU and the USA decreased by 1.8 and 0.9% yr^{-1}. These percentages correspond to 2.3, 0.91, 0.62, and 0.66 Gt CO_2, respectively (Le Quéré et al. 2018).

From 2016 to 2017, the estimated global fossil carbon dioxide emissions grew by about 1.6%, reaching an average of about 36.2 Gt CO_2 yr^{-1}, due mainly to coal (40%), oil (35%), gas (20%), and cement industry (4%). This EF increase in 2017 followed for three years with little or no emissions growth.

As shown in Fig. 8.6, the total amount of CO_2 emissions is quite asymmetric in relation to countries and economic areas and these distortions have been an element of disagreement between countries when it comes to sharing mitigation costs. In fact, four blocks emit more carbon dioxide than the rest of the world.

The differences between economic blocks and countries reflect the levels of economic development and industrialization almost always correlated with the intensity of the energy consumption, which ends up translating into the per capita values of CO_2 emission as shown in Fig. 8.7.

The average gross rate in atmospheric CO_2 concentration (ΔCAt) in 2017 was about 16.8 Gt CO_2y^{-1}, close to the 2008–2017 decade (17.2 Gt CO_2y^{-1}). The ocean sink (UCO) was about 9.1 Gt CO_2y^{-1} in 2017 compared with a yearly average of 8.8 Gt CO_2y^{-1} over the period 2008–2017. The terrestrial CO_2 sink (UCL) was 13.9 Gt CO_2 in 2017 compared with a yearly average of 11.7 Gt CO_2 over 2008–2017.

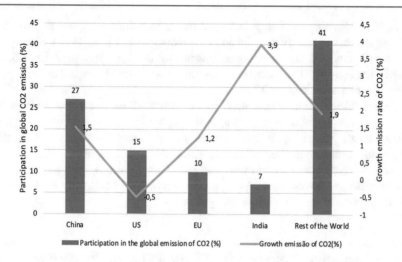

Fig. 8.6 Percentage distribution of CO_2 emissions and rate from 2016 to 2017 among the main polluting areas

Fig. 8.7 Global average emissions per capita in 2017 ($GtCO_2$-eq.)

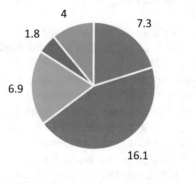

■ China   ■ US   ■ EU   ■ India   ■ Global average value

Despite the somewhat negative panorama at the global level, CO_2 emissions decreased significantly in a group of 19 countries within Europe and North America which were responsible for about 20% of the emissions. In 2018, the total CO_2 emissions were about 37.1 Gt, compared with 2290 Gt CO_2 for the accumulated emissions over the period 1870–2018. In that year, the atmospheric CO_2 growth concentration (ΔCAt) averaged around 18 Gt CO_2, corresponding to an average atmospheric concentration of about 407 ppm. Combined global ocean and land sinks were about 24 Gt CO_2. Assuming a probability of 66%, a net negative global GHG in 2060 and 2080 is forecasted, corresponding to global average temperature increases of around 1.7 and 1.3 °C by 2100 (Jackson et al. 2018).

In 2018, the biggest change in fossil fuel applications (EF) was due to a significant increase in energy demand and CO_2 emissions in China. In this country, the increases in coal, oil, and natural gas demands for the same year were 4.5, 3.6, and 17.7%, respectively. The increase in natural gas consumption was due to China´s policy for climate change mitigation.

Global carbon components need to be evaluated to assess the state of the environment, vulnerable to fast changes in biophysical and anthropogenic factors. These include variations in growth rates of fossil fuel emissions, environment temperatures, and dynamics of carbon sinks. Also, time series of data sets are essential to predict the carbon cycle, and many organizations, such as universities and laboratories, governments, private companies, and media, rely on these data sets to respond to climate change mitigation.

8.3   Alternative Scenarios for Climate Change Predictions

The AR4 and AR5 Reports describe sets of alternative GHG emissions scenarios linked with projections for atmospheric temperature increases. At present, the average global warming is 0.85 °C relative to the pre-industrial era (Fischer and Knutti 2015).

Under AR4, six climate change scenarios were proposed: B1, AIT, B2, A1B, A2, and A1FI, corresponding to atmospheric concentrations hypothesis of GHG (CO_2-eq.), in 2100 of 600, 700, 800, 850, 1250, and 1550 ppm, respectively.

The temperature and sea-level changes estimated under these scenarios, comparing the periods 2090–2099 and 1980–1999, are synthesized in Table 8.1.

The ranges reflect the limited understanding of physical mechanisms such as feedback from the carbon and carbon dioxide cycles (AR4 Report).

The six scenarios from four climate major trend scenarios (A1, A2, B1, and B2) assume alternative socioeconomic development roadmaps, encompassing a wide range of GHG emissions and demographic, economic, and technological driving forces which input into projections of climate change vulnerability. These six scenarios do not assume additional climate policies, besides the current ones.

The A1 scenario assumes high global economic growth, a global population peaking in the mid-twenty-first century, and a fast introduction of new efficient technologies. This scenario is divided into three which are A1FI for fossil intensive

Table 8.1 Characterization of climate change scenarios under the AR4 IPCC report

Climate scenarios	B1	AIT	B2	A1B	A2	A1F1
Temperature variations, °C	1.8	2.4	2.4	2.8	3.4	4.0
Temperature ranges, °C	1.1–2.9	1.4–3.8	1.4–3.8	1.7–4.4	2.0–5.4	2.4–6.4
Sea-level rise projections, m	0.18–0.38	0.20–0.45	0.20–0.43	0.21–0.48	0.23–0.51	0.26–0.59

technologies, A1T for non-fossil energy technologies, and A1B for an equilibrated forecast encompassing all the renewable and non-renewable sources.

Scenario B1 describes a world with the same global population as A1, converging fast toward a service and information economy, with rapid changes in economic structures.

The B2 scenario relates to a world with a medium-level population and economic growth, with emphasis on local policies for economic, social, and environmental sustainability.

The A2 scenario described a very heterogeneous world with high population growth, slow economic development, and slow technological change. No likelihood has been attached to any of these scenarios.

The AR5 Report assumed that anthropogenic GHG emissions are mainly driven by population size, economic activity, lifestyles, energy use, land-use patterns, technology, and climate policy. AR5 used improved knowledge to change the nomenclature of AR4, establishing for the twenty-first century a set of four Representative Concentration Pathways (RCPs) profiling for GHG emissions and atmospheric concentrations, air pollution emissions, and land use. The four RCPs are as follows: (i) RCP2.6, representative of a profile with likely global warming below 2 °C above pre-industrial temperatures that can be achieved through a package of rigorous mitigations policies; (ii) RCP4.5 and RCP6.0 are intermediate profiles; and (iii) RCP8.5 a profile with very high GHG emissions. Most models indicate that scenarios that meet forcing levels, like RCP2.6, are characterized by substantial net negative emissions by 2100 averaging about 2 Gt CO_2yr^{-1}.

RCP2.6 is the only climate profile where projections consider unlikely that global surface temperatures, in the period 2081–2100, exceed the average temperatures in the periods 1850–1900 and 1986–2005, by 2 °C and 0.3 °C to 1.7 °C, respectively. The baseline profiles corresponding to scenarios where no effort is made to control CGH emissions fall into the RCP6.0 and RCP8.5 classes.

Over the period 2010–2100, the RCP2.6 profile is within a predicted range, from 430 to 480 ppm ranges of atmospheric CO_2-eq. For the twenty-first century, the RCP4.5 profile corresponds to ranges from 430 to 480 ppm in the period 2010–2040, and from 530 to 580 ppm in the remaining decades. RCP6.0 falls in predicted ranges 480–530 ppm and 530 to 580 ppm over the periods 2010–2030 and 2030–2050, and over a predicted range of 720–1000 ppm from 2050 onwards. Finally, RCP8.5 corresponds to a predicted atmospheric CO_2-eq. concentration higher than 1000 ppm.

Overall, the main differences in the climate systems are expected to occur by the end of the twenty-first century over the period 2081–2100. For the period 2016–2035, the increase in global mean surface temperature is similar for the four RCP profiles and is expected to be in the range of 0.3–1.7 °C, relative to the period 1986–2005. This statement assumes that there are no drastic outliers such as major volcanic eruptions, dramatic changes in some natural sources, for example, CH_4 and N_2O, and big changes in total solar irradiance.

Over the period 2081–2100, significant differences in global surface temperatures, relative to the period 1986–2005, are forecasted among the 4 RCP profiles, with likely increases from 0.3 to 1.7 °C for RCP2.6, 1.1 to 2.6 °C for RCP4.5, 1.4 to 3.1 °C for RCP6.0, and 2.6 to 4.8 °C for RCP8.5.

If a comparison is made with the second half of the nineteenth century, the forecasted increases in global surface temperatures are higher than 1.5 °C for RCP4.5 and higher than 2 °C for RCP6.0 and RCP8.5. The average linear global trending combining land and ocean surface temperature data shows an average warming of 0.85 °C, over the period 1880–2012. In the Northern Hemisphere, the warmest 30-year period over the last 1400 years was likely the one found from 1983 to 2012, and each decade between 1984 and 2014 was successively warmer than any previous decade since 1850. Average Arctic temperatures have increased at almost twice the global average rate in the past 100 years. Also, the annual mean of sea ice Arctic extent decreased over the period 1979–2012 at a rate of 3.5–4.1% per decade.

Global temperatures show multi-decadal average warming, along with variability at smaller annual and decadal scales. Due to this pattern of variability, trends based on short records are very sensitive to periods selected for evaluations, not reflecting long-term climate trends generally (AR5 Report).

Predictions from global circulation models, showing a non-uniformity of global warming in the twenty-first century with changes in the hemisphere, season, and underlying surfaces, along with seasonal and inter-annual variability, were carried out by Boer et al. (2000). Land average warming will be higher than over oceans and, on a global scale, warming will be faster over the Arctic region. At daily and seasonal scales, frequent hot and fewer cold temperatures over most land areas are almost certain to occur along with an increase in the mean global temperature. Also, longer heatwaves, with higher frequency and sporadic winter extremes, will likely take place.

Ocean warming, especially near the surface, dominates the increase in energy stored in the climate system, accounting for more than 90% of the energy accumulated over the period 1971–2010, with only about 1% stored in the atmosphere. Over the same period, the upper 75-m depth of oceans on a global scale warmed at about 0.11 °C per decade, with the warming of the upper 700 m likely beginning in the 1870s.

The physical and geographical context for rainfall provides the essential framework for the analysis of patterns and intensity of precipitation (Stull 2000). Basically, the convergence of moisture-laden air masses leads to air uplift, cloud formation, and delivery of precipitation, snow, or hail. Atmospheric water vapor is central to this dynamic, insofar as the warm atmosphere can convey water vapor amounts as high as 6–7% per °C of warming near the Earth's surface, as expressed by the Clausius–Clapeyron principle. As water vapor also contributes to radiative cooling of the atmosphere, this later type of radiative forcing induces an additional in total global precipitation of only about 2–3%, per °C of warming. This radiative cooling of atmosphere is due to additional absorption and emission of infrared radiation (Allan 2011). Intense rainfall is mostly local, but it is fed by an atmospheric moisture supply that may deplete other regions from localized moderate rainfall. These are principles that must be throughout evaluated, because of the implications for flooding and drought in the future.

Global processes such as latent energy release during storms, or limitations of moisture sources, can be crucial to the development of distinct regional tendencies concerning the dependency of rainfall intensity with air temperature. Atmospheric aerosols modulate the energy balance in the atmosphere and the Earth's surface system, thereby influencing these processes (e.g., Allan 2011).

From 1900 to 2005, accumulated precipitation and frequency and intensity of heavy precipitation events increased significantly in the eastern parts of North and South America, Northern Europe, and Northern and Central Asia. On the other hand, precipitation declined in the Sahel, the Mediterranean, Southern Africa, and parts of Southern Asia. Globally, the area affected by drought has likely increased since the 1970s (e.g., AR4 and AR5 Reports).

The global water cycle is also likely affected by anthropogenic factors that influence atmospheric moisture content, changes in precipitation patterns over land, and the occurrence of heavy precipitation events. The global specific humidity has very likely increased in the near-surface boundary layer since the 1970s.

In the twenty-first century, the forecasted global changes in precipitation will not be uniform. The information available since 1901 and particularly 1951 onwards points to an increase in the average precipitation in mid-latitude land areas of the Northern Hemisphere. Fisher and Knutti (2015) mention available models predicting heavy precipitation days for Northern Europe and North America, in response to increasing global temperature.

High and equatorial latitudes, especially in the Pacific and Africa, are likely to face increases in average precipitation ranging from about 30 to 50% under the extreme RCP2.6 to RCP8.5 scenarios. In most dry regions in mid-latitudes, mean rainfall will likely decrease from 5 to 40% under the two extreme scenarios. In contrast, in mid-latitude wet regions, mean precipitation will likely increase by a similar range from 5 to 30% under the two extreme scenarios. Under profile RCP8.5, in dry regions, the frequency of droughts will likely increase toward the end of the twenty-first century.

Systems likely to be affected by climate change include, among others, the tundra, and the boreal forest due to the inability to adjust to warming. Negative impacts due to the lack of precipitation and water availability will cause hydrological negative impacts in areas as diverse as Mediterranean-type ecosystems, tropical rainforests, and in lower latitude agriculture areas. Also, sea-level rise and human health downgrading in populations with low adaptative capacity will be expected. Even in economically well-off areas, some activities, and vulnerable people, including the poor, young, and elderly, will be at risk.

8.4   Occurrence of Extreme Events

Climate change includes both variations in the mean climate variables as well as in weather extremes. Overall, by definition, an extreme event refers to an event that is rare in a specific place and time of the year. A rare phenomenon is one that may occur with probabilities equal to or less than the 10th percentile or equal to or greater than the 90th percentile, considering a probability density function derived from real data. If the pattern of extreme climate anomalies persists over a large period, such as a season, resulting in an extreme average, then it can be classified as an extreme climate event (AR5 Report).

Over the last 60 years, some global extreme events have increased in magnitude and frequency. These include heatwaves and intense precipitation, on different spatiotemporal scales, with unevenly distributed change profiles (e.g., Perkins et al. 2012; Coumou and Rahmstorf 2012).

The distribution of daily maximum and minimum temperatures has shifted to higher temperatures, with changes of variances in probability distribution functions showing a spatial heterogeneity, with wider distributions in the tropics. Global trends in night-time temperature events were stronger than those relative to daytime (Donat and Alexander 2012; Perkins et al. 2012; Donat et al. 2013).

The National Weather Service (NWS) in the USA defines, in a simple way (Robinson 2001), a heatwave as a period lasting at least 48 h wherein neither the overnight low nor the daytime high falls below the thresholds of 26.6 and 40.5 °C, respectively. When more than 1% of the recorded case values in automatic weather stations exceed NWS thresholds, alternative heatwave thresholds are defined as the values of these 1% cases. In general, a hot spell is a similar event referring to events ranging from the 1% values to the NWS thresholds, while a warm spell is an event occurring between the 1 and 2% values.

Extreme events, such as a decrease in cold temperature extremes, increase in warm temperature extremes, sea-level changes, discrete heavy precipitation events, or drought increase in dry areas, have been observed since the 1950s and many were linked to anthropogenic influences (AR5 Report). The frequency of heatwaves has increased in continental areas in Europe, Asia, and Australia, and there is a strong chance that human influence more than doubled the probability of heatwave events in some areas. Extreme diurnal patterns as warm days and nights increased, with high likelihood, at the expense of cold days and nights.

Global field observations over the second half of the twentieth century validated the likely human influence on the increase in heavy precipitation events. Costs related to flood damage worldwide have been increasing since the 1970s, with the risks of extreme events and/or related processes increasing with further warming, even if only by 1 °C.

Negative impacts of extreme climate events include disruption of food crops and water availability, deterioration of infrastructures, and human health and lives. The inability to cope with these drastic impacts applies to some extent to all countries regardless of their developmental status (e.g., Meehl et al. 2000; AR5 Report).

Some events such as extreme cold temperatures will decrease but overall, most events including heatwaves, precipitation episodes, and storms in urban areas are expected to increase along this century. There is an increasing trend for the occurrence of global warm spells of average, minimum, and maximum temperatures over a greater area and with higher magnitude than the typical summer heatwaves, contributing to changing the annual trends (Perkins et al. 2012).

Trends are generally more heterogeneous for daily precipitation extremes. For example, for the period 1900–2009, a modeling study showed that out of a global set of about 8320 weather stations, about two-thirds reported increases in extreme events, and these were mainly located in the tropics and higher latitudes (Westra et al. 2013; AR5 Report).

Reliable assertions of very heavy and extreme precipitation events are possible only in areas with dense networks of precipitation recording stations. In the mid-latitudes, a widespread increase in the frequency of the very heavy precipitation was seen over the twentieth century (Groisman et al. 2005). The later authors assume that an intense precipitation event has more than 0.3% of the upper daily rainfall, and that the projections for a greenhouse-enriched atmosphere reveal an increasing probability of these events. These extreme precipitation events show a return time of 3–5 years for a daily event for annual and 10–20 years for seasonal events, depending on the probability of local daily and seasonal rain events.

Other extreme phenomena such as droughts, that are not tightly linked to temperatures, are highly uncertain. These are highly complex processes caused by anthropogenic and hydrometeorological factors with impacts that vary greatly among locations. The common view of drought or dryness is the relationship with rainfall deficit. Additional complexity can be added by socioeconomic and/or atmospheric factors such as aerosols and land–atmosphere interactions which can induce or non-linear and self-enforcing events (e.g., Sippel and Otto 2014). Several studies also reported that in Europe, heatwave intensity, length, and frequency have increased over the last century and even more so in Southeast Europe over the last 50 years (Della-Marta et al. 2007; Kuglitsch et al. 2010).

It is very likely that several of the unprecedented events of the past decade would not have occurred without anthropogenic global warming (e.g., Coumou and Rahmstorf 2012). However, a clear deterministic cause-effect between human activity leading to warming and the occurrence of extreme events is not possible, because extreme events can occur by chance under strictly natural conditions (e.g., Sippel and Otto 2014; Stott et al. 2004). In this context, Coumou et al. (2013b) concluded that approximately 80% of recent monthly heat records would not have occurred without human influence. This share will increase to more than 90% by 2040 under a medium future global warming scenario.

Perkins-Kirkpatrick and Gibson (2017) studied the variation of characteristics of regional heatwave changes with mean global warming, assuming the 1.5–2 °C scenarios stipulated by the Paris Agreement as scenarios of warmer thresholds. Two global climate model ensembles, comprising 27 models, were applied. Basically, these models were the Coupled Model Inter-comparison Project Phase 5 (CMIP5) and a 21-member version of the Community Earth System Model (CESM), with daily data in the period 1861–2005 and from 2016 thereafter with projections until 2100 under profiles RCP8.5 and RCP4.5 of the AR5 Report. The modeling strategy considered distinct climate sensitivities, physical parameterization, and resolutions, through CMIP5, and the influence of internal variability, through CESM.

Four heatwave variables which were the sum of heatwave days, the total number of discrete events, length of the longest event, and peak wave intensity, were evaluated during a 5-month summer season. Results were obtained at the global level and for 21 land-based regions, for both global mean warming and heatwave thresholds relative to pre-industrial conditions. These results were assumed as applicable to heatwave and mitigation strategies.

The higher changes in heatwave days occurred in the tropics with 30 additional days per season and per °C of global temperature rise. These latter events occur over vast areas of Africa, Central and South America, and Southeast Asia. In mid to high latitudes, the variation was less drastic with 10–15 extra days expected over Northern Europe, North America, and Russia. Over South Australia and higher latitudes of South America, the correspondent value of extra days ranged from 4 to 8 per degree of global warming.

Also, over most regions, the number of heatwaves per season was expected to rise by about 1.5–2 events per degree of global warming, excepting Central and Southern Africa and Central Asia where an increase of 2.5 events per season was forecasted. The median change in the longest heatwave duration period ranged from 1 to 3 days with smaller ranges in higher latitudes, and estimated ranges of 4–6 days per degree of global warming over India, Southeast Asia, the US, and South America.

In Central America, African areas, and the Middle East, the longest heatwave period was projected to increase by 10–12 days per degree of global warming. About the peak wave intensity, it was concluded that the temperature of the hottest heatwave day per season was expected to rise between 1.2 and 1.5 °C per degree of global warming, by 1.8 °C across the US, south of Southern America, and some of the African areas, by 2 °C in Europe, and by 1.1 °C in Australia and Southeast Asia.

There is a huge variation of heat warming events and mean warming at regional levels, comparatively with global temperature increase. Over high latitudes in Alaska, Greenland, and North Asia, regional mean warming is twofold higher than global warming. In lower latitudes in the south of southern South America, south and Southeast Asia, and Australia, the regional mean warming is about the same order of magnitude as the global average. Variation in regional and seasonal changes in heatwave days is linear with a widespread. The increase in heatwave days with the global average temperature is much faster in tropical regions than in higher latitudes. Regions with larger overall maximum increases of around 150 days of heatwave days are those with average temperature change rates increasing between 2.5 and 3 °C.

The study of Perkins-Kirkpatrick and Gibson (2017) mentioned above, described the sensitivity of heatwave frequency in relation to the average global warming. Different thresholds of global warming imply different patterns in terms of regional changes in heatwave days. For example, with a global warming of 5 °C until 2100, the number of heatwaves per season is forecasted to increase from 60 days in south of Southern America to 120 days in Eastern Africa, the Amazon Basin, and Western Africa. Under such a scenario, heatwave conditions would become the new normal. For a global 1 °C increase in mean warming, a regional divergence prevails, with lower latitude regions showing faster increases in heatwave days. Lower levels of global warming will be reflected in minimal extremely intense heat events.

The regional variation of the number of seasonal heatwaves and global temperature increase is not linear. The number of events decreases with mean temperature increases of 1.5–2 °C in Eastern Africa and of 3 °C in Western Africa and Sahara regions. For many other regions considered, the predominant rate of increase of heatwave events peaks at about 3 °C of mean global warming. Some regions such as Southeast and South Asia and Mediterranean Basin show a peak at 4.5 °C mean global warming. Large increases in heatwave days result in no-lagging, long continuous events. The longest events are located over tropical areas, such as Sahara, the Amazon Basin, Central America, and Eastern Africa. In these regions, under a 5 °C mean global warming, heatwaves could last 80 days but under a scenario of 2–3 °C, the duration of heatwaves would be substantially shorter of about 20 to 40–50 days, respectively.

The median temperature regional peak heatwave increases linearly with global warming, with the largest increase occurring in the Mediterranean Basin, where heating intensity could be 9 °C under a mean global heating scenario of 5 °C in 2100, compared with pre-industrial values. Under mean global heating of 2.5 °C, heatwaves could peak between an additional 2.5 °C in Australia and Southern South America and 5 °C in the Mediterranean Basin and Central Asia. This increase is significantly higher when compared with the mean global 1.5 °C scenario, where heatwave additional peaking range is between 2 °C in East Asia and Southern America and 3 °C in the Mediterranean Basin and Central Asia.

Regional median return periods of heatwaves are forecasted to decrease at a non-linear rate with an increasing mean global threshold. At a 4 °C global warming, in almost all regions, a yearly intense heatwave could occur, compared with 1 equivalent event every 30 years, between 1861 and 1890. The return times are substantially different under 1.5 and 2.5 °C scenarios across regions. In some regions such as Tibet, central North America, eastern North America, and East Asia, a threefold higher return time can occur at scenarios of 1.5 and 2.5 °C mean global warming. The estimated return times were also about twofold higher in regions such as Greenland, Northern Europe, Australia, Central America, North Asia, and South Asia. In the Mediterranean Basin, it is about 1.5-fold higher.

The relationships detected by Perkins-Kirkpatrick and Gibson (2017), between regional heatwave characteristics and global mean warming, are model-dependent. For example, the overall spread of heatwave days per °C among the 27 models can be as large as 40. On the other hand, the lengthening of the longest event can also spread among the models across 30 days, under equal conditions. Tropical regions, very sensitive to average global temperature increases, displayed the highest differences among models. The spreads are reduced outside tropical regions, with heatwave days changing by 8–20 days, event number by 1–2.5 days, and the duration of the longest event between 4 and 12 days. The influence from internal climate variability evaluated through the CESM ensemble models was found as low, with very small intraregional variations of coefficients per °C of mean global warming, compared with using CIMP5. Thus, a conclusion to be drawn is that the overall variations in the rate of heatwave events are strictly dependent on differences in physical methodologies between climate models in terms of parameterization schemes, resolution, and overall climate sensitivity.

The role of internal climate variability becomes more visible when projecting changes in regional heatwaves per individual global mean warming threshold. Over the same regions and for the same heatwave characteristics, under a rate of expected heatwave changes per 0.5 °C of the mean global warming, internal variability accounted for at least 50% of these changes, and in most cases accounted for 20–30% of them.

The percentage P of internal variability for variables of heatwaves is given by

$$\%P = 100^*(CESM(99th - 1st)/CMIP(99th - 1st)) \qquad (8.1)$$

where the 99th and 1st refer to 99th and 1st percentiles of results from CESM and CIMP5 models.

Simulated internal climate variability can account for 21–70% of the number of events, for 12–35% of the duration of the longing event, and for 28–67% of peak heatwave intensity, depending on the region. Overall, that influence is greater over higher latitude regions such as Alaska, Central North America, and Northern Europe and in the tropics, e.g., in the Amazon Basin, Southeast Asia, and Western Africa.

Estimated internal variability over variation in 0.5 °C of global average heating, changes in a consistent way across regions and should be considered in evaluating the overall regional change for any specific global warming threshold. Only in the Amazon Basin did the simulated internal variability decrease with global temperature increases. In practice, the influence of internal climate variability remains unpredictable, and even if the anthropogenic forcing could be rigorously quantified, a range of variability of heatwave parameters per °C of global mean warming would be required for dealing with adaptation and mitigation policies. The future of climate variability remains uncertain and particularly in relation to the global climate response to anthropogenic forcing and in terms of the variability of regional heatwave profiles.

Fischer and Knutti (2015) studied the extent to which global warming can account for heavy precipitation and hot extremes. They concluded that about 75% of the moderate heat extremes, corresponding to the present day 0.85 °C average warming relative to the pre-industrial era, and about 18% of the global moderate precipitation extremes, are attributable to global warming, with the latter mainly due in a non-linear fashion to anthropogenic activities. Currently, the probability of a hot extreme event occurring is 1 in 1000 days, which is about five times higher than in pre-industrial conditions.

At 2 °C warming, the likelihood of most extreme cases (99.99% quantile) increases by a factor of 1.5–3 depending on the region and the model. Thus, an event that occurred once every 30 years under pre-industrial conditions is expected to occur every 10–20 years under a 2 °C warming scenario. The probability of extreme precipitation increases at the expense of days with moderate, low, or no rainfall. From this warming scenario, the fraction of precipitation episodes attributable to human influence rises to about 40%. The increased water vapor holding capacity of warmer air, could be linked to a decrease in the mean precipitation regimes giving way to the predominance of extreme rainfall events. Overall, more heavy rain events are expected in a warmer world due to the intensification of the hydrological cycle. Most models suggested a general increase of 10–30% in the precipitation intensity at most latitudes, for doubled carbon dioxide levels (Kettenberg et al. 1996).

The probability of extreme hot events under warming of 2 °C is double that for 1.5 °C and fivefold higher than that for the current 0.85 °C increase. This has strong implications in terms of mitigation targets in climate negotiations, because the differences of global average temperatures are small but large in terms of the probability of extreme events. For every single degree of warming, the likelihood of a rarest and extreme event occurring increases, along with a greater role for anthropogenic emissions, with higher socioeconomic and environmental impacts.

The anthropogenic contribution toward heatwave and heavy precipitation was assessed for the European heatwave of 2003 (Stott et al. 2004), for the Russian heatwave of 2010 (Otto et al. 2012), and for the Australia heatwave in the summer of 2013 (Lewis and Karoly 2013).

In 2003, the threshold of increase of annual temperatures of 1.6 °C in Europe was surpassed for the first time, by comparison with the period 1961–90, and 2001 showed the second warm European summer with a correspondent 1.5 °C threshold (Stott et al. 2004). This record-breaking heatwave in 2003, believed to be the hottest since AD 1500, arguably caused 70,000 excess deaths and damage to agriculture and forest amounting to more than €13.1 billion (e.g., Schiermeier 2010).

Stott et al. (2004) argued that asking whether external influences in climate, for example, increase of GHG emissions, are deterministic in this hot discrete event is not the right question to ask as such an event could happen by chance in an unmodified climate. A more relevant question would be whether the likelihood or the risk of the heatwave occurring would be higher with anthropogenic influence.

The same authors estimated, using a temperature threshold for mean summer which was surpassed in 2003, and before only once in 1851, that it is very likely that human activity at least doubled the odds of exceeding that threshold. Their calculations showed that with a chance higher than 90%, half of the risk of European summer temperatures to exceed a threshold of 1.6 °C was attributable to human influence on climate. The fraction attributable risk (FAR) is sometimes a parameter valid in establishing the liability for compensation for such events with a value of 0.5 corresponding to doubling the risk over natural conditions.

Otto et al. (2012) analyzed a heatwave in Russia over a month, from July to August 2010 that caused a high loss of lives, a reduction in crop production by about 25%, and a total economic loss of US$ 15 billion. This phenomenon was characterized by a level of air temperatures higher by more than 5 °C above the long-term mean for which natural causes cannot be discarded.

The same authors argue that two apparently distinct narratives for interpreting extreme events are complementary. These narratives are (i) that natural variability was the primary cause of this event and (ii) that with a probability of 80%, the 2010 July heat record would have not occurred without climate warming linked to an external trend that is the anthropogenic influence on greenhouse gas forcing.

In this case, the key point again posed for low probability events is that the probability of the event occurrence and the fraction of risk attributable to external forcing are two different questions. They reported that while in the 1960s a 2010-like event could have been expected every 99 years, in the 2000s the equivalent period was about 33 years.

Thus, the conjugation of the two narratives in the case of the heatwave in Russia is based on the principle that even a natural extreme event in 2010 would be more likely under the anthropogenic warming which occurred in this region since the 1960s. Indeed, the empirical analysis showed that under a stationary climate, without rising in average annual temperature, the observed monthly mean temperatures for July 2010 would be very unlikely, in relation to the distribution defined over the 1950–2009 period. This, because, in this period, return times ranged from 250 to 1000 years, implied that without an enhanced warming factor, the 2010 heatwave would have been a very unusual event.

The average increase in temperature is much smaller than the anomalies observed during the heatwave, although an increase occurred in a cascading non-linear effect. Meanwhile, the probability of a heatwave as large as the one observed in 2010 has increased three- to fourfold. The return time for the 2010 July temperature was estimated to be 250 years. Given that the area covers less than 1% of the global land area and was selected a posteriori, a 1/250-year event could eventually occur every few years. The heat event in Russia can be internally generated, but mainly externally driven in terms of probability of occurrence.

Lewis and Karoly (2013) analyzed the role of anthropogenic factors in the hottest Australian summer over the period where records were available, drawing conclusions like those of the studies already mentioned. Simulations of natural and anthropogenic forcings for the periods (i) 1976–2005 and (ii) 2006–2020, under an RCP8.5 scenario simulating a climate of high emission under the AR5 climate profile, found that the odds of extreme heat due to human influences during the two periods were 2.5- and five-fold higher. The natural "La Niña" events alone were unlikely to cause the extreme heatwave, compared with the RCP8.5 high emissions scenario. A significant decrease in return times of extremely hot summers was also evaluated for the periods 1976–2005 and 2006–2020 (RCP8.5). Beyond 2020, extremely hot summers were likely to occur and in the period 2080–2099, with

RCP8.5 simulations, at least 65% of seasons are projected to be extremely hot over all land areas, with severe implications for human health and natural systems.

Sippel and Otto (2014) described how the risk of hydrometeorological extreme events has changed with a warming climate. For Southeast Europe, these authors focused on the very hot and dry summer of 2012, which in combination with heatwaves reaching 40 °C affected the entire region. It was the hottest and the third driest season recorded in Serbia, due to the prevalence of high-pressure divergences and lead to an economic loss of about one billion Euros.

These authors explored the tendencies in inter-decadal changes obtained from the probabilistic analysis that can be used to evaluate a meteorological risk of extreme weather. The decade-long simulations smooth out the influence of natural climatic variability, allowing also evaluating the return times of meteorological events. Indexes for obtaining proxies for impact-relevant meteorological conditions were derived for combinations of climate variables. Such an index for seasonal/monthly precipitation deficit makes it possible to estimate probabilities for summer dryness. The likelihood of changes in heat and dryness in Southeast Europe using two-decade data series was also determined.

Seasonal temperature anomalies ranging from 2 to 5 °C, compared to those of the period of 1961–1990, showed that very low precipitation has occurred throughout South and Eastern Europe. Within this climatic scenario, there were multi-day heatwaves reflecting turbulent phenomena of higher frequency. During two heatwaves in August 2010, daily minimum and maximum temperatures ranged from values as high as 25–40 °C. The multi-day temperature events are relevant when compared with the full-year temperature increases.

For the evaluation of temperature impact events, a 5-day sampling period can be more important than, for example, monthly mean temperatures, for exampl, as a proxy for short-term heat stress during a summer heatwave. Sippel and Otto (2014) also showed that a 5-day mean wet-bulb global temperature in the summer works as a proxy for short-term heat stress for human health. When temperatures rise, evapotranspiration also increases causing an increase in water demand for agricultural crops, within monthly and seasonal time scales.

Sippel and Otto (2014) assumed thereby that general dryness increased in the given region, independently of the absence of pronounced changes in the precipitation regime. They showed also that a hypothetical water balance, defined as the difference precipitation and potential evapotranspiration (P-PET), could be used as a proxy for local dryness, with PET computed according to the Thornthwaite method:

$$PET = 16K(10T/I)^m \tag{8.2}$$

where K is a function of days in a given month and latitude, reflecting the average number of daytime hours in that month, T is the mean monthly air temperature (in ° C), and I is a location heat index based on monthly-mean temperatures defined as

$$I = \sum_{i=1}^{12} \left(\frac{T_i}{5}\right)^{1.514} \tag{8.3}$$

with i index referring to the month and m a coefficient defined as

$$m = 6.75 \times 10^{-7}I^{-3} - 7.71 \times 10^{-5}I^2 + 1.79 \times 10^{-2}I + 0.492 \tag{8.4}$$

The above water balance was shown to be a useful proxy for local dryness, despite the absence of computation of the dynamics of soil moisture and groundwater. The simulations enabled the summer monthly dynamics of the water balance from the 1960s to 2000–2010 to be analyzed. A handicap of this methodology is that precipitation data is very biased, so that the estimations are made mostly in inter-decade comparisons, rather than in absolute values of the water balance. Thus, this methodology only estimates the dryness risk relative variations instead of absolute risk.

Sippel and Otto (2014) also demonstrated a critical reduction in the return periods of multi-day heatwave events in the summer season. For example, a 5-day period above a given threshold of high air temperatures, could have returned on average once or twice per century in the 1960s, whereas in the 2000s it could return under ten year periods. These authors also showed that in Southeast Europe the inter-annual variability, dependent mainly on typical natural factors such as different sea surface temperatures and North Atlantic oscillations, is likely of minor importance relative to inter-decadal variability driven by greenhouse gases and aerosols and other factors.

Multivariate combinations of hydrometeorological variables, such as air temperature, environment humidity, and precipitation, are likely to be more relevant and complementary to the use of meteorological variables alone for assessing climate impacts. One example is the wet-bulb globe temperature (WBGT) which can be used as a proxy for heat stress and is used by weather services to issue health warnings. This variable can be given by the following expression:

$$\text{WBGT} = 0.567T_{air} + 0.393e + 3.94 \tag{8.5}$$

where T_{air} is the air temperature and e is the water vapor pressure (Fischer and Knutti 2015).

The abovementioned expedite indexes of dryness and heat stress that can provide quantitative evaluations into changes are extreme weather risk and impacts of joint variables. This information can assist stakeholders in planning across sectors such as health, water availability, and agriculture in Southeast Europe. Sippel and Otto (2014) refer also that although changes both in natural and in anthropogenic forcing contributed to weather events and exerted influence in the global climate, the bulk of global mean warming in the last 50 years has been due to anthropogenic causes.

8.5   Long-Term Climate Change

The impacts of actual CO_2 emissions will last for centuries even if the anthropogenic emissions of greenhouse gases are stopped. Global warming will remain after 2100, except for the RCP2.6 scenario of the AR5 Report. The forecasts of AR5 are that global temperature will be stationary, remaining high long after an eventual termination of anthropogenic CO_2 emissions. To revert the current forecasts of climate change until 2100, there must be continuous extraction of atmospheric CO_2 during a significant period. Also, the direct impacts of climate change, for example, soil carbon and ecosystems re-equilibrium, ice sheets, and sea-level rise, have long-term dynamics of their own, so that their effects will be felt even after the global temperatures stabilize. Mitigation may, however, strongly reduce the number of heat extremes by the second half of the twenty-first century (Coumou and Robinson 2013).

The global mean sea level will go on rising for centuries beyond 2100, and the same can be said about ocean acidification if CO_2 emissions go unchecked. The RCP2.6 projections for the sea-level increase until 2300, indicate that sea rise can be lower than the 1 m above the pre-industrial era, if the GHG concentrations are kept below the 500 ppm CO_2-eq. The rise could be higher than 3 m if the GHG concentrations range from 700 ppm CO_2-eq. to 1500 ppm CO_2-eq., under scenario RCP8.5.

It should be noted that the Antarctica solid ice discharge contribution is underestimated, so that the sea-level rise in 2100 could be higher. In this context, sustained ice mass loss could cause a larger sea-level rise, under an irreversible scenario. For Greenland, continued global warming ranging from 1 °C to about 4 °C would deliver a drastic mass of ice during a millennium time scale, which would eventually cause a sea-level rise as high as 7 m.

A scenario of a sudden drastic instability in response to climate forcing is possible, although the evidence available is not enough to carry out an accurate quantitative assessment. The RCP4.5, RCP6.0, and RCP8.5 profiles, corresponding to medium to high GHG emissions scenarios, could abruptly disrupt regional scale terrestrial, marine, and freshwater systems, both in terms of composition and structure.

An example of an impact condition with global consequences is the reduction in permafrost areas which will occur if the global temperatures rise further. Under the RCP8.5 climate profile, net carbon emissions in the twenty-first century from permafrost could range from 180 Gt CO_2 to 920 Gt CO_2 adding to the 2040 Gt CO_2 already accumulated carbon emissions, and almost reaching the 2900 Gt CO_2 threshold critical to attaining a likely dangerous climate change.

8.6   A Brief Analysis of the Mediterranean Case Study

8.6.1   Introduction

The Mediterranean and its southern and eastern rims are regions where climate change will be felt strongly over the twenty-first century. The direct impacts of phenomena, such as global temperature rise, precipitation decrease, sea-level rise, and other extreme climatic events, would be reinforced with the effects of anthropogenic activities on the natural systems. These impacts include the increasing water scarcity, and negative effects on sectors such as agriculture, fishery, infrastructures, energy, and hydropower production.

The impact of climate change on the environment is already noticeable in the Mediterranean and is producing observable effects on human activity. The Mediterranean Basin incorporates a multitude of sub-areas with complex interactions between the physical systems that define the climate and biosphere, making it a climate "hot spot" for climate change (e.g., Giorgi 2006). This region can also be seen as a natural laboratory for assessing vulnerability and climate change impacts and for introducing adaptation and emission-reducing measures (Grunderbeeck and Tourre 2008). A geographical division commonly used for the sub-regions of the Mediterranean Basin (e.g., Adloff et al. 2015) are the seven sub-basins ordered from East to West: Levantine, Aegean, Ionian, Adriatic, Gulf of Lions, Western Mediterranean, and Atlantic.

The climate projections forecasted for the Mediterranean fit into the profiles of IPCC, characterized by an increase in air temperature of 2.2 and 5.1 °C (scenario A1B of the AR4 Report) comparing 2080–2099 with 1980–1999, with differences in sub-regions. The projections also point to (i) a decrease in rainfall from 4 to 27%, (ii) an average increase of sea-level-rise of 0.35 m, (iii) an increase in drought days evaluated by the higher yearly number of days with temperature exceeding 30 °C (Giannakopoulos et al. 2005), and (iv) higher frequency and intensity of extreme climatic events, such as heatwaves and floods.

Further predicted negative consequences are soil desertification, water scarcity, rural wildfires, or forest biotic plagues. North African regions will be among the Mediterranean areas more prone to drastic effects of climate change, in comparison with the Northern Mediterranean.

In 2025, the GHG emissions in the Mediterranean Basin are predictably to be twice as high as they were in 1990, largely due to the energy demand. The growth rate in emissions of carbon dioxide will tend to be higher in Southern and Eastern Mediterranean regions, by comparison with the Northern Mediterranean. The GHG emissions in some regions increased between 1990 and 2004 by about 58%, arguably a considerable emission rate on the global scale. While in the Northern Mediterranean countries, the transport sectors ranked first as GHG emitter between 1990 and 2004, in the Southern and Eastern that ranking was for electricity and heating.

About 6000 years ago, the climate was warmer in the Mediterranean Basin, with temperatures in winter higher by 1–3 °C than in the present day, and about more than 8–15% water was available for plant growth (Cheddadi et al. 1998). A similar pattern was established for a period 18000 years ago (Doumenge 1997).

Evidence suggests that the seasonal amplitude used to be higher than in the present day, with average temperatures in hotter summers higher by 2 °C than in the present, and colder winters with average temperatures lower by 1 and 2 °C than in the present. A 1 °C temperature increase would provoke a 100 km vegetation shift to the north. Simulations based on a 2 °C temperature increase and doubling CO_2 concentrations, delivered the development of deciduous forest areas in the northern part of the Mediterranean Basin instead of an increase of the arid zones. Anthropogenic actions supposedly cause climate changes comparable to natural causes at a much faster rate (e.g., Cheddadi et al. 1998, 2008; Grunderbeeck and Tourre 2008).

8.6.2   Changes in Temperature

The climate change projections to the Mediterranean region are built on the numerical regional climate models, adding to information of CGMs, with resolutions up to tens of kilometres, e.g., 50 × 50 km or 10 × 10 km (Grunderbeeck and Tourre 2008). Modeling studies have shown that in the South-Central European region, extreme temperatures are increasing faster than mean temperatures indicating that heatwave frequency, amplitude, and length will become more drastic in the twenty-first century. Drought indexes and proxies also show an increased tendency for droughts in Southern and Southern-central Europe (Sippel and Otto 2014).

During the twentieth century, average air temperature in the Mediterranean region rose within the range of 1.5–4 °C, depending on the sub-region. In South-western Europe, (Portugal, Spain, and Southern France) over the same period and with a clear increase since 1970, air temperatures rose by almost 2 °C. In North Africa, a similar warming tendency was harder to quantify largely due to an incomplete measurement network.

Even if the EU objective of a 2 °C average temperature is achieved until the end of this century, the temperature increases in the Mediterranean will exceed that objective and due to environmental and socioeconomic specific conditions, the impact will be more incisive than in other regions of the world. The predicted increase of global average temperature in the Mediterranean Basin is correlated with changes in precipitation and drought at local and regional levels, as it is in all regions across the globe.

The average Mediterranean temperature has regional specificities based on its geography, giving rise to seasonal variations. During winter, regions where the average temperature is larger than the global regional average, abridge areas of the Iberian Peninsula and Eastern Mediterranean. During spring, the opposite situation prevails, and in the autumn a large area displaying strong sensitivity to global climate change is seen. It includes parts of Southern France, Spain, and Northwest Africa with a 1.3 °C temperature increase, compared to a 1 °C mean global temperature increase (Grunderbeeck and Tourre 2008).

The rising concentrations of greenhouse gases, excluding any other forcings, could cause warming over the Mediterranean region in a magnitude higher than the global increase (Karas 2006). One forecast (Wigley 1992) indicated that average temperatures over the region could rise by about 3.5 °C between 2000 and the latter half of the twenty-first century. Other estimates indicate that about half of this rise —between 1.4 and 2.6 °C—could occur by the 2020s (Rosenzweig and Tubiello 1997). Parallel evidence from numerical modeling (Kattenberg et al. 1996) for the South Mediterranean Basin pointed to temperature increases of 1–4.5 °C (with a mid-point of about 2.5 °C) by the latter half of the twenty-first century. Even if emissions of greenhouse gases were stabilized by then, temperatures would continue to rise for several decades due to a time lag in the response by the oceans (Karas 2006). Cubasch et al. (1996) estimated by modeling that until 2100, average temperatures could rise by 2.5–3 °C across the Mediterranean Sea, 3–4 °C on coastal areas, 4–4.5 °C over most inland areas, and up to 5.5 °C across Morocco. There was also a similarity among the profiles of maximum summer warming obtained by climate simulations in the Balkans and Iberian Peninsula (Giannakopoulos et al. 2009).

These results reflect a spatial pattern of temperature distribution, wherein the general tendency that warming over the sea lags that over the inland areas. The twenty-first century warming tendency is also higher in inland areas compared with coastal areas. Strong seasonal dependence is seen in the Mediterranean area with average warming up to 4 °C in summer, above 2 °C in autumn, and below 2 °C during spring and winter (Giannakopoulos et al. 2009). Parallel simulations of impacts of doubling of CO_2 concentration in relation to Europe gave similar tendencies.

Possible increases in aerosol emissions could disguise some of this warming. For example, Mitchell et al. (1995) showed that aerosols may reduce warming over the Mediterranean region by 1–2 °C for the period 2030–2050, relatively to the nineteenth century. Hasselmann et al. (1995) reported that the net effect of aerosols could even show a hypothetical cooling over the central Mediterranean in summer over the next few decades. Indeed, the presence of aerosols may locally counter the greenhouse effects but, unlike GHGs, their effects are transient, lasting weeks or months. Overall, aerosols could exert some influence, and long-term climate projections cannot ignore their effects (Karas 2006).

Giannakopoulos et al. (2005) formulated climate predictions over the period 2025–2050 (average values under A1B, B1, and A2 scenarios of the AR4 Report), with different uncertainty ranges, and assuming a conservative global 2 °C temperature increase during the twenty-first century. In the case of the Iberian Peninsula, an increase of 14–42 days in the summer, with temperatures higher than 30 ° C, was projected.

Modeling projections between now and the second half of the twenty-first century in European areas, of days per year with maximum temperatures higher than 30 °C, pointed to a significant increase of the areas corresponding to the range of 100–200 days. This heating effect is particularly accentuated throughout the

Mediterranean Basin. The Iberian Peninsula is particularly affected with a significant area presumably included in the 100–200-day class (ACACIA Project 2000).

According to the AR4 Report, doubling of CO_2 atmospheric concentration will induce a temperature variation ranging from −5 to 2.5 °C in the Mediterranean Basin, with the Iberian Peninsula being one of the areas with the largest temperature increases (Grunderbeeck and Tourre 2008). Even if the concentration of GHG was maintained at the 2000 levels, an increase of 0.1 °C in air temperature per decade would be expected.

Global and regional circulation models (RCM), show uncertainties that can be parameterized and that might be expected to decrease over time, in the light of new knowledge on climate. Among the main sources of uncertainty are the incomplete knowledge of the physical mechanisms and processes dealt with numerical models for regional climate, the socioeconomic impact, and on the human demographic and technological evolution. This latter aspect is important for assessing specific impacts on climate change of agriculture, industry, water resources, and energy.

Parameterization schemes in the RCMs are based on the knowledge of observed values that might be exceeded in the future. For example, some projections of temperature ranges in the Mediterranean Basin are based on average air temperature increases which may not be representative of current daily average temperature change (Grunderbeeck and Tourre 2008).

8.6.3   Changes in Precipitation and Moisture Availability

Global warming has been linked with increases in average daily precipitation, followed by increased water vapor in a warmer atmosphere. Increases in annual rainfall were recorded in many mid- to upper latitudes, except for the Mediterranean Basin where a decreasing trend has been observed (e.g., Alpert et al. 2002). Since 1900, precipitation decreased by over 5% over much of the land bordering the Mediterranean Sea, apart from the contiguous coastland extending from Tunisia to Libya where it increased slightly (Karas 2006).

In the Central-Western Mediterranean Basin encompassing Italy and Spain, the total precipitation decreased by 10–20%. In the southern Mediterranean strip, including Southern Italy, Southern Spain, and Tunisia, a higher global decrease of 26% occurred, and in the Eastern Mediterranean mixed tendencies of rainfall with mean precipitation did not change, and heavy precipitation events increased (Groisman et al. 2005). Other evidence clearly shows that rainfall and specific and relative atmospheric humidity, predictors of precipitation, have decreased in the Iberian Peninsula and Greece over the second half of the twentieth century, and this correlated positively with decreased rainfall (e.g., Grunderbeeck and Tourre 2008).

For Central and Southern Israel, an increasing trend in total rainfall was recorded over the period 1951–1990 and is likely to be associated with large modifications in the land cover and land use. Alpert et al. (2002) noted that in the Eastern Mediterranean region, mean precipitation during the rainy season did not change noticeably in opposition to episodes of heavy precipitation which showed change patterns. This general temporal tendency contained discernible regular time cycles between wetter and drier periods.

In the western Mediterranean and the Balkans, the available recorded data shows major moist cycles during the periods of 1900–1920, 1930–1956, and 1968–1980 with interim dry periods (Maheras 1987; Maheras and Kolyva-Machera 1990).

The forecasts from 22 models for Southern Europe and the Mediterranean region, for the period 2050–2100, point to a significant decrease in rainfall, ranging from 30 to 50%, while for countries of Northern Europe the previsions are opposite for a rise between 30 and 50% (scenario A1B, from the AR4 Report). For most of the Iberian Peninsula, there is a trend toward more rainy days excepting southeast Spain where the reverse is expected with high rainfall amounts.

In all this context, the established consensus from climate projections is for decreased precipitations from 4 to 27% in Southern Europe and Mediterranean regions, whereas in Northern Europe it will increase between 0 and 16%. An increase in drought periods associated with land degradation, along with a high number of days with temperatures higher than 30 °C, is also expected (Giannakopoulos et al. 2005).

If the numerical models are broadly correct about precipitation changes in response to increases in greenhouse gases, the droughts over the western Mediterranean Basin could be symptomatic of growing human influence on regional climate. Wetter conditions in the east might reflect the stronger influence of aerosols in this area (Karas 2006). Giannakopoulos et al. (2009) cite Portugal as the Mediterranean-influenced country where the modeling results of precipitation profiles, under different emission scenarios, differ greatly, mainly due to intra- and inter-annual rainfall variability. Apart from this source of variability, the remaining differences in precipitation profiles are not statistically significant.

The characterization trends of precipitation in the Mediterranean Basin, should consider the complexity of physical processes underlying precipitation. The use of high-resolution updated models at smaller scales should be considered to take into account the specificities of the Mediterranean Sea with uneven and rugged coastlines and smaller seas. Hence, the regional trends in the Mediterranean Basin typically distort the scale of changes in precipitation at the local level. For example, in much of North-Western Africa, Spain, Italy, and Greece for the period 1975–1994, average precipitation was lower by 17% than during the preceding 20 years. This contrasts with the conditions elsewhere in northern Africa and the eastern basin, where average rainfall was higher over the same period (Karas 2006).

In any case, a long-term model for the Mediterranean region suggests that from 2050 onwards, precipitation will decrease markedly as the relative influence of greenhouse gases grows (Karas 2006). However, some models are found to underestimate year-to-year variability, with problems in reproducing the observed trends when Spain and Greece are compared. This means that different underlying physical processes are at stake, and further investigation on downscaled processes is required.

In the short term, aerosol effects may counter the effect of rising concentrations of greenhouse gases in some areas. Results from transient experiments for around the middle of the twenty-first century suggest that once aerosol effects are considered, precipitation over Southern Europe and Turkey may increase slightly (Kattenberg et al. 1996). These changes are far from certain, as they depend on both the aerosol scenario used and how aerosols are represented in the models. In any case, the long prospect remains of hotter, drier conditions throughout the Mediterranean region, as the relative influence of greenhouse gases increases over time.

In addition to the tendency for reduced rainfall, a reduction in moisture availability is also predicted. Indeed, moisture availability is determined both by water gains from precipitation and water losses through runoff and evapotranspiration. The future global temperature increase will cause an increase in evapotranspiration, and therefore loss of available water.

The projected seasonal precipitation variations are also likely to reduce water availability during the plant growth season (Kattenberg et al. 1996; Wigley 1992), due to water runoff in the winter. This implies that even when precipitation is forecasted to increase, available water can decrease if the gains are surpassed by losses. On the other hand, if rainfall peaking episodes increase, the likelihood is that water will not be stored in the soil, but lost as runoff (Segal et al. 1994). The evaluation of moisture availability by GCMs can be complex due to the difficulty in assessing potential evapotranspiration and the hydrological cycle (Rind et al. 1990). Anyway, the global influence of GHGs is likely to grow and the long-term prospect is of a tendency for dryness throughout the Mediterranean region.

8.6.4 Changes in Extreme Events

The Mediterranean Basin as aforementioned, is a region particularly vulnerable to global warming with specificities at the local level. In addition to changes in the mean climate, more fluctuations in extreme events, which are a component of natural climate change variability, were recorded both in terms of frequency and intensity. Presently, about 75% of the daily hot extremes over land is mainly due to climate change, and the frequency of very hot days, exceeding the 99th percentile of the daily maximum temperature, has tripled during the twentieth century (e.g., Fischer and Knutti 2015).

Attention should also be paid by all stakeholders to changes in seasonal shifts and the timing of climate change events. Based on phenology as well as climate data, seasonal shifts have been reported in different regions, mainly in early spring and shortened winters. Phenomena including extended warm periods over the years, and shifting in hot weather profiles, have multidisciplinary impacts on distinct but interconnected issues such as the environment, human health, agriculture, and energy demand (Founda et al. 2019). For example, Fontana et al. (2015) showed that heat stress in Italy had a strong impact on wheat yield.

As the global climate changes, the frequency of extreme events in the Mediterranean region will change both in relation to the average climate as to climate flutuations. Warmer conditions over the Mediterranean region should lead as to an increase of extremely high temperatures as to a decrease in extremely low-temperature events. An increase in drought periods is related to the previously mentioned high frequency of days during which the temperature exceeds 30 °C (Giannakopoulos et al. 2005). In areas experiencing a general decrease in precipitation, droughts are likely to become more frequent.

Heatwaves, droughts, or floods are also likely to be more frequent and violent. This tendency for the higher occurrence of extreme events has been noted since the mid of the twentieth century, with decreases in global frequency of cold days and nights and increases of warm days and nights (Hartmann et al. 2013). Likewise, heatwave frequency has increased in large parts of Europe, Asia, and Australia during this period. In the same way, it is likely that from about 1950 onwards, the number of heavy precipitation events over land has increased in more regions than it has decreased (AR5, Report).

Similarly, in areas experiencing a general decrease in precipitation, droughts are likely to become more frequent, as the probability of dry days and the length of dry spells increases. The converse is true for areas where precipitation increases. Precipitation events changed abruptly in 1996, with drying areas suddenly experiencing extreme wet conditions and vice versa (WMO 1997). It was also reported that the probability of a dry spell lasting more than 30 days in summer in Southern Europe would increase with the doubling of atmospheric carbon dioxide concentration, by a factor between two- and fivefold (Kattemberg et al. 1996). Extreme precipitation events over most of the mid-latitude land masses and over wet tropical regions will likely become also more intense and more frequent.

Alpert et al. (2002) identified a paradoxical increase in Mediterranean extreme daily rainfall, in the period 1951–1995, despite a decrease in the total precipitation. This so-called paradox reflects a presumable scenario of the existence of a substantial change in the rainfall distribution over the Mediterranean area wherein the "increase of variance of the rainfall distribution overcomes the reduction of the mean."

The same author proposed 6 rainfall daily categories as powers of 2, from (A) light: 0–4 mm/day; (B) light–moderate: 4–16 mm/day; (C1) moderate–heavy: 16–32 mm/day; (C2) heavy–torrential: 32–64 mm/day; (D1) 64–128 mm/day; and (D2) Torrential: ≥ 128 mm/day up. This enables the evaluation of several rainfall categories relative to rainfall totals and how these rainfall classes vary with time. An alternative non-discrete approach would be the use of gamma distribution and the analysis of changes in the shape and scale parameters (e.g., Groisman et al. 1999). The rainfall trends were analyzed in 265 Mediterranean stations distributed across the following countries: Spain (182 stations), Italy (42 stations), Cyprus (3), and Israel (38). It was found that, for example, in Spain, the torrential category contribution toward the total annual rainfall increased significantly over the 45-year period. On the other hand, the classes C1 and C2, contributing to about half of the total rainfall, decreased significantly, dropping from about 49% in the early 1950s to about 43% in the 1990s. The rate of decreasing rainfall was about 1.52 mmy^{-1} and the contribution of the lower extremes to the annual rainfall also increased, although not significantly.

Likewise, in Italy, there was a clear distinction between the heavy torrential categories (C2, D1, D2) showing an increasing trend, and the lighter classes (A, B, C1) showing a significant lowering trend. In this country, torrential rainfall, above 128 mmd^{-1}, contributed to about 4–5% of the total rainfall in the 1990s compared with only 1% in the 1950s. In contrast, in the 1950s heavy torrential rainfall classes above 32 mmd^{-1} accounted for only 23% of rainfall versus 32% in the 1990s. Italian weather stations showed that heavy precipitation episodes contributed mainly to rainfall in the summer and transition seasons. In Italy, torrential classes D1 and D2 exhibit higher inter-annual variability among all classes.

In Israel and Cyprus, no significant trends were detected, although as in the western Mediterranean there exist heavy categories which are C1 and C2 in Israel and C2 and D1 in Cyprus that augment. Lighter categories like A in Israel, and B and C1 in Cyprus show, a decreasing trend. In Israel, all classes of discrete rainfall events have high inter-annual variability, probably due to the semi-aridity in the south. Annual precipitation averages ranged from 776 mmy^{-1} in the north to 105 mmy^{-1} in the south, across a mere distance of 220 km. To some extent this may be due to the influence of the Red Sea, which contributes to significant rainfall in the southern region of Israel.

Alpert et al. (2002) also reported that torrential rainfall classes tend to peak in El Nino years such as 1953, 1965, 1982/3, and 1986/7. For example, in the latter two years, the D2 class accounted for 15% rainfall compared with the typical annual average of 1–4%. This relationship between discrete torrential events and El Nino has been more pronounced in recent decades as experienced in northern Israel and Turkey. This reflects the rainfall paradox referred to above, giving high variation in precipitation patterns in the Mediterranean Basin which may be related to greenhouse gas warming. The rain distribution is also characterized by the fact that the increase in variance overcomes the reduction of the mean (Meehl et al. 2000).

For the period 2025–2050, matrix analysis of modeling results of temperature and precipitation extremes was carried out by Giannakopoulos et al. (2005). It consisted of average values under A1B, B1, and A2 scenarios of the AR4 Report, under a conservative scenario of 2 °C average temperature increase during the twenty-first century. The matrix showed that the east Mediterranean region (Mashriq) was clearly the hottest and driest area, with prevailing one-month continuous period and of 2–3 continuous weeks with high summer temperature increases of 3 and 2 °C, respectively, and with one-month continuous periods with night temperatures increasing also by 3 °C. In Mashriq, the impact of the 2 °C increase scenario in daily precipitation is not so drastic, with prevailing periods of 2–3 weeks and one week corresponding, respectively, to increases of 2 dry days and decreases in 2 mm of daily precipitation.

Forecasts for North Iberian, Southern France, Corsica, Sardinia, and Sicilia showed increased precipitation of 3 mmd^{-1} in prevailing one-month continuous periods and to rainfall decreases of 1 and 2 mmd^{-1} in predominant continuous

1-week and 2–3-weeks periods, respectively. In those regions, moderate temperature increase prevails mainly in 1 month and 1 and 2-3 weeks continuous periods.

Founda et al. (2019) also pointed that for the Eastern Mediterranean region, under scenario RCP8.5 (AR5 Report), a marked expansion of hot extreme seasons is a key feature of climatic change. This expansion at both ends was not symmetrical with advancements prevailing over the delay of the hot extremes. There was also an earlier beginning of tropical nights, defined as nights when daily minimum temperatures are higher than 20 °C, and of longer duration. The averaged simulation across the regions indicated a likely extension in the hot extreme season by about 1 month between now and 2050 and by about 2 or more months by the end of this century. The projections showed higher increases for night-time hot temperature extremes, due to the higher impact of greenhouse gases under conditions where the infrared radiation re-emitted by the atmosphere is the only energy source.

8.7   The Potential of Biochar in the Modulation of Carbon Budget and of Climate Change Mitigation

Biochar is a low-cost carbonaceous material obtained when biomass, such as wood, leaves, straw of cereals, and manure, is pyrolyzed or charred at temperatures below 700 °C in a closed reactor under a low supply of oxygen. By mirroring charcoal, one of the most ancient technologies, biochar can also be defined as "charcoal, from biomass that has been pyrolyzed in a zero or low oxygen environment, for which, owing to its inherent properties, scientific consensus exists that added to soil at a specific site is expected to sustainably sequester carbon and concurrently improve soil functions, under current and future management, while avoiding short- and long-term effects to the wider environment as well as human and animal health" (Verheijen et al. 2010). Biochar as a feedstock can also be defined as "charcoal for application in soils", being also conceptualized as the solid product of pyrolysis and gasification of biomass and denominated as charcoal if derived from woody biomass.

Biochar has been produced and utilized for thousands of years in soil management practiced by ancient Amerindian populations before the arrival of Europeans and to the development of complex civilizations in the Amazon Basin. Certain dark Earth in the Basin or "terra preta" are associated with enhanced sustained soil fertility (Lehmann and Joseph 2009; Lehmann et al. 2006).

In the Amazon Basin, large amounts of biochar-derived C stocks remain in these soils today after hundreds or thousand of years passed after their abandonment, and slash-and-burn practices have ceased in the region. In Amazonian soils, the total soil C storage is about 250 $MgCha^{-1}m^{-1}$ compared with typical values of 110 $MgCha^{-1}m^{-1}$ in soils derived from similar parent material (Verheijen et al. 2010). This kind of information allowed us to conclude positively about the huge potential of biochar as an efficient tool for environmental management, which can be nowadays directed to four main objectives which can be achieved solely or jointly: climate change mitigation along with energy production, soil improvement for productivity, and waste management.

A wide scope of biochar systems with distinct strategies and socioeconomic benefits can thereby be designed. Yet, the knowledge of biochar by today's living farmers is still very scarce, and political debates about climate change continue ignoring this thematic, while industries that could benefit immensely have barely considered it. In all this context, biochar production and its land application have been proposed as a possible strategy for climate change mitigation (e.g., Lehmann et al. 2006; Lehmann and Joseph 2009).

The main influence of biochar in the mitigation of climate change is related to the achievement of the much higher stability of organic matter or biomass. This stability is achieved by the thermochemical conversion of fresh organic materials, which mineralize comparatively quickly, into biochar which mineralizes much more slowly. The difference between the mineralization rates of uncharred soil organic matter and charred biochar materials results in a greater amount of carbon sequestration and storage in soils and a lower amount of carbon dioxide released in the atmosphere. The biomass carbon is transformed in a more stable form in biochar which remains in the soil for hundreds of thousands of years.

Globally, soils are the largest terrestrial organic C pool with estimations of soil organic carbon amounts of 1500 GtC at 1 m depth, or 2400 GtC at 2 m depth, much higher than the atmosphere with 750 GtC, or than vegetation with 560 GtC. The estimated C pool in soils is also 240-fold higher than annual fossil fuel emissions of about 10 GtC (e.g., Paustian et al. 2016). These are gross modeling estimates reflecting the preliminary status of fundamental knowledge of attainable soil organic carbon contents relative to variation with environmental factors.

The principle for using biochar for carbon sequestration is related to the contribution of soil in the C-cycle, for example, with the yearly global upward flux of carbon dioxide from the soil of 60 Gt. This CO_2 results from the decomposition of soil organic matter (SOM). Also, more than 80% of the terrestrial organic carbon is contained in the soil. However, the potential for the accumulation of carbon in soil is low insofar that the respective carbon sinks have low permanency, can be depleted with land-use change, and are often offset by greenhouse gas emissions. The carbon amount in the soil through biochar input will increase in comparison with a microbial decomposition of SOM, with a negative carbon balance which is a significant contribution to climate change mitigation.

The application of biomass-derived black carbon or biochar to soil, provides a large and long-term carbon sink and minimizes the leakage phenomena associated with additions of non-charred organics to soil. Leakage is the denomination used for accounting with default emissions, which occur on the source locations from where the added organic matter was removed for application in soil sink sites. In this circumstance, the net carbon release reduction in the whole system may not have been achieved, insofar that the carbon increase in soil sinks could be offset by carbon removal in source locations. Indeed, only if the delivered organic matter resulted from an increase in biomass productivity, would the net carbon balance in the system boundaries be possibly negative. With biochar, not only the soil carbon is stored for decades, but also biomass productivity in the ecosystems increases, due to soil quality improvement, with a secondary cascading effect of additional carbon added in soils (e.g., Lehmann and Joseph 2009; Lehmann et al. 2006; Gaunt and Cowie 2009).

Chemically, biochar has a high content of fixed carbon, ranging from 50 to 90% with volatiles and ash amounts ranging from 0–40% to 0.5 and 5%, respectively (e.g., Verheijen et al. 2010; Weber and Quicker 2018). The high carbon amount and the existence of aromatic rigid structural compounds, which are hexagonal compact carbon units linked together and deprived of oxygen and hydrogen, are the common features of recalcitrant biochar feedstocks which can derive from a wide range of biomasses, through a scope of thermochemical pyrolytic and gasifying conversions.

The residence time of biochar versus its mineralization rate in the soil is a key variable for the establishment of scenarios of the global reduction in carbon emissions. Under a conservative perspective, the composition of biochar is about 80% of stable aromatic recalcitrant carbon and 20% labile mineralizable carbon which can be released into the atmosphere as carbon dioxide in the first years of application in soil. This release is very dependent on the weight of these fractions in the biochar composition.

Modeling analysis shows that a maximum reduction in carbon emissions would require a mean residence of biochar in soil higher than 100 years, considering a system of 50% conversion rate for biochar production from slow biomass pyrolysis or carbonization. A mean residence time of 50 years returns 96% of carbon to the atmosphere over a period of 200 years, whereas a return of 28% would correspond to a residence time of 500 years (e.g., Lehmann et al. 2010).

The concretization of the whole potential of biochar feedstock in environmental management, depends essentially on a sustainable development of a paradigm of slash-and-char. The later consists of biomass conversion through, for example, technologies of carbonization, torrefaction, pyrolysis, or gasification of biomass feedstocks, such as from agricultural and forestry wastes such as forest residues, mill residues, field crop residues, and urban wastes. Those slash-and-char practices should thereby replace traditional slash-and-burn practices of biomass residues in the field.

A myriad of conversion technologies has in common the heating and/or the oxidation of biomass feedstock under the partial or total absence of oxygen. The yields of biochar in these processes range from 20 to 60% and carbon contents of biochar higher than 50%, depending on operative variables and feedstock (e.g., Schenkel et al. 1999; Lehmann 2007; Meyer et al. 2011; Enders et al. 2012; Lee et al. 2013).

The stability or longevity of carbon in/and biochar is another important matter, insofar that only a long residence time ensures relevant carbon sequestration. Over century lifecycles, biomass mineralization dynamics is more sensitive to variations in the amount in labile carbon, say from 0 to 5%, than to decrease, for example, from 1000 to 500 years in mean residence times. As aforementioned, most carbon in biochar is highly stable and has an average time of 1000 years or longer at 10 °C mean annual temperature (e.g., Roberts et al. 2010).

The other big issue to address concerns the soil loading capacity for biochar (BLC) or in another way, the question of how much biochar soil can tolerate. Soils with high biochar concentrations corresponding to about 40% of their soil organic carbon were found in the Amazonian Basin. Furthermore, after hundreds or thousands of years, the productivity of these soils did not decrease or even increase. The evidence available points out, for example, that in a weathered tropic soil, crop yields kept increasing with biochar loadings up to 140 t Cha^{-1}. On the other hand, other trials showed that some crops delivered positive effects concerning biomass and yields, at much lower applications. For example, experiments with bean (*Phaseolus vulgaris* L.) cultivation showed positive yield effects with application rates up to 50 t Cha^{-1}, which disappeared with a rate of 60 t Cha^{-1}, along with yield decreases at application rates of 150 t Cha^{-1}. This is indicative that BLC is likely to be dependent on crops, on soil and climate, and on rates of application and total accumulated amounts as well. Overall, it is likely that crops respond positively to biochar additions up to 50 t Cha^{-1} with possible growth reductions at very high rates of application.

Smoldering combustion is another variable interfering with BLC, insofar that this kind of combustion can be supported for years or long time periods in organic soils that dry out sufficiently. Thus, soils with extreme loadings of biochar, subjected to drastic drying conditions, could eventually withstand smoldering fires ignited by natural or anthropogenic factors. This is an issue also requiring long-term research.

The sum of downward CO_2 from the atmosphere to the ocean, soil, and vegetation is of the order of 213.35 GtC, and the sum of upward fluxes to the atmosphere of non-anthropogenic sources which were soil, vegetation, oceans, and forests accounts for about 211.6 GtC, so that e.g. in 2004 a net loss of carbon from the atmosphere was of around 1.75 GtC. This loss was however more than compensated by anthropogenic sources of 5.5 GtC, delivering to a net gain of around 3.75 GtC, for which a significant contribution for mitigation could be achieved through biochar addition in soil. The hypothetical equilibrium of anthropogenic carbon emissions should be thereby the main objective to achieve with biochar in soil (e.g., Verheijen et al. 2010). As aforementioned, the carbon dioxide emissions were higher of around 9.9 GtC in 2017 and averaging 4.7 GtC in the period 2008–2017.

Also, global emissions of greenhouse gases were reported as increasing by 35% from 1990 to 2010 reaching about 13.6 GtC. These later data ranges from 0.4 to 2.5 orders of magnitude from the value of 5.5 GtC of 2004, a variation which is not incompatible with the hypothetical equilibrium role of compensation, already mentioned, are attributed to biochar. Estimations were carried out indicating that storing carbon in biochar would compensate global carbon yearly emissions of about 0.1–0.3 Gt CO_2.

Other projections mention that carbon sequestration of biochar, resulting from forestry and agricultural residues and urban wastes, would reach by 2100 yearly values ranging from 5.5 to 9.5 Gt C, values of the same order of magnitude of current fossils fuel emissions. Also, an application of 10–100 $Mgha^{-1}$ of biochar, with carbon concentration around 50–78% considering a global cropped area of 1411 Mha, would determine a sequestering of carbon of 7–110 GtC. The assumed range of biochar application in soils of 10–100 Mgha^{-1} is within the range of values for soil biochar loading capacity (Oliveira et al. 2017).

Where biomass is converted to biochar, the potential net C storage is 20 percent of the C captured through photosynthesis (gross primary productivity, or GPP). As the global gross primary productivity (GPP) flux is 120 GtC, this would equate to a theoretical annual potential of 24 GtC (88 GtCO$_2$ eq.), though much of this biomass is not available for conversion to biochar. The actual potential will depend upon our ability to access biomass feedstocks in an economically viable and environmentally responsible manner (Lehmann 2007; Gaunt and Cowie 2009).

Biochar has also a relevant role in reducing emissions of nitrous oxide and methane from arable soils. These gases are about 310 times and 21 times more potent as greenhouse gases than CO$_2$. Evidence points that biochar complemented with other soil management practices could reduce N$_2$O emissions by as much as 80%. The results concerning methane emissions are mixed with positive results obtained, for example, with near-complete suppression of emissions with maize and forages in tropical soils and with no impact on methane emissions by rice paddy fields. Carbon dioxide represents about 83% of the total GHG emissions while CH$_4$ and N$_2$O constitute about 8 and 5% of that total (e.g., Verheijen et al. 2010; Roberts et al. 2010; Gaunt and Cowie 2009). The annual emission of NO$_2$ from cultivated soils with are of around 4.2 Mt., higher than those from any other anthropogenic source, for a total global emission of 8.1 Mt NO$_2$ (e.g., Paustian et al. 2016).

For methane, the total global emissions are about 200 Mt, with more than one-third occurring from microbial anaerobic breakdown of organic compounds in soil. Wetlands and rice cultivations, with annual estimates ranging from 177–284 Mt to 33–40 Mt, respectively, are the largest soil sources of CH$_4$. In contrast, well-aerated soil is potential sinks for methane with estimates of 30 Gt mainly in forests and uplands via methane oxidation. Alternative estimates point to amounts of 80 and 50% contributions to global emissions of N$_2$O and CH$_4$, respectively. In this context, a vast scope exists for the mitigation of emissions of these two gases, through soil management with biochar applications among a set of 10 alternatives. Indeed, studies about the effects of biochar on the mechanisms for the reduction of methane and nitrous oxide emissions under different scenarios are needed for a real quantification and design of strategies of these mitigations (e.g., Paustian et al. 2016).

References

ACACIA Project. (2000). In Parry, M. (Ed.), *Assessment of potential effects and adaptations for climate change in Europe: The Europe Acacia Project* (p. 324). Norwich, University of East Anglia.

Adloff, F., Somot, S., Sevault, F., Jordà, G., Aznar, R., Déqué, M., et al. (2015). Mediterranean sea response to climate change in an ensemble of twenty first century scenarios. *Climate Dynamics, 45,* 2775–2802.

Allan, R. (2011). Human influence on rainfall. *Nature, 470,* 344–345.

Alpert, P., Ben-Gai, T., Baharad, A., Benjamini, Y., Yekutieli, D., Colacino, M., et al. (2002). The paradoxical increase of mediterranean extreme daily rainfall in spite of decrease in total values. *Geophysical Research Letters, 29*(11), 2–4.

AR4 IPCC Report. (2008). In R. K. Pachauri & A. Reisinger (Eds.) *Climate change 2007: Synthesis report. Contribution of working groups I, II and III to the fourth assessment report of the intergovernmental panel on climate change* (112 pp.). Geneva, Switzerland: Fourth Assessment Report, IPCC.

AR5 IPCC Report, 2015. *Climate Change 2014: Synthesis Report. Contribution of Working Groups I, II and III to the Fifth Assessment Report of the Intergovernmental Panel on Climate Change* [Core Writing Team, R. K. Pachauri and L. A. Meyer (Eds.)] (151 pp.). Geneva, Switzerland: IPCC. ISBN 978-92-9169-143-2.

Boer, G., Flato, G., & Ramsden, D. (2000). A transient climate change simulation with greenhouse gas and serosol forcing: Projected climate for the 21st century. *Climate Dynamics, 16,* 427–450.

Cheddadi, R., Lamb, H. F., Guiot, J., & van der Kaars, S. (1998). Holocene climatic change in morocco: A quantitative reconstruction from pollen data. *Climate Dynamics, 14,* 883–890.

Coumou, D., & Rahmstorf, S. (2012). A decade of weather extremes. *Nature climate change| Advance Online Publication* (6 pp.). Macmillan Publisher Ltd. Retrieved from www.nature.com/natureclimatechange.

Coumou, D., & Robinson, A. (2013). Historic and future increase in the global land area affected by monthly heat extremes. *Environmental Research Letters, 8,* 03401.

Coumou, D., Robinson, A., & Rahmstorf, S. (2013). Global increase in record-breaking monthly-mean temperatures. *Climatic Change, 118,* 771–782.

Cubasch, U., von Storch, H., Waszkewitzl, J., & Zorita, E. (1996). Estimates of climate change in southern Europe derived from dynamical climate model output. *Climate Research, 7,* 129–149.

Della-Marta, P., Haylock, M., Luterbacher, J., & Wanner, H. (2007). Doubled length of western European summer heat waves since 1880. *Journal of Geophysical Research, 112,* D15103.

Donat, M., & Alexander, L. (2012). The shifting probability distribution of global daytime and night-time temperatures. *Geophysical Research Letters, 39,* L14707.

Donat, M., Alexander, L., Yang, H., Durre, I., Vose, R., Dunn, R., et al. (2013). Updated analyses of temperature and precipitation extreme indices since the beginning of the twentieth century: The HadEX2 dataset. *Journal of Geophysical Research: Atmospheres, 118,* 2098–2118.

Doumenge, F. (1997). Environmental Change and the Mediterranean. (UNU Lectures, 16, 17).

Enders, A., Hanley, K., Whitman, T., Joseph, S., & Lehmann, J. (2012). Characterization of biochars to evaluate recalcitrance and agronomic performance. *Bioresource Technology, 1114,* 644–653.

Fischer, E., and Knutti, E. (2015). Anthropogenic contribution to global occurrence of heavy-precipitation and high-temperature extremes. *Nature Climate Change, 5,* 560–565.

Fontana, G., Toreti, A., Ceglar, A., & De Sanctis, G. (2015). Early heat waves over Italy and their impacts on durum wheat yields. *Natural Hazards Earth Systems Science, 15,* 1631–1637.

Founda, D., Varotsos, K., Pierros, F., & Giannakopoulos, C. (2019). Observed and projected shifts in hot extremes' season in the eastern mediterranean global and planetary. *Change Global and Planetary Change, 175,* 190–200.

Gaunt, J., & Cowie, A. (2009). Biochar, greenhouse gas accounting and emissions trading, Chapter 18. In J. Lehmann & S. Joseph (Eds.), *Biochar for environmental management* (405 pp.). Earthscan. ISBN 978-1-84407-658-1.

Giannakopoulos, C. M., Bindi, M., Moriondo, M., LaSager, P., & Tin, T. (2005). *Climate change impacts in the mediterranean resulting for 2 °C global temperature rise. A report for the WWF for a living planet.* Gand, Switzerland: The Global Conservation Organization, WWF.

Giannakopoulos, C., Le Sager, P., Bindi, M., Moriondo, M., Kostopoulou, E., & Goodess, C. M. (2009). Climatic Changes and Associated Impacts in the Mediterranean Resulting from a 2 °C global warming. *Glob Planet Change, 68*(3), 209–224.

Giorgi, F. (2006). Climate Change Hot-spots. *Geophysical Research Letters, 33*, L08707. Global Warming Potential, Wikipedia. Retrieved April, 2019, from https://en.wikipedia.org/wiki/Global_warming_potential.

Groisman, P. Y., Karl, T. R., Easterling, D. R., Knight R. W., Jamason, P. F., Hennessy, K. J. et al. (1999). Changes in the probability of heavy precipitation: Important indicators of climatic change. In: T. R. Karl, N. Nicholls & A. Ghazi (Eds.), *Weather and Climate Extremes*. Dordech: Springer. https://doi.org/10.1007/978-94-015-9265-9_15.

Groisman, P. Y., Knight, R., Easterling, D., Karl, T., Hegerl, G., & Razuvaev, V. (2005). Trends in intense precipitation in the climate record. *Journal of Climate, 18*, 1326–1350.

Grunderbeek, P., Tourre, Y. (2008). Mediterranean basin: Climate change and impacts during the 21st century. Part I, Chapter 1, pp. 1–64. In: Climate Change and Energy in the Mediterranean. Plan Bleu, Regional Activity Center, Sophia Antipolis. (https://www.eib.org/attachments/country/climate_energy_mediterranean_en.pdf) (accessed in April 2019).

Hansen, J., Ruedy, R., & Sato, M. (2001). A closer look at United States and global surface temperature change. *Journal of Geophysical Research, 106*, 23947–23963.

Hartmann, D. L., Klein, A. M., Tank, Rusticucci, M., Alexander, Brönnimann, S., et al. (2013). Observations: Atmosphere and surface supplementary material. In Climate change 2013: The physical science basis. Contribution of working group i to the fifth assessment report of the intergovernmental panel on climate change [T. F. Stocker, D. Qin, G.-K. Plattner, M. Tignor, S. K. Allen, J. Boschung, A. Nauels, Y. Xia, V. Bex & P.M. Midgley (Eds.)]. Retrieved from www.climatechange2013.org.

Hasselmann, K. F., Bengtsson, L., Cubasch, U., Hegerl, G. C., Rodhe, H., Roeckner, E. et al. (1995). Detection of anthropogenic climate change using a fingerprint method. *Report/Max-Planck-Institute für Meteorologie, 168*. http://hdl.handle.net/21.11116/0000-0000-3F03-7.

Jackson, R., Le Quéré, C., Andrew, R., Canadell, J., Korsbakken, J., Liu, Z., et al. (2018). Global energy growth is outpacing decarbonization. *Environmental Research Letters, 13*(120401), 1–7.

Karas, J. (2006). *Climate Change and the Mediterranean Region*. Report, Greenpeace (34 pp.). Retrieved April, 2019, from http://www.greenpeace.org/international/Global/international/planet-2/report/2006/3/climate-change-and-the-mediter.pdf).

Katttemberg, A. et al. (1996). Climate models—Projections of future climate. In J. T. Houghton et al. (Eds.), *Climate change 1995: The science of climate change*. Report of IPCC Workin g Group I (pp. 289–357). Cambridge University Press.

Kuglitsch, F., Toreti, F., Xoplaki, E., Della-Marta, M., Zerefos, F., Türkeş, M., et al. (2010). Heat Wave Changes in the Eastern Mediterranean since 1960. *Geophysical Research Letters, 37*, 1–5.

Le Quéré, C., Andrew, R., Friedlingstein, P., Sitch, S., Hauck, J., Pongratz, J., et al. (2018). Global carbon budget. *Earth System Science Data, 10*, 1–54.

Lee, Y., Park, J., Ryu, C., Gang, K., Yang, W., Park, Y.-K., et al. (2013). Comparison of biochar properties from biomass residues produced by slow pyrolysis at 500 & #xB0;C. *Bioresource Technology, 148*, 196–201.

Lehmann, J., Gaunt, J., & Rondon, M. (2006). Bio-char sequestration in terrestrial ecosystems—A review. *Mitigation and Adaptation Strategies for Climate Change, 11*, 403–427.

Lehmann, J. (2007). Bio-energy in the black. *Frontiers in Ecology and the Environment, 5*, 381–387.

Lehmann, J., & Joseph, S. (2009). Biochar for environmental management: An introduction, Chapter 1. In J. Lehmann, & S. Joseph, (Eds.), *Biochar for environmental management* (405 pp.) Earthscan. ISBN 978-1-84407-658-1.

Lehmann, J., Amonette, J., & Roberts, K. (2010). Role of biochar in mitigation of climate change, Chapter 17. In D. Hillel & C. Rosenzweig, C. (Eds.), *Handbook of Climate Change and Agroecosystems Impacts, Adaptation, and Mitigation, Series on Climate Change Impacts, Adaptation and Mitigation* (Vol. 1, 452 pp.).

Lewis, C., & Karoly, D. (2013). Anthropogenic contributions to Australia's record summer temperatures of 2013. *Geophysical Research Letters, 40,* 3705–3709.

Maheras, P. (1987). Temporal Fluctuations of annual precipitation in Palma and Thessaloniki. *Journal Meteorology, 12,* 305–308.

Maheras, P., & Kolyva-Machera, F. (1990). Temporal and spatial characteristics of annual precipitation over the balkans in the twentieth century. *International Journal of Climatology, 10,* 495–504.

Mann, M., Bradley, R., & Hughes, M. (1998). Global-scale temperature patterns and climate forcing over the past six centuries. *Nature, 392,* 779–787.

Meehl, G., Karl, T., Easterling, D., Changnon, S., Pielke, Jr. R., Evans, J. et al. (2000). An introduction to trends in extreme weather and climate events: Observations, socioeconomic impacts, terrestrial ecological impacts and model projections. *Bulletin of the American Meteorological Society, 81*(3), 413–416.

Meyer, S., Glaser, B., & Quicker, P. (2011). Technical, economical, and climate-related aspects of biochar production technologies: A literature review. *Environmental Science and Technology, 45,* 1473–1479.

Mitchell, J., Davis, R., Ingram, W., & Senior, C. (1995). On surface temperatures, greenhouse gases and aerosols: Models and observation. *Journal of Climate, 8,* 2364–2386.

Mitra, A. (2020). The blue carbon concept. In: S.C. Patra & Mitra (Eds.), *Webinar articles on Blue Carbon Domain: A Potential Regulator of Climate Change.* University of Calculta.

Oliveira, F., Patel, A., Jaisi, D., Adhikari, S., & Lu, H. (2017). Environmental application of biochar: Current status and perspectives. *Bioresource Technology, 246,* 110–122.

Otto, F., Massey, G., van Oldenborgh, G., & Jones, R. (2012). Reconciling two approaches to attribution of the 2010 Russian heat wave. *Geophysical Research Letters, 39,* L04702.

Paustian, K., Lehmann, J., Ogle, S., Reay, D., Robertson, G., & Smith, P. (2016). Climate-Smart soils. *Nature, 532,* 49–57. https://doi.org/10.1038/nature17174.

Perkins-Kirkpatric, S., & Gibson, P. (2017). *Changes in regional heatwave characteristics as a function of increasing global temperature, scientific reports,* 7: 12256. Retrieved December, 2019, from www.nature.con/scientificreports.

Perkins, S., Alexander, L., & Nairn, J. (2012). Increasing frequency, intensity and duration of observed global heatwaves and warm spells. *Geophysical Research Letters, 39,* L20714.

Quiggin, J. (2019). *Opportunity costs: Can carbon taxing become a positive-sum game?.* Retrieved April, 2019, from https://aeon.co/ideas/opportunity-costs-can-carbon-taxing-become-a-positive-sum-game.

Rind, D., Goldberg, R., Hansen, J., Rosenzweig, C., & Ruedy, R. (1990). Potential evapotranspiration and the likelihood of future drought. *Journal of Geophysical Research, 95,* 9983–10004.

Roberts, K., Gloy, B., Joseph, S., Scott, R., & Lehman, J. (2010). Life cycle assessment of biochar systems: Estimating the energetic. *Economic and Climate Change Potential, 44,* 827–833.

Robinson, P. (2001). On the definition of heat have. *Journal of Applied Meteorology, 40,* 762–775.

Rosenzwieg, C., & Tubiello, F. (1997). Impacts of global climate change on mediterranean agriculture: Current methodologies and future directions. An introductory essay. *Mitigation and Adaptation Strategies for Global Change, 1*(3), 219–232.

Schenkel, Y., Temmerman, M., Van Belle, J.-F., & Vanker kove, R. (1999). New Indicator for the evaluation of the wood carbonization process. *Energy Sources, 21,* 935–943.

Schiermeier, Q. (2010). Mediterranean most at risk from European heatwaves. *Nature News.* https://doi.org/10.1038/news.2010.238.

Segal, B., Mandel, M., Alpert, P., Stein, U. & Mitchell, M. J. (1994). Some assessment of potential $2 \times CO_2$ climatic effects on water balance components in the eastern mediterranean. *Climate Change, 27,* 351–371.

Sippel, S., & Otto, F. (2014). Beyond climatological extremes—Assessing how the odds of hydrometeorological extreme events in south-east Europe change in a warming climate. *Climatic Change, 125,* 381–398.

Stott, P., Stone, D., & Allen, M. (2004). Human contribution to the European heatwave of 2003. *Nature, 432*, 610–614.

Stull, R. (2000). *Meteorology for scientist and engineers* (2nd ed., p. 502). Thomson Learning: Brooks/Cole.

Verheijen, F., Jeffrey, S., Bastos, A., van der Velde, M., & Diafas, I. (2010). *Biochar application to soils-A critical scientific review on soil properties, processes and functions.* European Commission, Institute for Environment and Sustainability, 166 pp. ISBN 978-02-79-14293-2.

Westra, S., Alexander, L., & Zwiers, F. (2013). Global increasing trends in annual maximum daily precipitation. *Journal of Climate, 26*, 3904–3918.

Wigley, T. (1992). Future climate of the mediterranean basin with particular emphasis on changes in precipitation. In L. Jeftic, J. D. Milliman, & G. Sestini (Eds.), *Climatic change and the mediterranean* (pp. 15–44). London: Edward Arnold.

WMO. (1997). WMO Statement on the Status of the Global Climate in 1996.

Annex A1: Instrumentation in Environmental Physics

Abstract

Annex A1 aimed to present fundamentals of environmental instrumentation beginning with damping factors and dynamic response of zeroth, first and second-order sensors, to step functions. General concepts of Fourier and Laplace transforms, transfer functions and transient processes follows. General properties of sensors, e.g. accuracy, sensitivity, linearity, or repeatability are defined with further characterization of transducers for measurement of temperatures, radiation, humidity, airspeed, wind direction, precipitation, and data acquisition. Temperature transducers with significant relevance in environmental physics are based on electric, expansion or thermodynamic principles. The examples of sensors discussed are thermocouples with Seeback electronic principles in bimetallic junctions, isolated or in thermopiles, thermistors, aspirated and condensation hygrometers or pyrradiometers.

A1.1 Introduction

Quantification of mass and energy flow in the surface layer is one of the main aims of environmental physics. Such quantification is done by previously described methods such as the eddy covariance, aerodynamic and Bowen methods. The application of these methods involves carrying out measurements on the ground, some of which have been presented in relation to the eddy covariance method.

A sensor, detector or transducer is a device that converts a form of energy or physical quantity into a distinct form, that is, a device that converts an input signal into an output signal of another type. Typically, the sensor generates a signal (electrical, mechanical, etc.) corresponding to the quantity to be measured. A recording device processes this signal and registers it in internal or external memory, on paper, or for example on a monitor. Instrumentation analysis involves the study and operation of sensors, requiring a recorder and/or visualization device.

Instrumentation plays an important role in our high-tech world. For example, in the automobile industry, instrumentation is crucial in creating and improving vehicle efficiency in terms of fuel consumption and emissions. Use of robotics in vehicle manufacture involves adequate instrumentation and quality control feedback.

The Editor(s) (if applicable) and The Author(s), under exclusive license to Springer Nature Switzerland AG 2021
A. Rodrigues et al., *Fundamental Principles of Environmental Physics*,
https://doi.org/10.1007/978-3-030-69025-0

Sensors have a wide range of different features. For example, analogue sensors are those where the input and output signals are continuous functions of time. The amplitudes of the signals may have any value constrained by the physical limits of the system. Digital sensors have discrete output signals. In an analogue-digital sensor, the input signal is a continuous function of time and the output signal is a quantified signal that can only have discrete values. The combination of a computer or a data logging system, with an analogue system produces digital signals, typically as binary numbers so that it becomes a combined digital-analogue system.

Converting an analogue signal to a digital one involves an approximation because the continuous analogue signal can take on an infinite number of values, while the variety of different numbers that can result from a finite set of digits is limited. For example, in a 14-*bit* data logging system, the analogue values detected by the sensor (e.g. mV) thermocouple measurement) are digitized in binary form (digits 0 and 1) where any 14-digit sequence will be storage as whole numbers between 0 (lowest binary number comprising 14 0s) and $2^{13}-1 = 8191$ (largest binary number constituted by a 14-digit sequence of 1). That is, the data acquisition unit can only store integers between 0 and 8191, that is, provide only 8192 memory locations for storing of analogue information.

The resolution of the acquired measurement can follow the same principle. If the data acquisition apparatus has a 3-bit capacity only, integers that can be stored vary between 0 and 7. If the analogue value to be stored is 4 mV, then this integer must be distributed over 8 positions between 0 and 4 mV at 0.5 mV ranges (or 4/8) so that this will be the level of precision.

Transfer functions (Chap. 3) establish the relationship between the value of the measured quantity and the signal in the sensor output. This function may be linear or not or may be obtained from a conversion table of discrete values. For example, for a platinum resistance thermometer, the transfer function may be of the type:

$$R_T = R_o\left(1 + aT - bT^2 - cT^2(T - 100)\right) \tag{A1.1}$$

where R_T and R_o are resistances to temperature T °C and to 0 °C and a, b, and c are constants given in tables. Typically, the transfer functions are integrated into the data logging system.

The dynamic response of a sensor is its response to the input variation over time. The dynamic responses of the sensors, with thermal or mechanical inertia, are often described by equations linear ordinary differential. In a mechanical system, a moving mass stores kinetic energy and can store potential energy due to its position in the force field. The order of the differential equation that describes the response will always equal the number of energy storage reservoirs (Shaw 1995).

A linear zero-order instrument has an output proportional to the input at all time instants, according to the equation:

$$y(t) = kx(t) \tag{A1.2}$$

where k is the gain constant or instrument's static sensitivity. The output signal follows the input signal without distortion or delay.

The potentiometer is an example of a zero-order, linear instrument used for measuring displacement, where the change in the potential difference across the element is proportional to the displacement. The measuring instrument zero-order displays ideal dynamic behaviour without any phasing. The instrument is zero-order when static output is produced in response to a static entry.

A linear first-order instrument has an output given by a first-order differential equation, as follows:

$$\tau \frac{d(y)}{dt} + y(t) = kx(t) \tag{A1.3}$$

where τ is the time constant (time dimensions) and k the gain constant or static sensitivity of the instrument (output/input dimensions). Thermometers are also first-order instruments. The time constant of temperature measurements depends on the thermal capacity and the interaction between the thermometer and the contact surface.

Cup anemometers for wind velocity measurements are also first-order instruments where the time constant depends on the inertia momentum of the sensor (Kristensen 1993). The reaction of this type of equipment to variable continuous inputs is exemplified by their response to sine wave input functions at various frequencies. If the angular frequency of breakpoint ω_b, is defined as the inverse of the time constant τ and a dimensionless frequency $\alpha = \omega_i/\omega_b$, as the ratio between the angular frequency of the input and the angular frequency of the breakpoint, it can be shown that the output is a modified sine wave with reduced amplitude $A(\alpha)$ and a phase angle $\phi(\alpha)$ given by:

$$A(\alpha) = \frac{1}{\sqrt{(1+\alpha^2)}} \tag{A1.4}$$

and:

$$\phi(\alpha) = tan^{-1}(\alpha) \tag{A1.5}$$

A linear second-order instrument has an output given by a second-order differential equation:

$$\tau \frac{d^2 y(t)}{dt^2} + 2\delta\omega \frac{dy(t)}{dt} + \omega^2 y(t) = k\omega^2 x(t) \tag{A1.6}$$

where δ and ω are constants for the damping factor with a maximum unit value and natural frequency of the undampened instrument, respectively.

A second-order system, subject to a static entry, tends to oscillate around its equilibrium position (or response). The natural frequency of the instrument is defined as the frequency of these oscillations. The internal friction of the instrument opposes these natural fluctuations, with intensity proportional to the rate of change of output signal. The damping factor is a measure of the opposition to the natural oscillations. Depending on the extent of damping, a second-order system can stabilize in the form of undampened oscillations when subjected to input in the form of a step function. Wind direction sensors are an example of such systems.

If the input function $x(t)$ varies continuously with time, the output of a zero-order apparatus varies the same way, being multiplied by a static gain k. Conversely, outputs of first and second-order instruments do not vary in the same way. In these instruments, the configuration of the output function $y(t)$, is different from that of the input function $x(t)$.

It follows from the above that the response of various systems (zero, one or two order) to a step function, characterized by a quick change in the input value such as $x(t) = 0$ for $t \leq 0$ and $x(t) = 1$ for $t > 0$ is:

(i) for the zero-order systems a step function of the type $x(t) = 0$ for $t \leq 0$ and x $(t) = k$ for $t > 0$

(ii) for first order systems (Eq. A1.3 and Fig. A1.2), where $y(0) = 0$, the response will be: (Fig. A1.1)

$$y(t) = K[1 - exp(-t/\tau)] \tag{A1.7}$$

The initial rate of change of response $y(t)$ at time instants near $t = 0$ is of the order of k/τ. At instant $t = \tau$ (time constant) the value of $y(t)$ is about 0.632 K. For longer periods, $y(t)$ tends asymptotically to K. The sensor response velocity in reaching K depends on the time constant value.

(iii) for second-order sensors, the solution of Eq. (A1.6) to step function input and considering $y(0) = 0$ and $dy(0)/dt = 0$, depends on the damping factor, δ. If the damping factor is greater than 1, over-damping (Fig. A1.3) can occur, and $y(t)$ reaches the K threshold slowly.

If δ is 1, the response is:

Fig. A1.1 Schematic representation of a zero-order system

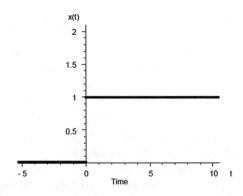

Fig. A1.2 Representation of
a 1st order system

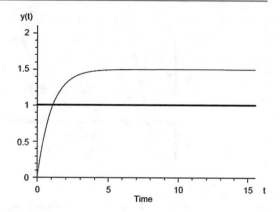

$$y(t) = K[1 - (1 + \omega t)exp(-\omega t)] \tag{A1.8}$$

and the condition of critical damping is reached (Fig. A1.3). Such systems are like
to first-order systems as the response approaches a K threshold.

If the damping factor is between 0 and 1 (Fig. A1.4), there will be
sub-dampening, so that the response $y(t)$ oscillates about the steady-state value
K, with an amplitude which decreases with time. Finally, if the damping factor is
zero (Fig. A1.4) there is no damping and the response oscillates regularly around
the constant K, with amplitude K and oscillation period of $2\pi\omega$.

It can be shown that a damping factor of 0.69 allows the desired measurement
value K to be reached more quickly (Fig. A1.5). The design of the second-order
instrument is thus delineated for such optimum dampening. The response of a
second-order sensor, with this damping factor, is characterized by an initial
oscillatory peak that overshoots K before reaching a constant value.

Fig. A1.3 Representation of
a 2nd order system ($\delta \geq 1$)

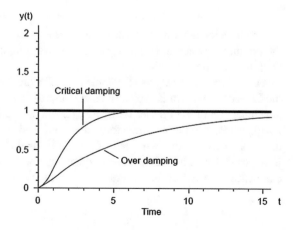

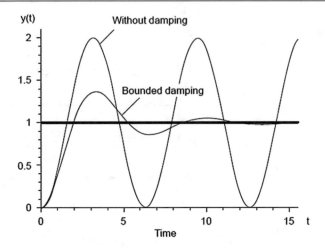

Fig. A1.4 Representation of a first order system (curves correspond to δ between 0 and 1)

Fig. A1.5 Representation of a first order system with $\delta = 0.69$

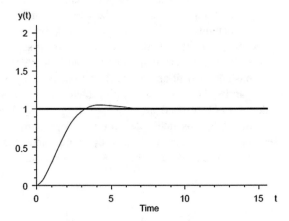

The transfer function of a measuring system, mentioned above, is indicative of the phasing between the input signal and sensor output, as well as damping of the amplitude of the output signal.

For a simplified analysis of complex transfer functions in the frequency domain, the Laplace transform is typically used (Connor 1978). The Laplace transform is a generalization of the Fourier transform, (Chap. 3) applicable in situations at instant $t = 0$, where the Fourier transform does not lead to a finite solution. The main difference between the two types of transforms is that while the Fourier transforms uses positive and negative wave frequencies, the Laplace transforms uses waves dampened by a $e^{-\sigma}$ factor where σ is a positive number.

As referred to in Chap. 3, a Fourier transform is defined as:

$$F(i\omega) = \int_{-\infty}^{\infty} f(t)e^{-i\omega t}dt \tag{3.123}$$

so that the transform converges the integral of Eq. (3.123). This happens when the integral:

$$I = \int_{-\infty}^{\infty} |f(t)|dt \tag{A1.9}$$

converges as $e^{-i\omega t}$ has unit magnitude (Husch and Stearns 1990). If $f(t)$ does not decrease significantly with increasing t, the integral of Eq. (3.123) does not exist, then it becomes necessary to use different techniques to overcome this situation.

In this context, the Laplace transform can be seen as a modification of the Fourier transform, to enable processing of a function $f(t)$ that does not disappear with increasing time dependent variable, t.

If the Fourier integral, Eq. (3.123), does not converge to a specific $f(t)$, the integration can be carried out by multiplying it with a decreasing exponential function $e^{-\delta|t|}$ with $\delta > 0$ so that the product $f(t)e^{-\delta|t|}$ will decrease to zero, with increasing t. Also, physical problems begin to occur at $t = 0$, and the t signal module can be removed. The Fourier transform becomes:

$$F(\delta + i\omega) = \int_0^{\infty} f(t)e^{-(\delta + i\omega)t}dt \tag{A1.10}$$

Putting the complex independent variable, $(\delta + i\omega)$, in Eq. (A1.10) as s, the Laplace transform becomes:

$$F(s) = \int_0^{+\infty} f(t)e^{-st}dt \tag{A1.11}$$

Equation (A1.11) shows that the function $f(t)$ is represented by an infinite set of terms e^{st}, with complex s. These terms give the sines and cosines of the Fourier transform, as well as sines, cosines and increasing and decreasing exponentials, depending on the damping factor δ (Ventura 1985). On the other hand, if $f(t) = 0$ for $t < 0$ and if the Fourier transform $F(i\omega)$ exists, then $F = (i\omega) = F(s)$ for $s = i\omega$ with $\delta = 0$, where δ is the real part of the complex variable s. The $F(s)$ transform of $f(t)$ is also the Fourier transform of $f(t)e^{-\delta t}$, where $f(t) = 0$, for $t < 0$ (Husch and Stearns 1990).

The inverse Laplace transform, considering δ constant is:

$$f(t) = \frac{1}{2\pi i} \int\limits_{\delta-i\infty}^{\delta+i\infty} F(s)e^{st}ds \qquad (A1.12)$$

This transfer function concept is useful in analysing systems in which one (or more) input to the sensor produces one (or more) output function.

Linear systems with variable order are described by linear differential equations with constant coefficients:

$$\left[A_n \frac{d^n}{dt^n} + A_{n-1} \frac{d^{n-1}}{dt^{n-1}} + \ldots\ldots + A_0 \right] g(t) = \\ \left[B_n \frac{d^n}{dt^n} + B_{n-1} \frac{d^{n-1}}{dt^{n-1}} + \ldots\ldots + B_0 \right] f(t) \qquad (A1.13)$$

where for each term the differential operator is in square brackets, multiplied by a constant coefficient, and $g(t)$ and $f(t)$ are output and input system functions, respectively. The Fourier transform of the two equality terms becomes (Stearns and Husch 1990):

$$[A_n(i\omega)^n + A_{n-1}(i\omega)^n + \ldots\ldots + A_0]G(i\omega) = \\ [B_n(i\omega)^n + B_{n-1}(i\omega)^n + \ldots\ldots + B_0]F(i\omega) \qquad (A1.14)$$

and the transfer function $H(i\omega)$, describing the system in terms of ratios of the output and input transforms will be given by:

$$H(s) = H(i\omega) = \frac{G(i\omega)}{F(i\omega)} = \frac{\sum_{i=0}^{n} B_i(i\omega)^i}{\sum_{i=0}^{n} A_i(i\omega)^i} \qquad (A1.15)$$

Laplace transforms of Eq. (A1.14) are equal to the Fourier transforms if the initial values are assumed to be zero. This is a quick way for the analysis of practical applications for measurement systems. For example, in Chap. 3, for the eddy covariance method, it was noted that the empirical transfer functions relating to the corrections applied, is multiplied by co-spectral or spectral densities relative to flows or variances to be determined (Eqs. 3.203 and 3.204).

A1.2 General Properties of Sensors

The static response is the observed output of a given instrument for a given stationary input. Some features included in this type of response are:

  i. Sensitivity is defined as the slope of the input/output curve, which may be an input value of the function if the curve is non-linear. It may also be defined as the minimum variation of the input parameter that will create a detectable change in the output. The sensitivity error is the deviation from the ideal slope of the sensor response curve.

ii. Sensor range corresponds to the maximum and minimum values that can be measured. For example, a thermometer can operate in the range −40° to 60 °C, which gives a span of 100 °C.

iii. Accuracy of a sensor is the smallest possible deviation from the most probable or exact value of the measured quantity and refers to the degree of reproducibility of a measurement. Thus, if a quantity is measured repeatedly, an ideal sensor would yield the same output each time. But in effect, sensors produce a range of output values distributed around the presumptive true value.

iv. A sensor's stability is its capacity to function according to the calibration curve for a reasonable period.

v. Sensor offset error is the deviation that occurs when the output should be zero or alternatively, the difference between the true and observed output values, under specified conditions. The offset, such as nonlinearity, is readily corrected for by digitalization.

vi. Linearity or nonlinearity of the calibration curve, dependent on the environmental conditions, affects the sensitivity of the measuring equipment, even though linearization can be done digitally. The linearity of the transducer is the degree of deviation from the true curve, measured by a sensor in relation to a straight line.

vii. Resolution of a sensor is the lowest input variation which causes a detectable change in the output, which can be related to the input value. The resolution can be expressed either in relation to the percentage of the entire scale or in absolute terms.

viii. Sensor accuracy is the maximum difference between the true value (which should be measured by a standard primary or a good secondary standard) and the output value. Accuracy may be expressed as a percentage of full scale, or in absolute terms.

ix. Hysteresis is due to different sensor output depending on the input value be achieved by an increase or decrease of the measured quantity.

x. Sensor threshold is the lowest measurable quantity above zero and is usually the result of friction in mechanical devices, for example, in the case of anemometers cups.

xi. Repeatability is the degree of closeness between consecutive output measurements for a given input value under identical operating conditions. Generally, this feature is quantified as a percentage of non-repeatability in the measurement range.

xii. Measurement errors are either systematic or random. With systematic errors, the error does not change between successive measurements. With random errors, the errors vary between successive measurements, for example, due to electronic noise, temperature fluctuation, operational error, environmental disturbances, etc.

An error is a difference between the measured value and its true value, expressed in relative or absolute terms. The total error in measurement is a combination of errors such as errors due to an improper exposure, exposure to a contaminant influence (e.g. a temperature sensor exposed to a radiative source), electronic noise, uncertainty or fluctuation in calibration or improper protrusion of the device in the environment or signal loss during the transmission process.

When checking the magnitude of a combination of errors, dependence or independence of errors should be considered. In the case of independence, the square root of the sum of squares of individual errors must be less than the sum of the absolute values of the individual errors (Shaw 1995).

A1.3 Temperature Measurements

Environmental temperature is a fundamental variable for the study of environmental physical processes in the constant flux layer, insofar that as aforementioned temperature profiles are essential factors for the dynamics of atmospheric processes. The physical fundamentals underlying temperature measurement are based on electric, expansion or thermodynamic principles. Temperature is an intensive property common to all points of a microsystem and delivering a sensation of heat or cold located in its boundaries. The temperature measurement is common in everyday procedures such as meteorological previsions, control, and regulation in rooms and apartments, healthcare amongst many other practical situations. The principles of measuring air temperature and sensible heat fluxes by sonic anemometry were presented in Chap. 3.

Thermometers are very common instruments for measuring temperatures of microsystems, by recording e.g. the variation of volume of a gas or liquid with temperature. The air temperature measured by a thermometer is the result of its energy budget determined by the heat exchanges between the environment and the device through the processes of radiation, convection, and conduction. The usual rational is the use of a substance with a property which changes with temperature in a regular way. This process can be evaluated through a linear relationship such as:

$$t(x) = ax + b \qquad (A1.16)$$

where t is the temperature of the used substance with the property x varying with that temperature. The constants a and b are dependent on the used substance and can be calculated specifying two points in the temperature scale. Thermometers of alcohol and mercury are based on this principle. As air temperature increases the liquid swells and the dilatation is recorded in a small tube connected with the liquid reservoir. For thermographs, the liquid reservoirs are connected to mechanical devices for data recording which allow the continuous recording of temperature evolution. The bimetallic thermometers operative principle is based on the fact the distinct metals show different expansion coefficients. Two strips of distinct metals are joined and as they are heated or cooled the differential expansion causes a

Fig. A1.6 Example of an
electronic thermometer

bending of the strip, which allows quantifying the temperature. The bimetallic thermometers may be incorporated in thermostats wherein the bending metallic strip activates a mechanism for environmental temperature control. The classical mercury thermometers and bimetal thermometers have been replaced by transducers with electrical measurement principles.

The electronic thermometers are the most used devices for temperature measurements in environmental applications (Fig. A1.6).

The electronic transducers of temperature measurement must be of dimensions and geometrical forms compatible with any type of utilization allowing for digital or analogue outputs, which may be saved or processed in an expedite way. Four main types of electronic sensors are thermocouples, thermistors, diodes, and platinum resistance thermometers (PRT).

Thermocouples are common sensors based the *Seeback* effect that which if two metals are connected to establish an electric circuit and if the two junctions are at two distinct temperature, an electromotive force (emf) will be generated (ΔV) which is proportional to temperature difference (ΔT) so that (e.g, Oke 1992):

$$\Delta V = a_1 \Delta T + a_2 \Delta T^2 \tag{A1.17}$$

In real conditions, under the temperature ranges prevailing in the surface layer, the second term on the right side can be cancelled. The sensitivity of thermocouples, evaluated through the constant a_1 , is dependent on the metals used. For the thermocouples of combination of copper and constantan (Fig. A1.7), which can be used at temperatures ranging between −200 and 350 °C, sensitivity ranges between 40 and 60 μV °C^{-1}.

For the thermocouples using platinum and rhodium, usable at temperatures ranging between 0 and 1600 °C the sensitivity is lower of around 10 μV °C^{-1}. Thermocouples of iron and constantan can be used at temperatures ranging between −40 750 °C with a sensitivity of 55 μV °C^{-1}.

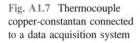

Fig. A1.7 Thermocouple copper-constantan connected to a data acquisition system

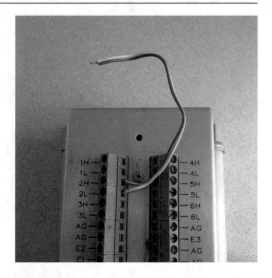

The commonest thermocouples are based on chromel (90% of nickel plus 10% of chromium) and alumel (95% of nickel plus 2% of manganese plus 2% of aluminium and 1% of silica), being used in a large scope of objectives with a sensitivity of 41 μV $°C^{-1}$ and applied under a range of temperatures between -200 1300 °C.

Thermocouples are robust and cheap sensors, of easy use and manufacturing, and not needing of external electric power for the measurements, because they produce a voltage by themselves. Usually, thermocouples don't require internal calibration, except for very precise work, and deliver an output voltage which has a very stable linear variation with temperature.

For obtaining absolute temperature, one of the junctions must be referenced to equilibrium with known constant temperature, e.g. ice. For example, if for a thermocouple copper-constantan, with a sensitivity of 40 μV $°C^{-1}$, the reference junction is at 0 °C for measurement of 1000 μV the junction temperature will be of 0 °C + 1000/45 $\approx$ 22 °C. Thus, for the temperature measurement with a thermocouple, using Eq. A1.16, it is necessary to know the value of a_1 constant and measure ΔV with a device e.g. voltmeter.

The low values of ΔV signals, around 10^{-3} and 10^{-6}, require equipment of high quality for data acquisition. The problem can be simplified by connecting the number of the junction is series forming a so-called thermopile, so that the outputs can be summed. Thermopiles can be used in applications wherein the differences in air temperature, through differential measurements, are more relevant than the absolute values. This is the case e.g. in the quantification of vertical fluxes of sensible heat via flux-gradient methodologies.

The modern data acquisition devices measure the temperature of the connection between thermocouples and thermistors or PRT and add this temperature by digital or analogue means to the temperature reported from voltage measurements.

The main errors for temperature measurements with thermocouples are due conditions of turbulence and radiation, being indeed much greater, ranging between 401 K and 0.05 K, than the most recent electrical measurement techniques (−0.001 K) (e.g., Foken 2017).

The radiation error is due to additional heating by the absorption of radiation by the sensor and is a function of the radiation balance at the sensor surface, and of the heat transfer properties. The principles of heat transfer through conduction, forced convection and radiation, presented in Chap. 6, are therefore involved. Radiation errors are minimal for very thin thermocouples. For thin platinum wires, under forced convection and incident solar radiation of 800 Wm^{-2}, radiation errors of below 0.1 K are possible for wire diameters lower than 20 μm. The sensitivity drastically improves, to values below 0.05 K, for increasing wind velocities in the range between 1 ms^{-1} and 10 ms^{-1} (Foken 2017).

Thermistors are sensors whose electric resistance changes with temperature. Their resistance can be determined for a range of temperatures for calibration and operative temperature measurement. The curve of variation of resistance, although nonlinear can be written as follows:

$$ln(R) = \alpha + \frac{\beta}{T} \tag{A1.18}$$

with α and β being constant for each thermistor and T the air temperature in K.

Thermistors are manufactured from sintered semiconductors usually ceramic with manganese, nickel, copper, iron, cobalt, and uranium oxides. In data acquisition devices, thermistors are connected to a source of electric voltage in series, with the voltage drop in the internal resistance in the internal resistance used to determine temperature. Thermistors show an advantage in delivering a high electrical output, by temperature units, thus more easily quantifiable. Among the disadvantages, are the fact that these sensors are more expensive than thermocouples, and require more calibration.

The voltage drops through a junction PN of a semiconductor diode under constant electric current, is linear dependent on the junction temperature, in specific temperature ranges. The diode can be therefore used as a thermometer if inserted in a proper integrated circuit.

The thermometers of platinum resistance (PRT) are very precise instruments. The electric resistance of platinum increases by around 0.4% $°C^{-1}$ and thus these devices can be inserted in an electric circuit with the changes in voltage indicative of temperature changes. Although the electrical outputs of PRT are much smaller than that from thermistors, the former has the advantage of keeping the calibration active for a longer period and being so potentially used as patterns or in applications with high precision requirements (Campbell 1997; Fritchen and Gay 1979). The use of physical white shelters for temperature measurements aims to minimize the heterogenous environmental conditions associated with radiation and airflow. Those conditions delivered changes in the energy budget of sensor surfaces linked with microvariations of heat transfer. The insulation against short wavelength radiation is guaranteed by the reflection of the white colour of the shelter surface which is usually doubled. Inconvenient convection exchanges between environment and sensor device are controlled e.g. by forced ventilation or by minimization of sensor dimension.

The several types of equipment for temperature measurement have different kinds of performance. A thermocouple, for instance, with a very small junction will detect very fast fluctuations of temperature. Alternatively, thermometers with higher dimension will allow an expedite evaluation of average temperatures and simplified outputs which are easier to process. In many situations temperature differences in air are more relevant than the absolute values (Oke 1992) e.g. for applications of aerodynamic methods. As aforementioned data acquisition of temperature differences of about 0.01 is easily achievable by differential measurements ascribing a fixe level of measurements to an absolute temperature threshold.

The measurements of soil temperature are also subjected to errors, mainly due to the heat conduction between soil and wires and cables of temperature sensors. Soil temperature change at a slower rate than air temperatures and heat conduction prevail, since radiative and convective exchanges are minimal in soil. Sensors used in measurements in air temperature are overall applied for soil temperature measurements (e.g., Oke 1992). Special attention should be paid to changes in soil structure when installing the instruments. Ideally, the installation should be horizontal, from a pit excavated at an intended deepness, so that longitudinal conduction and soil moisture flow as well are minimized.

Measurements of surface temperatures (e.g., leaves) can be carried out with very thin thermocouples, although in practice the design of a representative temperature sampling can be a very complex ask. The use of a radiation thermometer is an expedite methodology for surface temperature measurements. This kind of device is based on the measurement of longwave radiation, ranging between 8 μm–14 μm, emitted by the surfaces and detected in its hemispherical view. The radiation detected by the device is composed by longwave radiation emitted from the surface and downward solar radiation which is reflected from the ground. The reflected component can be neglected and only considered the emitted longwave radiation component as follows:

$$T_0 \sim T_k = (L \uparrow / \sigma)^{1/4} \tag{A1.19}$$

where T_o is the real surface temperature, T_k is the measured radiative temperature, $L\uparrow$ the longwave radiation emitted by surface and detected by the instrument and σ the Steffan-Boltzman constant. This methodology has the advantages of avoiding the intrusive contact with the surface and of resulting from an integration of the view area.

The measurement of heat flux in the soil is carried out with a flux plate which is basically a thermopile connected with a fine plaque whose material has a known thermal conductivity. The thermopile records the temperature difference between the two plaque surfaces followed by the application of Fourier Law (Chap. 6) for calculation of soil heat flux. The measurement of soil heat conductivity can also be carried out under Fourier Law principles, by insertion of a soil sample in a heated connected probe subjected to electric heating with the transversal temperature gradient measured with a set of thermocouples.

A temperature sensor provides a response to temperature change which is an exponential time function. If a thermometer at a temperature T_i is immersed in an environment with a temperature T_f, under a step variation, the time variation of measured temperature will be a first-order response system, like:

$$T = T_f + (T_i - T_f)exp(-t/\tau) \tag{A1.20}$$

where T is the thermometer temperature, t is the time instant and τ the time constant defined here as the time necessary for the measured temperature changes 63%, comparatively with the initial temperature. This time constant is dependent on the calorific capacity of the sensor and of the heat transfer rate. Response time wherein the thermometer at a given initial temperature measures the temperature of the adjacent environment can be estimated through Eq. A1.20. For example, if the calorific capacity is high, considering sensors of higher dimensions, the time constant is high and the rate of response adaption to environmental temperature variation is low. The time constant allows also to estimate the attenuation and phasing of temperature fluctuations e.g. through a sine function. If the temperature, T, fluctuations are sinusoidal with amplitude A and frequency w the temperature measured will be given by:

$$T = T_{mes} + \frac{Asin(wt - \phi)}{\sqrt{1 + w^2\tau^2}} \tag{A1.21}$$

where T_{mes} is the average temperature measured by the thermometer and ϕ is the phase angle given by:

$$\phi = tan^{-1}(w\tau) \tag{A1.22}$$

Equation (A1.21) show that although the average measured temperature is the correct average environmental temperature, the measured values are affected by an error dependent of the time constant and frequency of environmental temperature oscillations, increasing with both.

A1.4 Measurement of Radiative Fluxes

The radiation measurement sensors are based on the principle of heating and temperature increase of the receiving surface, caused by radiation (Foken 2017). For absolute measurement sensors, used for calibration, the temperature is measured directly on a dark radiation-receiving surface from radiation irradiated from the sun without any filters. Selective measurement of direct sun radiation allows neglecting longwave radiation.

The radiation measuring devices, globally called radiometers, are generally based on the measurement of the differential heating of an instrument surface exposed to radiation, compared to the part of the instrument that is not exposed (Paw 1995). Another option is based on the temperature difference between two irradiated surfaces, one black and the other white (Foken 2017).

The use of thermopiles by multijunction of thermocouples in series or parallel, in several types of radiometers (pyranometers, pyrradiometers, etc.) is a possibility of transduction or conversion of the radiative flow, in a thermal response and hence in a voltage signal suitable for readable electronics (Oke 1992). The receiving surface of the thermopile is usually coated with a glass, quartz, or polyethylene coating, which works as a protective agent against atmospheric conditions, as spectral filters for the separation of long and short wavelength radiation and as a means of normalizing the transfer of convective heat to the surface of the thermopile, to minimize the effects of wind speed on the energy balance of the instrument. Quartz coverings are only permeable to short-wavelength radiation in the range between 0.3 and 3 μm. Special polyethylene covers (Lupolen) are used for measurements of short and long wavelength radiation (ranges between 0.29 and 100 μm) and silicon covers are used for measurements in the ranges of long wavelengths (between 4 and 100 μm) (Foken 2017).

Radiometers can also be used to measure radiation incident hemispheric radiation at very narrow angles. Radiometers can measure the period in which direct sunlight occurs, for example by differential reading, between a photocell exposed to direct and diffuse solar radiation and another exposed to diffuse solar radiation. When the differences between the readings are significant, a switch is activated that closes a circuit, indicating the presence of direct solar radiation (Paw 1995). Radiometers, which operate according to the heating principle, are based on short and long-wavelength radiation energy balances to which they are subjected (Paw 1995). Basically, the energy balance of the sensor surface, Ri, is:

$$R_i - \varepsilon \sigma T_s^4 = h_c(T_s - T_a) + k(T_s - T_{est}) \qquad (A1.23)$$

where T_s is the sensor temperature, T_a the air temperature, h_c the convective heat transfer coefficient, σ the Stefan-Boltzmann constant, ε the emissivity of the surface, k the thermal conductivity and T_{est} the body temperature of the device. If the sensor has two exposed zones (as in the case of the net radiometer) or two zones on the same surface, with different absorbances, then the energy balance of the Eq. (A1.23) is modified, to obtain the differential temperature of the two different surfaces or zones.

Pyranometers, (Fig. A1.8) measure the total radiation of short wavelength incident on a surface. Pyranometers are devices used to measure solar radiation of short wavelength, incident on a horizontal surface. The respective thermopile is covered by double glass coatings, whose radiative properties are such that, as mentioned above, they only allow the passage of radiation from the range 0.3–3 μm, to the incidence surface.

The incidence surface may be painted with black paint for increasing absorbance. Half of the thermal junctions are connected to small bands whose

Fig. A1.8 Pyranometer for measurement of PAR radiation (400–700 nm)

temperatures fluctuate rapidly according to the variation of incident solar radiation. The other half is connected to the sensor structure, whose temperature varies slowly. The difference between the potentials of the two groups of junctions is related to the reception of short wavelength radiation.

Another possible design of pyranometers is based on alternating contact with black and white painted surfaces (Oke 1992). An inverted pyranometer samples the short-length radiation reflected from the soil surface, thus allowing the surface albedo to be obtained. A pyrradiometer or solarimeter, measures the incident solar radiation, in the hemispheric volume contiguous to the flat surface.

A pyranometer can be transformed into a device for measuring diffuse radiation, by including a ring to induce a shadow on the incident surface by permanent interposition to direct solar radiation. The device will now measure diffuse solar radiation after correcting the diffuse radiation flux intercepted by the ring. Under these conditions, direct solar radiation can be obtained by the difference between total solar radiation, obtained with another instrument without shadow, and diffuse solar radiation. Direct solar radiation can also be measured directly using pyrheliometers focused on the solar disk and measuring solar radiation on a surface perpendicular to the radiative beam (Oke 1992). This value must be multiplied by the cosine of the zenith angle for conversion to the direct solar radiation incident on a horizontal surface (cosine law in Chap. 6). In vegetation canopies, where radiative fluxes vary in space, radiation sampling can be optimized by sensor movement systems (Oke 1992).

The net pyrradiometers (also called net-radiometers) (Fig. A1.9) are the instruments used to measure the balance of total radiation of short and long wavelengths. Its radiation-receiving surface is a dark-flat plate, across which there is a thermopile with two sets of joints connected to the upper and lower surfaces. With the plate aligned in parallel to the surface, the output of the thermopile is related to the temperature difference between the two surfaces, and, in turn, the latter is proportional to the difference between the incident radiation and the radiation reflected by the soil surface at all wavelengths. Convective exchanges, between both surfaces and the environment, also contribute to the observed temperature differences.

Fig. A1.9 Net-radiometers

To cancel the effects of wind differences on the two surfaces, the radiation receiving plate is ventilated at a constant air speed, with a protection of a hemispheric polyethylene shield transparent at wavelengths between 0.3 and 100 μm. Ventilation allows a cancellation of the effects of the differencies of airflow velocity in convected heat exchange between the upper and lower surfaces of the instrument. Ventilation also avoid dew formation in the hemispherical cover which is also protected with a hemispherical polyethylene cover. It is also necessary to remove dust or particles outside the protective cover, as they absorb radiation, reducing the transparency of the polyethylene covers. This type of apparatus is not reliable for measurements in rainy conditions (Oke 1992).

A net-pyrradiometer can be adapted to estimate the radiative balance of long wavelengths. At night, these devices measure the long-wavelength radiative balance becoming a net-pyrgeometer. By day, if the short-wavelength solar radiation balance is obtained from a pyranometer, the long-wavelength radiation balance can be obtained by difference.

Pyrgeometers are radiometers that measure the balance of long-wavelength radiation on a surface (e.g., Paw 1995; Oke 1992; Foken 2017). Pyrgeometers have a silicon coating, due to the transmissivity only of radiation with wavelengths between 4 and 100 μm. The discrimination of incident radiation of long wavelength, in relation to the radiation emitted by the detector, is achieved through an electronic compensation circuit (Oke 1992).

Other pyranometers have been developed using silicon photocells, which are sensitive to photons, with wavelengths in the range of 0.4 μm to 1.1 μm. Silicon photocells, due to their stability and characteristics of the PN junction, are suitable for measurements of solar radiation in the visible range.

Photovoltaic sensors produce a voltage or current, which is a function of the incident radiation flow. Longer wavelengths do not have enough energy to establish

Table A1.1 Accuracy of radiation measuring instruments (Ohmura et al. 1998)

Parameter	Device	Accuracy in 1990	Accuracy in 1990
Global solar radiation	Pyranometer	15	5
Diffuse radiation	Pyranometer with shadow ring	10	5
Long wavelength descending radiation	Pyrgeometer	30	10

Table A1.2 Precision of radiation measurement instruments (ISSO 1990: WMO 2008)

Property	Secondary standard	First class	Second class
Time constant	<15 s	<30 s	<60 s
offset	±10 Wm^{-2}	±15 Wm^{-2}	±40 Wm^{-2}
Resolution	±1 Wm^{-2}	±5 Wm^{-2}	±10 Wm^{-2}
Long term stability	$\pm1\%$	$\pm2\%$	$\pm5\%$
Non-linearity	$\pm0,5\%$	$\pm2\%$	$\pm5\%$
Spectral sensitivity	$\pm2\%$	$\pm5\%$	$\pm10\%$
Temperature response	$\pm1\%$	$\pm2\%$	$\pm5\%$

electron transport in the photocell. The photovoltaic effect is proportional to the number of photons absorbed by the photocell (Paw 1985). Diffusion plates or disks are added to the photovoltaic device so that it is sensitive to desirable radiation ranges (e.g., visible). This type of pyranometer shows substantial errors in cloudy sky conditions, due to calibration needs (Paw 1985; Campbell 1997). A parallel resistance, (shunt) can be integrated into the sensor structure, for the purpose of optimizing its linear response to the incident radiation regardless of its temperature.

In the past 10–15 years, significant progress has been made on the accuracy of radiation sensors. This progress was based on the classification of sensors by the World Meteorological Organization (WMO), standard indicators of error limits (Foken 2017), as well as the publication of the Basic Surface Radiation Network (BSRN) manual with a quality control code for these instruments. In Table A1.1, precision values of different radiation measurement instruments are indicated according to the OMM standards. Table A1.2 shows the quality requirements for pyranometers.

A1.5 Measurement of Relative Humidity

The measurement of atmospheric humidity is another major objective of environmental physics study, with a wide variety of sensors, with different operating principles. Measurement principles are based, for example, on changes in the electrical properties of materials, changes in the physical dimensions of substances due to moisture absorption, cooling of wet surfaces by evaporation, absorption of radiation by water or when measuring the temperature of a surface when dew forms on it. The principles of measuring atmospheric humidity, for the application of the turbulent covariance method, based on the use of infrared analyzers and detectors for analyzing the absorption of radiation by water vapor, were developed in Chap. 3.

Hair hygrometers are based on the principle that moisture causes changes in the length of human hair strands. Human hair varies by about 2.5% in length, when the relative humidity ranges from 0% to 100%. Changes in the dimensions of human hair and other materials are magnified mechanically and transmitted in digital or analog terms. Air length also varies with temperature and fatigue, so calibration is critical with this type of equipment. The time constant for this type of hygrometer is high, in the order of tens of minutes (Campbell 1997).

Methods based on psychrometric relationships are very common. These methods, consisting of the use of psychrometers are fundamentally based on the evaporative cooling of a water surface and measurement of the consequent lowering of temperature, from which the humidity is calculated. In this context, the aspiration ventilated psychrometer is based on the temperature measurements of the dry and wet bulb thermometers. The temperature of the humid bulb thermometer, in conditions of non-saturated air, will always be lower than that of the dry thermometer and by assuring adequate conditions of ventilation and radiation shielding, the calculation of partial pressure of water vapor and other moisture properties, according to the principles mentioned in Annex A2.

For high accuracy demanding operations, such as comparison experiments and calibrations, dew point and frost point hygrometers are commonly used. Dew point hygrometers are expensive devices based on the water condensation on a mirrored surface for direct measurement of the dew point temperature.

The condensation hygrometers use a thermoelectric cooling system to cool the surface until condensation occurs. A radiative beam, reflected from the mirrored surface to a photocell, is indicative of the beginning of the condensation process. The photocell re-emits the signal to the controller of the surface cooling process, so that the mirrored surface remains at the dew point temperature (Campbell 1997). In meteorological networks, capacitive sensors are widely used. Hair hygrometers are still used for temperatures below $0°$ to which other psychrometers do not have enough precision (Foken 2017).

Another very common category of humidity sensors is electrical sensors. These instruments are based on absorption principles, through a volume of material or surface adsorption, which causes variations in electrical properties such as electrical resistance or capacity (Fig. A1.10). The resistance units can be coated with materials such as lithium chloride or aluminum oxide. The capacitive units have a polymer coating with a mesh, which measures the change in electrical capacity, with the adsorption or desorption of water, and are also radiation-shielded. The measurement of variations in resistance or capacity of these sensors, in general,

Fig. A1.10 Capacitive sensor for measurement of air relative humidity and with a radiation shield in installation

Table A1.3 Time constant of temperature and humidity measurement systems

Sensor	Time constant (seconds)
Sonic thermometer	<0.01
Optical humidity measurement system	<0.01
Thermocouples	<0.01
Thermistors	0.1–1
Resistance thermometers	10–30
Liquid in glass-thermometers	80–150

requires a high-frequency excitation in alternating current (Campbell 1997) which can be provided by modern data acquisition devices.

The time constant of temperature and moisture sensors is given in Table A1.3 (Foken 2017).

A1.6 Air Speed Measurement

Air velocity profiles are, mentioned in Chaps. 2 and 3, crucial for the processes of heat transfer by convection and for the development of heat and mass flows in environmental systems. Anemometers are the instruments that are used to measure air velocity, with different types of anemometers, such as cup anemometers (Fig. A1.11), helical anemometers, hot wire anemometers, anemometers of dynamic pressure and, for the application of the turbulent covariance method, the

Fig. A1.11 Cup anemometer

sonic anemometers used mainly for analyzing turbulence structure instead of average winds.

The functioning of the latter has already been analyzed in Chap. 3. Cup anemometers are possibly the most usual for measuring the average horizontal speed (Campbell 1997). The most common versions are based on sets of three lightweight cups, which are mounted on shafts with low friction connections. The wind causes the horizontal rotation of the axis that supports the cups. This rotation can be used to generate discrete voltage pulses or to provide a continuous voltage signal. The rotation speed of cup anemometers, under ideal operating conditions, is linearly related to the wind speed. The friction at the junctions of the instrument induces inertia that stops the rotation of the shaft, at average speeds of the order 0.2–0.3 ms^{-1}, in anticipation of the cancellation of the air velocity (Oke 1992).

The response of the anemometer is measured in terms of distance constant, rather than the time constant (Foken 2008). The distance constant, on the order of 1 m to 2 m, is a measure of the inertia of the cup anemometer. This constant is equal to the wind travel distance downstream of the instrument, necessary for the measured speed value to be within a maximum range of 37%, relative to the airspeed, when there is a step variation of the speed of the ambient air. It can be measured by inserting the immobile cup anemometer in a wind tunnel at a constant speed, (situation representing the step function), with a temporal record of the increase in the speed of the cups. Under these conditions, the distance constant is equal to the product of the period necessary for the cups to reach 63% of their final speed, by the constant airspeed (Campbell 1997).

Propeller anemometers can be used to measure the horizontal air velocity if the device is continuously oriented by a pinwheel in the direction of the prevailing winds. These anemometers are configured in sets of three propellers to obtain measurements of the three wind speed components. Hotwire anemometers are based on an operational device which is a heated wire or a junction, the temperature of which will be influenced by heat exchanges by convection, thus enabling the calculation of wind speed (Oke 1992). These instruments are more useful in

Fig. A1.12 Precipitation measuring device

confined spaces where cups or propellers cannot rotate, such as within crop canopy. These devices are more suitable for measuring the total air velocity than their components.

Wind direction sensors (pinwheels) are second-order sensors, coupled with air velocity measuring devices, capable of detecting variations in wind direction, at speeds of the order of 0.6 ms^{-1}, with a precision of $\pm$ 2°. Modern dataloggers for data acquisition, already allow the necessary connections with sensors with complex electronic circuitry.

A1.7 Precipitation Measurement

Precipitation is another important micrometeorological variable, necessary for accounting for water inputs and comparison with evapotranspiration. Its measurement is performed with buckets (Fig. A1.12), with water collection areas ranging from 200 cm^2 to 500 cm^2 (Foken 2017). A possible operating option is the filling and emptying of the bucket, a rain gauge, and the consequent generation of electrical impulses, which are counted by an appropriate data acquisition device.

A1.8 Data Acquisition Devices

Many sensors are designed for spot and real-time measurements of physical quantities, and the readings of these measures are adapted to the intended purposes. A multimeter is a particular case of an independent reader, prepared to measure only electrical quantities. Independent readings of quantities such as air temperature and humidity or soil respiration are portable solutions in which the measurement processes, signal conversion, possible memory storage with the indication of the measurement time, and reading are integrated, for possible real-time usage. The

Fig. A1.13 Data acquisition equipment

registered information can be viewed either on the reader itself or transmitted to a computer through specific computer links and packages.

The data acquisition equipment (dataloggers) (Fig. A1.13) are complementary equipment to the measurement sensors, configured to allow digitization on a data acquisition board and storage in internal memory, for later reading on a computer reading software. These devices also allow computer programming of measurement rates in predetermined periods of time.

The data acquisition devices allow the processing of all types of electrical signals, such as voltage, current intensity, resistance, pulse count and frequency reading. The reading ranges are the most diverse, depending on the needs of the sensors, and the power supply can be guaranteed by batteries with solar panels in the field or by alternating current. These devices allow reading electrical voltage signals in differential or absolute mode, which are thereafter subjected to processing operations such as amplification, linearization, digitization or storage. The processed information can be sent remotely, or be transfered to a computer. The data acquisition devices also allow the electrical supply of the sensors or the emission of excitation pulses, necessary for their operation (e.g., wind direction sensor).

The equipment for application to the turbulent covariance method (Chap. 3) has its own memory systems for digitization and data storage for post-processing of information. These processes are regulated using specific software for each system.

Annex A2: Basic Topics on Laws of Motion and Evaporation

Abstract

Annex A2 showed topics on laws of movement and evaporation physics intended to guide the reader on embedding in matters presented in the book. Momentum and mass conservation under discrete and infinitesimal approaches, Bernoulli equation and fluid pressure and velocity, buoyancy, boundary layer flows with stagnation points and wake drag around circle sections, streamlined or bluff bodies include the issues discussed. Further analysis is performed on dynamics of evaporation involving concepts such as dry, wet or dew point temperatures, relative and absolute humidity, or interpretation of psychometric charts.

A2.1 Newton's Laws

An elementary approach to the physical principles of Newtonian mechanics, valid both for the movement of bodies and of fluid particles, is useful for understanding common phenomena in environmental physics, such as the atmospheric processes inherent to the vertical flows of mass and energy in the atmospheric boundary layer. The principles of classical mechanics, based on Newton's three laws, are quite acceptable for the normal ranges of velocity and distance despite the most recent and correct approach to Einstein's theory of relativity.

Newton's First Law of motion is as follows: the whole body remains in a state of rest or uniform rectilinear motion unless it is compelled to change that state by external forces applied to it. The tendency of a body to maintain its state of rest or uniform rectilinear movement translates the principle of inertia, so Newton's First Law is called the Law of Inertia.

Newton's First Law is only valid in the so-called inertial reference points. In practical terms, fixed Earth references are considered as inertial references. Strictly speaking, the Earth's rotational and translational movements have small accelerations ($3,4*10^{-2}ms^{-2}$ and $0,6*10^{-2}ms^{-2}$, respectively) whose effect is negligible in short duration movements, and hence Earth can be considered a reference of inertia. Any inertial reference that moves at a constant speed relative to

The Editor(s) (if applicable) and The Author(s), under exclusive license
to Springer Nature Switzerland AG 2021
A. Rodrigues et al., *Fundamental Principles of Environmental Physics*,
https://doi.org/10.1007/978-3-030-69025-0

an inertial reference (e.g. a car or an airplane) is also considered as an inertial reference. The reference systems where the Law of Inertia is not valid are referred to as non-inertial references. An example is that of a speeding car in which a body in its interior such as a cup on a tray, at rest while the car kept constant speed, shifted because of the increase of speed without any force exerting on the car cup (Giancoli 2000).

The concept of relative velocity concerning the motion velocity experienced by the observer, is dependent on his reference frame. The frame usually considered is the earth's ground. If the observer is moving, the relative motion of the observer relative to a given object will be given by the difference between the velocities of the observer relative to the object and of the object relative to the reference frame.

From common experience, we know that a moving body tends to stop, because of the retarding effect of surface friction, even when there is apparently no force inducing such a result. On the other hand, a body that is supported and dropped in free fall moves with increasing speed, due to the force of gravity, even in the absence of any known force that induces such movement. Such phenomena can be explained by the introduction into the balance of external forces of more subtle forces whose existence is not obvious, such as surface friction, gravity, or air resistance. The Law of Inertia applicable as referred to in inertial frames is only strictly valid in an ideal imaginary world without any forces, (e.g., Asimov 1993) even those as indicated, e.g., friction forces, not directly obvious.

The force is a vector quantity and if the addition of the force vectors defining their balance is non-zero then the body starts moving at variable speed in the direction of the resulting force. If this result is null, by Newton's First Law, the body or particle remains in rest or in motion condition with uniform velocity. In the real world, there are always acting forces, at least the force of gravity, and a given body, e.g. a particle, can remain at rest, if the vector sum of the applied forces is null.

Newton's First Law explains the concept of force, but it is not enough to quantify it. For this, Newton introduced the notion of mass to quantify the amount of inertia possessed by the body in question. A larger mass body has more inertia than a smaller mass body. In this context, Newton's Second Law states that the acceleration of a body is equal to the ratio of the balance of forces acting on it and its mass, or rate of change of movement. The direction of movement is in the direction of the resultant of forces.

Newton's Second Law can therefore be written as:

$$a = \frac{\sum F}{m} \Rightarrow \sum F = ma \tag{A2.1}$$

The concepts of mass and weight, although distinct, are sometimes confused in practice in the sense that it can be said that a heavy body has a large mass and vice versa.

The weight of a body represents the force (gravitational) with which the body is attracted to the surface (by the Earth). Mass is a property of the body, its quantity of matter, quantifying its inertia. In the International System units, the mass is expressed in kg and force is expressed in Newtons. The unit of force, Newton, is defined as the force required for causing an acceleration of 1 ms^{-1} to a body with a mass of 1 kg.

If we consider in Eq. (A2.1) a body subjected to a zero balance of forces, one can deduce that the acceleration is null, which translates Newton's First Law of Motion. Newton's First Law may thus be considered as a particular case of the second. Newton's Second Law quantifies the effect of force systems in motion. The next question concerns the provenance of a force. In fact, the existence of one force implies its application by one body against another body. A charging animal pulls a sleigh, a hammer nails a nail, a magnetic body attracts a metal clip, or in the biosphere, a gust of wind causes a ripple of the vegetable coverings. Strictly speaking, the force application of a body or system of fluid particles is not unilateral. In the case of the hammer, it is evident that the hammer exerts a force on the nail, but also the nail exerts a counterforce on the hammer, as the hammer speed cancels out because of the contact with the nail. By Newton's Second Law this hammer braking effect can only be caused by an enough force emitted by the nail.

Newton thus considered that the two bodies should be treated equally, formulating his Third Law of Motion: whenever one body exerts a force on another body, the second body exerts a force, on the former body, which is equal and opposite to the first force.

Newton's Third Law can be enunciated according to the principle of action-reaction: each force of action always corresponds to an equal and opposite force of reaction. The fundamental assumption is that the forces of action and reaction are acting on different bodies. Newton's Third Law explains processes such as that of the movement of a person on the ground in which their movement is initiated by the force exerted by the foot on the ground and the ground exerts an opposite reaction force with opposite sign (friction) on the person, which causes the movement. In fact, a person could not move on ice without friction. Likewise, a bird flies because of the force of reaction of the air on its wings, equal and of opposite signal of the force exerted by the bird on the air. Another example is the movement of an automobile. The automobile motor causes rotation of the wheels and the car's displacement is caused by the friction of the ground in the tires, being the reaction force in opposition to the force of action of the tires in the ground.

The normal tendency is to associate forces with active bodies such as people, animals, motors or a hammer-like moving object, and it is more difficult to imagine how an inanimate object at rest like a wall, a desk, or the ground can exert the same forces. The explanation lies in the occurrence of forces less apparent such as the elasticity of solid materials, existing in a greater or lesser degree, the superficial friction, or the impulsion of fluids.

The Third Law of Motion may lead to the following misconception: if the two forces are equal although of opposite signs, then they cancel each other, and no acceleration of the bodies occurs. In fact, two equal forces of opposite signs cancel each other out when applied to the same body. For example, if two equal forces of opposite directions were applied to the same rock, there would never be any movement of the rock, regardless of the force applied: the forces could cause rupture or pulverization of the rock, without however displacing it, (e.g., Asimov 1993).

The Law of Interaction implies, however, the existence of two equal forces of opposite signs, applied to two separate bodies. If a person exerts a force on a stone, the stone exerts an equal force in opposite direction on the person. In this case, neither force is compensated for: the rock is accelerated in the direction of the force applied by the person, and the person is accelerated in the direction of the force of equal magnitude applied by the rock on the person. If the launching of the stone happens on rough ground, the friction developed between the shoes and the ground introduces new forces to the system that prevent the person's movement. The acceleration is then cancelled, so the true effect of the Law of Interaction is no longer visible. However, if the throw occurs on an icy surface without friction, we will see the person sliding in the opposite direction to that of the throw.

By the same token, the gases formed by combustion in a rocket motor expand, exerting a force against the respective inner walls as they exert an opposite and equal force against the gases, expelling them. The gases are forced to a downward acceleration and by the principle of action-reaction they exert an opposite force in the structure of the rocket, impelling it to an ascending acceleration.

In these examples, the two bodies involved are physically separated or separable, insofar that one of them can accelerate in one direction and the other can accelerate in the opposite direction. If the two bodies are coupled, e.g. a horse pulling a wagonette, all this analysis is apparently more difficult to accomplish.

According to the Law of Interaction, the sleigh will also pull the horse in the opposite direction with equal speed, although it is visible that these two bodies do not accelerate in opposite directions but are together safely evolving in the same direction. If the forces connecting the horse and the wagonette were the only ones existing, there would be no movement. The frictional force exerted by the ground surface (exerted by the Earth) induces the progressive movement of the horse-sleight system. On an ice surface, without any frictional force, neither the horse nor the sleigh would progress.

In real conditions, and in the presence of frictional forces, the sleigh system exerts a frictional force on Earth and this, in turn, causes a frictional force on the sleight system. As a result of this interaction, the sleight system is displaced and progresses in a certain direction, while Earth is displaced in the opposite direction. As the planet has much more mass than the system, we only notice the movement in progression of the system.

These concepts of various forces in interaction in a system are exemplified in Fig. A2.1 exemplified by the effort made by a person to move a sleigh. In fact, it can be verified by the figure that the two forces mainly responsible for the progressive movement of the person-sleigh system are the friction force F_{ps} exerted by the ground on the person and the force F_{pt} exerted by the sleigh on the person. When the ground exerts a force F_{ps} on the person oriented in direction of movement

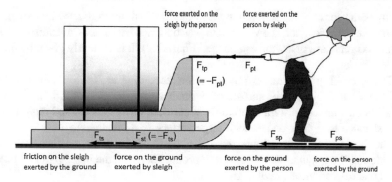

force exerted on the sleigh by the person

force exerted on the person by sleigh

F_{tp}
$(=-F_{pt})$

F_{pt}

F_{ts}

$F_{st} (=-F_{ts})$

F_{sp}

F_{ps}

friction on the sleigh exerted by the ground

force on the ground exerted by sleigh

force on the ground exerted by the person

force on the person exerted by the ground

Fig. A2.1 Balance of forces in action in a system in which a person moves a sleigh (adapted from Giancoli 2000)

of progress, is superior to the opposite force exerted by the sleigh on the person, F_{pt} this moves forward. As for the sleigh, one may also say that it is pulled by the person in the progressive direction of its movement when the force exerted on the sleigh, F_{tp}, is higher than the frictional retardation force, F_{ts}, exerted by the ground on the sleigh. In the case of an absence of friction, e.g. on an ice surface, there would be no displacement of the system (person-sleigh).

A2.2 Force of Gravity

Newton was not only the inventor of the Three Laws of the Movement that support the study of the Body Dynamics, but also elaborated the Law of Universal Gravitation, applied to describe one of the basic forces of nature. This law was further extended to interpret the motion of the Moon around the Earth and the planets in general.

The Law of Universal Gravitation can be stated as follows:

"Each particle in the Universe attracts any other particle with a force that is proportional to the product of its masses and inversely proportional to the square of the distance between them". The force acts along the straight line joining the two particles. The intensity of the gravitational force, $\left|\overrightarrow{F}\right| = F$, can be calculated as (e.g., Giancoli 2000):

$$F = G\frac{m_1 m_2}{r^2} \tag{A2.2}$$

with m_1 and m_2 the mass of the two particles, r being the distance between them and G a universal constant that is obtained by experimental measurement, and the same for all objects and conditions. The distance r is roughly given as the distance between the center of the bodies in question. The value of G must be necessarily small, as the force of attraction between objects of common use, while existing, is in practice undetectable. The value of G was obtained by Henry Cavendish in 1798, about 100 years after Newton established Eq. (A2.2). It is currently given by 6,67 X 10^{-11} Nm^2kg^{-2}.

The force of gravity imposes acceleration, according to Newton's Second Law, of 9.8 ms^{-2} to bodies at the surface of Earth. This acceleration of 9.8 ms^{-2} is valid for all bodies under vacuum conditions, regardless of its masses, and regardless of air resistance to lighter bodies with large surface (e.g., feathers).

The Eq. (A2.2) also allows the calculation of the mass of the Earth, m_T, considering the gravitational force as the product between the mass m of any object, and the acceleration of gravity, g:

$$F = mg = G\frac{mm_T}{r_T^2} \Rightarrow g = G\frac{m_T}{r_T^2}$$ (A2.3)

being r_T the average radius of the Earth, 6.38×10^3 km, giving to the mass of the Earth, m_T, the value of 5.98×10^{24} kg (e.g., Giancoli 2000).

The application of the product between mass and acceleration of gravity is enough to calculate the gravitational force (weight, as abovementioned) exerted on a body at the surface of the Earth. To calculate the force of gravity exerted on a space body far from the planet Earth, or the gravitational force due to that body, we can calculate the effective value of g by applying the appropriate values of r and m. By Eq. (A2.3) the weight of 1 kg of mass on Earth will be 1 kg $\times$ 9.8 ms^{-2} = 9.8 N. The same body will weigh about 1.64 N on the Moon, since its force of gravity is approximately six times smaller.

The force of gravity is a very weak natural force. An Earth-sized body is needed to produce a gravitational force enough to induce an acceleration of 9.8 ms^{-2} along its surface. In practice, we find that relatively weak forces are enough to counteract the force of gravity (e.g., push-ups, high jump, or mountain climbing, to name only cases of physical exercise). For planetary bodies with large masses, compared to Earth, the decline in gravitational force has noticeable effects. While in the case of Earth, the gravitational force is enough to attract the gaseous atmosphere, as we know it, in the case of Mars, a planet with about 1/10 of Earth mass, the respective gravitational force can only attract a much less thick atmosphere. The Moon with 1/81 of Earth's mass does not have enough gravitational force to attract any atmosphere (e.g., Asimov 1993).

Two natural bodies at the surface of the Earth will not exert between each other a significant and visible gravitational force since the product between their masses is an infinitesimal fraction of the product between the mass of the Earth and the mass of any of these objects. On the other hand, we can deduce, applying Newton's Third Law of gravitational interaction, that the Earth has an upward movement relative to the descending body. For practical purposes, and given the gigantic mass of Earth, this movement could be considered null. Eq. (A2.2) allow us to explain that the enormous mass of Earth cancels the effects of forces of reciprocal attraction between bodies at its surface. That is, the Earth attracts all bodies to its surface, without the gravitational interactions between them being noticed, since the Earth is the only present body capable of producing a significant gravitational force to be noticeable.

A2.3 Conservation of Linear Momentum

The concept of momentum or linear momentum is fundamental in environmental physics. The Law of Conservation of Momentum, being strictly associated with Newton's Laws of Motion, is fundamental to approaching the interaction of bodies, such as collisions

The amount of linear motion, $\vec{p}$, or linear momentum of a particle is defined as the product of its mass considered constant by the velocity of the particle:

$$\vec{p} = m\vec{v} \tag{A2.4}$$

and considering Newton's Second Law in the following formula:

$$\vec{f} = m\vec{a} = m\frac{\vec{dv}}{dt} = \frac{\vec{dp}}{dt} \tag{A2.5}$$

we can define Newton's Second Law in an alternative formula: the temporal rate of change of the momentum of a particle is equal to the balance of forces, $\vec{f}$, which act on the particle. By Newton's Second Law, a force acting on a body or particle at rest induces its motion. If the force is constant, then the velocity at a given instant is proportional to the strength of the force multiplied by the length of time during which it is applied.

The impulse designation of a force, I, (units SI Ns^{-1}) is attributed to the product between the force $\vec{f}$, and the time Δt during which the force is applied.

$$\vec{I} = \vec{f}\,\Delta t \tag{A2.6}$$

or by Newton's Second Law:

$$\vec{I} = m\vec{a}\,\Delta t \tag{A2.7}$$

and whereas the velocity variation during the time the force acts, $\Delta v = v_f - v_i$, is the product of the acceleration $\vec{a}$ by the time Δt, it will arrive to the momentum of the force:

$$\vec{I} = m\Delta\vec{v} = \Delta\vec{p} \Leftrightarrow \vec{I} = p_2 - p_1 = m\vec{v}_2 - m\vec{v}_1 \tag{A2.8}$$

i.e. the impulse force on a particle is equal to the change in the momentum of the particle from the end instant t_2 and initial t_1 from the actuation force. On the other hand, from Eq. (A2.5) it follows that:

$$\vec{f}\, dt = \vec{d}\, p \Leftrightarrow d\vec{I} = \vec{d}\, p \qquad (A2.9)$$

allowing to conclude that the infinitesimal impulse of a force is equal to the infinitesimal variation of the linear momentum of the body that underwent this impulse.

According to the principle of variation in the amount of motion of a particle, according to which the impulse of a force in a body (or particle) in a finite interval Δt, is equal to the variation of the particle's linear moment. Generalizing, it can also be said that in a body already in motion, the application of an impulse induces a change in movement equal to the impulse (for example, Asimov 1993).

The linear momentum of a body is a true measure of its motion, insofar as it reflects the principle that it depends on both the mass and velocity of the body in question. The effort required to stop a fast body will be greater than the effort to stop a slow body with the same mass. Likewise, the effort required to stop a heavy body will be greater than the effort to stop a light body at the same speed.

The law of conservation of linear momentum stipulates that the total linear momentum of an isolated system of bodies, that is, a set of bodies not subject to external forces, remains constant. By isolated system we mean a system in which external forces do not act (e.g., Giancoli 2000).

Let us consider a system made up of two bodies (particles) that collide with masses m_1 and m_2 and with linear momentum p_1 and p_2 and p'_1 and p'_2 before and after the collision, respectively. During the collision, assume that the instantaneous force exerted by the body 1 in the body 2 is $\vec{f}$. Simultaneously the force exerted by the body 2 in the body 1, by Newton's Third Law, is $-\vec{f}$. During the collision it is assumed that no other forces exist, or that in this case these forces are negligible. From Eq. (A2.9), relative to Newton's Second Law, and integrating the two members of equality between instants t_1 and t_2, we obtain an equivalent equation relative to a finite time interval:

$$\int_{t1}^{t2} \vec{dp} = \int_{t1}^{t2} \vec{f}\, dt \qquad (A2.10)$$

Applying this equation firstly to body 2, where the force $-\vec{f}$ acts, we have:

$$\int_{t1}^{t2} \vec{dp} = \Delta p = p'_2 - p_2 = \int_{t1}^{t2} \vec{f}\, dt \qquad (A2.11)$$

and then to the body 1, under the action of the force $-\overrightarrow{f}$:

$$\int_{t1}^{t2} \overrightarrow{dp} = \Delta p = p'_1 - p_1 = -\int_{t1}^{t2} \overrightarrow{f} \, dt \qquad (A2.12)$$

so:

$$\Delta p_1 = -\Delta p_2 \qquad (A2.13)$$

and:

$$p'_1 - p_1 = -\left(p'_2 - p_2\right) \qquad (A2.14)$$

or:

$$p_1 + p_2 = p'_1 + p'_2 \qquad (A2.15)$$

confirming the conservation of momentum before and after the collision.

This derivation of conservation of momentum can be generalized to a set of any number of n bodies interacting:

$$\overrightarrow{p} = m_1 \overrightarrow{v}_1 + m_2 \overrightarrow{v}_2 + m_3 \overrightarrow{v}_3 + \ldots\ldots + m_n \overrightarrow{v}_n = \sum p_i \qquad (A2.16)$$

deriving in relation to time:

$$\frac{dP}{dt} = \sum \frac{dp_i}{dt} = \sum f_i \qquad (A2.17)$$

being f_i the balance of forces in the ith body. The forces in presence can be external, exerted by bodies outside the system, and internal forces exerted by bodies, (particles) within the system, between or among themselves. By Newton's Third Law, such forces occur in pairs of equal action-reaction forces and opposite signs, canceling themselves out. Eq. (A2.17) may then be then written as:

$$\frac{dP}{dt} = \sum f_{ext} \qquad (A2.18)$$

where $\sum f_{ext}$ is indicative of the sum of all external forces acting on the system. If this sum is null, then:

$$\frac{dP}{dt} = 0 \qquad (A2.19)$$

and $\Delta P = 0$ or P constant, thus verifying the linear momentum conservation law.

Given the occurrence of outside forces, in practice the Law of Conservation of Linear Momentum does not rigorously occur. The boundary of the system can, however, be modified to include these forces. For example, if we consider as a system a vertical falling body, there is no conservation of momentum, since the force of gravity exerted by the Earth, considered as external, acts on the body causing change in its linear momentum. The system can then be altered to include the Earth so that the total momentum of the new system (Earth + body) remains unchanged, with gravity being considered as the inner force. In this case, we assume a movement of the Earth towards the falling body, whose velocity, as mentioned above, is practically null, given the enormous mass of the Earth.

A2.4 Topics on Fluid Mechanics

A2.4.1 Fundamental Principles

Fluids are substances, liquid or gaseous, which do not maintain a fixed shape, and can flow with greater or lesser ease, because their particles do not occupy fixed positions. Contrary to solids, liquids do not maintain a fixed shape assuming the shape of their container. A liquid such as a solid is not readily compressible and its volume can only be significantly modified by a very significant force. The liquid particles move adjacent to each other, with some internal friction due to their viscosity. For its part a gas does not have a fixed shape or volume, expanding to fill in full and simultaneously, the volume of its container and not from the bottom, as in the case of liquids. Under normal conditions, the gas molecules are very far apart, in the order of about one hundred diameters, moving randomly and quickly within their container space.

Density and relative density are two known physical properties fundamental to the characterization of fluids. The density of a substance, ρ is defined as the mass, m, per unit volume, V, given by the ratio m/V. The density is expressed in kgm^{-3} (units IS). For example, density values (in gcm^{-3}) of liquids such as water at 4 °C or ethyl alcohol are 1 and 0.79, respectively. The values of density of gases such as carbon dioxide or air are 1.98 and 1.29 (in kgm^{-3}) and the values of solids density (in gcm^{-3}) as wood, cork, or steel are of the order of 0.3–0.9, 0.24 and 7.8, respectively. In turn, the relative density of a substance relative to a standard substance is a dimensionless quantity defined as the ratio between the masses of these substances, occupying the same volume. The standard substance normally considered for solids and liquids is water at 4 °C. The standard substance usually considered for gases is air, at the same pressure and temperature as the gas whose relative density is to be determined.

A fluid is in hydrostatic equilibrium when, in macroscopic terms, at any point in its container space there is no accumulation or decrease of particles in the fluid. This situation of macroscopic equilibrium does not occur at the level of atoms and

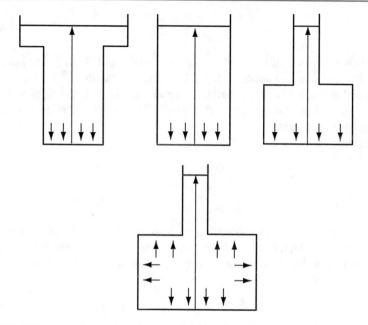

Fig. A2.2 Diagram representative of the distribution of forces inside containers with various configurations (Adapt of Asimov 1993)

molecules wherein rest does not occur. Static equilibrium liquids and gases obey the same basic laws, notwithstanding differences in behavior when the pressure and temperature that affect the gases vary more widely.

A fundamental variable for the characterization of fluids is their pressure, P, defined as the ratio of the total force to the surface of the fluid, on which the force acts perpendicularly. The SI unit of pressure is Pascal, defined in terms of Nm^{-2}. The fluids exert pressure in all directions and the permanence of a fluid at rest implies the equality and symmetry of the forces applied in its interior. This pressure exerted by a fluid at rest is called hydrostatic pressure.

Another important property of fluids at rest is that the force due to the pressure of a fluid acts perpendicularly to a surface in contact with the fluid (such as the wall of its container). Under these conditions the pressure exerted by a liquid depends only on its height level and density, being independent of the shape of the container (Fig. A2.2).

If there is a component of tangential force parallel to the contact surface, considering Newton's Third Law, the surface would exert a reaction force also parallel to the contact surface, resulting in the fluid flow that thus loses their static condition.

According to the Basic Law of Hydrostatics, the hydrostatic pressure in any fluid in a closed container varies according to the height of the fluid point considered:

$$\frac{dP}{dy} = -\rho g \tag{A2.20}$$

where ρ is the density of the fluid at height y. The Eq. (A2.20) is indicative of how the hydrostatic pressure varies with the height within the fluid. The negative sign indicates that the hydrostatic pressure decreases with increasing height in the fluid or increases with increasing depth. This relation is valid for situations in which the density varies with depth.

Integrating Eq. (A2.20) we will have:

$$\int_{P_1}^{P_2} dP = -\int_{y_1}^{y_1} \rho g dy \Rightarrow P_2 - P_1 = -\int_{y_1}^{y_1} \rho g dy \tag{A2.21}$$

Two particular cases for calculating the hydrostatic pressure variation are those corresponding to liquids of uniform density and pressure variability in Earth's atmosphere. For liquids in which density variation can be neglect the integral of Eq. (A1.22) will be:

$$P_2 - P_1 = -\rho g(y_2 - y_1) \tag{A2.22}$$

For the ordinary situation of a liquid in an open container, for example water in a glass or a lake, there is an uncovered surface on top, making it necessary to measure the depth of the liquid from that surface and to consider that the pressure on that surface is the atmospheric pressure P_o. So, we have for Eq. (A2.22), the following expression:

$$P = P_o + \rho g h \tag{A2.23}$$

where the height h is equal to $yy_2 - y_1$, $P = P_1$ and P_2 is replaced by P_o. Figure A2.3 is indicative of the stated principle that the pressure of a liquid (water) in a container is independent of height, so its height will be equal in the four tubes of the left side container, apart from the effects of capillarity (e.g., Asimov 1993). In the right-side

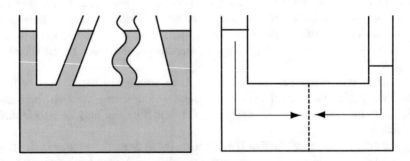

Fig. A2.3 Representative diagram of the pressure distribution of a liquid (e.g., water) in a container with multiple tubes on the left and two tubes separated by a porous membrane on the right-side (adpt. Asimov 1993)

container is an explanation for this principle, according to which the two tubes with distinct heights of liquid are separated by a vertical porous membrane. Consequently, the pressure of the liquid in the left-hand tube will be greater. The pressure gradient, oriented from left to right, will induce the liquid to move through the porous membrane, until the pressure and height of the liquid on both sides are equal.

Equations (A2.2) and (A2.3) are applicable to gases. Since gas density is very small, the pressure differences may be ignored if the measurement levels are not too high. However, if the height value is high, the gas pressure will be of a different order of magnitude and should, therefore, be considered. An interesting example is that of the terrestrial atmosphere, whose pressure at sea level is of the order of 1.013×10^5 Nm^{-2}, or 101.3 kPa, designated as an atmosphere, gradually decreasing with altitude. Considering that the density ρ is proportional to P, it can be written:

$$\frac{\rho}{\rho_o} = \frac{P}{P_o} \tag{A2.24}$$

where P_o is equal to 1.013×10^5 Nm^{-2}, and $\rho_a = 1.29$ kgm^{-3} is the air density at air level at 0 °C.

From Eqs. (A2.20) and (A2.24) we get:

$$\frac{dP}{dy} = -\rho g = -P\left(\frac{\rho_0}{P_0}\right) g \tag{A2.25}$$

$$\frac{dP}{P} = -\left(\frac{\rho_0}{P_0}\right) g dy \tag{A2.26}$$

which gives:

$$P = P_0 e^{-(\rho_0 g / P_0)y} \tag{A2.27}$$

so, that the air pressure in Earth's atmosphere decreases exponentially with height. From Eq. (A2.27), it can also be deduced that the atmospheric pressure decreases by half at 5550 m (Giancoli 2000). In general, high hydrostatic pressure values are referred to the excess pressure, relative to the atmospheric pressure, denominated as relative or gauge pressure. The pressure due to the weight of the Earth' atmosphere is exerted on all bodies present on the earth's surface, so that the bodies present on the earth's surface must withstand the pressure due to the huge atmospheric mass. In the case of a human being the pressure of his/her cells is equal to the atmospheric pressure. In the cases of a balloon or of a tire, the respective internal pressures must, respectively, be a little and a lot (3.2 atm.) higher than the atmospheric pressure.

Earth's atmosphere puts pressure on all the objects it contacts with, including other fluids. The external pressure applied in a fluid is transmitted through it, in addition to the actual weight of the fluid which is transmitted to the rest of the fluid at lower levels, as additional pressure force. In this context, the Pascal principle

Fig. A2.4 Diagram represen-
tative of the application of the
Pascal principle (Adapt de
Giancoli 2000)

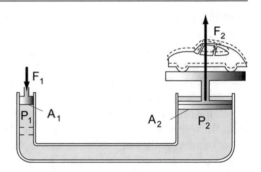

stipulates that the pressure applied to a confined fluid is transmitted integrally to the
entire volume of fluid and perpendicularly to the walls of the container as well.

For example, the pressure due to water at a depth of 200 m below the surface of a
lake, considering P_o as the atmospheric pressure at the lake surface, is by
Eq. (A2.24):

$$P = P_o + (1000 \text{ kgm}^{-3}) (9.8\text{ms}^{-2}) (200\text{m}) = P_o + 19.6 * 10^5 \text{ Nm}^{-2} = P_o + 19.4$$
atm = 20.4 atm.

Pascal's principle is susceptible of many practical applications, of which the
hydraulic jack is an example. In this mechanical device, a small force exerted on a
small section inlet piston is converted into a larger force exerted on the larger
section outlet piston (Fig. A2.4). If the two pistons are at the same height, by the
Pascal principle the force applied on the intake piston induces an increase in
pressure which is propagated homogeneously throughout the system, so that:

$$P_{out} = P_{in} \Leftrightarrow \frac{F_{out}}{A_{out}} = \rightleftarrows \frac{F \text{ in}}{A_{in}} \qquad (A2.28)$$

or:

$$\frac{F_{out}}{F_{in}} = \rightleftarrows \frac{A_{out}}{A_{in}} \qquad (A2.29)$$

The fraction on the left side of the equality of Eq. (A2.29) is called the
mechanical advantage of the hydraulic jack, being equal to the ratio of the sections.
If the sectional area of the outlet piston is, for example, 10 times greater than the
sectional area of the inlet piston, the outlet force is 10 times greater than the input
force. For example, an input force of 150 N can raise a body of 1500 N. This is the
effect by which the larger weights can be lifted by a much smaller force.

In this hydraulic system the work of the input and output forces remains
constant. To understand this proposition, suppose a small piston with a section of 1
cm^2 to which a force capable of causing a displacement of 1 cm of the level of
hydraulic fluid is applied. The volume of fluid displaced is therefore 1 cm^3. The
output piston, whose section is assumed to be 10 cm^2 can only move upwardly over

a distance enough to allocate a volume of 1 cm^3 of fluid. The distance travelled will therefore be 1 cm^3/10 cm^2 = 0.1 cm.

The force on the output piston was increased 10 times, but the path through which this force was applied was reduced by 1/10. The total work (given by the product of force by distance) obtained from the hydraulic system remained constant (e.g., Asimov 1993).

A2.4.2 Buoyancy

Bodies submerged in a fluid appear to weigh less than outside it. A huge rock, which under normal conditions can hardly be moved, is more easily lifted if it is underwater. When the moving rock reaches the water surface it appears to weigh more. Other materials, such as wood, may float on liquid's surface. In each of these situations, there is an upward force, buoyancy, inherent to the liquid, which opposes the downward gravity force.

Buoyancy occurs because the pressure of a fluid at rest increases with depth. In this way, an upward pressure force exerted by the fluid on the lower surface of a submerged solid body is greater than the downward force exerted on the upper body surface. The buoyancy is the upward force resulting from the balance of these two forces. For example, for a cylindrical body of height h, with the upper and lower heights h_1 and h_2, a cross sectional area A and subject to pressures P_1 and P_2, the resulting buoyancy F_I, from $F_1 = P_1A$ and $F_2 = P_2A$, on the upper and lower surfaces, respectively, becomes:

$$F = F_2 - F_1 = \rho_f g A(h_2 - h_1) = \rho_f g A h = \rho_f g V = m_f V \qquad (A2.30)$$

where V is the volume of the cylinder. The term $m_f V$ corresponds to the weight of the fluid with a volume equal to the volume of the cylinder. In this way, buoyancy is equal to the weight of fluid displaced by the immersed body.

Generally, submersion of any irregular solid body in a liquid contained in a vessel causes a displacement of an equal volume of liquid that rises to a level suitable with the displaced volume. It follows that the immersed body exerts a downward force enough to compensate for the weight of the displaced liquid and that, by Newton's third law, the liquid will exert an upward reaction, equivalent to the weight of the same volume of liquid. The force of impulsion, exerted on a body submerged in a fluid, is equal to the weight of the fluid, displaced by that body (Archimedes Principle).

The weight of the submerged body is equal to the product of its volume V and density D. The weight of the displaced liquid is equal to the product between its volume (equal to that of the submerged body) and its density d_1.

The weight of the body after submersion, W, is equal to the original weight, minus the weight of the displaced water:

$$W = VD - Vd_1 \qquad (A2.31)$$

whence the density of the submerged body D, is given by:

$$D = (W + Vd_1)/V \qquad (A2.32)$$

Using Eq. (A2.32), the density of the body can be calculated as both weight of the submerged body and the volume of liquid displaced as well as the density of the liquid. If an immersed body has a density greater than that of the fluid in which it is immersed, then D is greater than d_1 and VD greater than Vd_1, so it submerges in the liquid which it is immersed.

Figure A2.5 is representative of buoyant force in a hypothetical fluid cube, where it is shown that buoyance is manifested when the vertical force F_1, exerted on the lower horizontal surface is greater than the vertical force F_2, exerted on the lower horizontal surface. The difference in the intensity of these forces is because the height of liquid above the lower horizontal surface is greater than the height of liquid above the upper horizontal surface.

If an immersed body has a lower density than the fluid in which it is immersed, then D is less than d_1 and VD less than Vd_1, the body will float in the liquid. A solid body less dense than the surrounding fluid will float partially submerged on the liquid's surface under conditions where the weight of the displaced fluid is equal to its original weight: its weight in water is then zero and the body neither rises nor falls. An object only floats in a fluid if its density is lower than that of the fluid.

Air is also a fluid that exerts buoyancy, although due to its low density, this effect on solid bodies is small. Normal bodies weigh less in the air then in vacuum. Helium balloons float in the air because the density of helium is lower than the air.

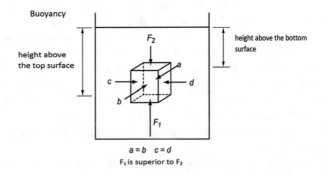

Fig. A2.5 Representative diagram of the balance of forces on the surfaces of a hypothetical liquid cube in a container. (a, b, c, and d, are pressure forces exerted on the vertical surfaces) (after Asimov 1993)

A2.4.3 Conservation of Mass and Motion Quantity

Fluid motion, or fluid dynamics, is a complex process that can be simplified by assuming the incompressibility of the fluid and the stationarity of the flow.

Two key forms of fluid movement are laminar and turbulent flow. Under laminar flow, the fluid layers slide smoothly over each other, the fluid particles move along streamlines, in an orderly manner, without overlapping. Under such flow, there is some energy dissipation due to the internal friction arising from viscosity. Turbulent flow, common in natural environments, is characterized by erratic, random, and circular motion in the form of swirls or turbulent vortices. These vortices dissipate energy in a far greater amounts than viscous dissipation in laminar regime.

A key principle in fluid dynamics calculations is that of mass conservation, according to which the rate of temporal accumulation of fluid mass within a control volume is given by the difference between the mass entering and leaving from that volume per time unit.

When the incompressible laminar flow is stationary through a tube of variable size (characteristic dimension = Δl_1), the mass flow rate $\Delta m_1/\Delta t$, in the larger input section A_1, is:

$$\frac{\Delta m_1}{\Delta t} = \frac{\rho \Delta V_1}{\Delta t} = \frac{\rho A_1 \Delta l_1}{\Delta t} = \rho A_1 v_1 \tag{A2.33}$$

where the volume $\Delta V_1 = A_1 \Delta l_1$ is the volume of the mass Δm_1, ρ is the density of the fluid and v_1 is the velocity of the fluid in the section A_1. As there is no lateral loss of fluid and the flow occurs at constant density (incompressible flow), the mass flow rate in the outlet A_2, is equal to the outlet mass flow rate:

$$A_1 v_1 = A_2 v_2 \tag{A2.34}$$

This equation shows that when the sectional area is large, the velocity is small and when the sectional area is small the velocity of the fluid is greater.

Since:

$$Av = A\Delta l/\Delta t = \Delta V/\Delta t \tag{A2.35}$$

where Av represents the volumetric flow rate (expressed in m^3/s).

The principle of mass conservation can be formulated generically by differentiation:

$$\left.\frac{dM}{dt}\right)_{sistema} = 0 \tag{A2.36}$$

where $M_{system} = \int\limits_{\substack{massa \\ (system)}} dm = \int\limits_{\substack{volume \\ (system)}} \rho dV$

applying the equations to the system and control volume (e.g. Fox and McDonald 1985) an integral form for a control volume comes as:

$$\left.\frac{dM}{dt}\right)_{system} = 0 = \frac{\partial}{\partial t} \int_{CV} \rho d\Lambda + \int_{CS} \rho \vec{V} \cdot \vec{d}A \qquad (A2.37)$$

where CV is the designative term of the control volume delimited by the control surface CS, $\vec{V}$ the velocity vector, $d\Lambda$ an infinitesimal element of the control volume, and $\vec{d}A$ the infinitesimal vector perpendicular to an arbitrary infinitesimal area A, sampled on the control surface. The first term of the equation, on the right side, represents the rate of change of flow within the control volume, and the second term, on the right-hand side, quantifies the balance between the input and output of mass of the system on the control surface considered, expressed in terms of vectoral product.

According to the principle of mass conservation, the sum of the rate of change of mass within the control volume of the system be equal and of a sign contrary to the balance of exchanges through that volume. That is, the rate of change of mass within the control volume of the system is equal to the time variation of the input-output balance across the control surface.

Assuming an incompressible and stationary flow, Eq. (A2.37) can be simplified to:

$$\int_{CS} \rho \vec{V} \cdot \vec{d}A = \rho \int_{CS} \vec{V} \cdot \vec{d}A = \int_{CS} \vec{V} \cdot \vec{d}A = 0 \qquad (A2.38)$$

Equation (A2.38) is the so called the continuity equation. Eq. (A2.37) is a case of a more general equation related with the change of an arbitrary extensive property N, such as linear momentum, within a control volume in the form:

$$\left.\frac{dN}{dt}\right)_{system} = \frac{\partial}{\partial t} \int_{VC} \eta \rho dV + \int_{CS} \eta \rho \vec{V} \cdot \vec{dA} \qquad (A2.39)$$

where the term in the left side is the total change of any extensive property N of the system, the first term of the right side is the time rate change of N within the control volume with η being the value of N per unit of mass, and the second term of the right side is the rate of the efflux of the extensive property through the control surface.

The principle of conservation of momentum is also applied to fluid dynamics. From a theoretical point of view, a fluid occupying a continuous volume is subject to surface forces acting at the boundary of the surface by direct contact and volume forces, distributed throughout the volume. Examples of volume forces are gravitational and electromagnetic forces.

As we mentioned earlier, Newton's 2nd Law establishes that in a moving system, the sum of all forces acting on a system is equal to the rate of time variation of the linear momentum of the system. The principle of conservation of the quantity

Fig. A2.6 Schematic show-
ing static pressure distribution
and flow velocity in horizontal pipes of homogeneous
section (upper) and variable
(lower) (after Asimov 1993)

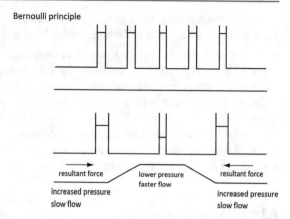

Bernoulli principle

resultant force | lower pressure | resultant force
faster flow

increased pressure | increased pressure
slow flow | slow flow

of linear momentum can be formulated, in a generic way, on a differential basis
identical to that presented for the conservation of mass, that is:

$$\overrightarrow{F} = \left.\frac{d\overrightarrow{P}}{dt}\right)_{system} \tag{A2.40}$$

where the linear moment, or quantity of movement of the system $\overrightarrow{P}$, is given by:

$$\overrightarrow{P}_{system} = \int_{\substack{mass \\ (system)}} \overrightarrow{V}\,dm \tag{A2.41}$$

where in the resulting force $\overrightarrow{F}$, includes all forces and surface $\overrightarrow{F}_S$ and volume, $\overrightarrow{F}_V$:
Whence,

$$\overrightarrow{F} = \overrightarrow{F}_S + \overrightarrow{F}_V \tag{A2.42}$$

For an infinitesimal mass system dm, Newton's 2nd Law can be expressed by:

$$\overrightarrow{d}F = dm\left.\frac{d\overrightarrow{V}}{dt}\right)_{system} \tag{A2.43}$$

Equation (A2.43) can be generically developed for a fluid particle of infinitesimal
mass, dm:

$$d\overrightarrow{F} = dm\frac{d\overrightarrow{V}}{dt} = dm\left[u\frac{\partial\overrightarrow{V}}{\partial x} + v\frac{\partial\overrightarrow{V}}{\partial y} + \omega\frac{\partial\overrightarrow{V}}{\partial z} + \frac{\partial\overrightarrow{V}}{\partial t}\right] \tag{A2.44}$$

with the summation term in the ssquare brackets being relative to the particle acceleration, and u, v and ω the time derivatives, according to the three coordinate axes, x, y, and z of the components of the velocity vector, $\vec{V}(x, y, z, t)$. The term included in the right parenthesis relating to acceleration of the fluid particle includes a convective component corresponding to the first three terms on the left side of the summation and a local acceleration component, as the velocity field may also be a function of time. A particle can thus be accelerated by convective transport or by local effects because the flow may not be stationary. Equation (A2.44) is the basis of the Navier-Stokes equations needed to describe Newtonian fluid movement in the laminar regime, referred to in Chap. 3.

A2.4.4 Bernoulli Equation

The Bernoulli principle refers to steady flow in ideal (non-viscous and incompressible) fluids, basically stating that if the fluid velocity is low, its pressure is high, whereas if its velocity is high, the pressure is low. It can be written as:

$$P + \frac{1}{2}\rho v^2 + \rho g h = \text{constant} \tag{A2.45}$$

where h, v and ρ are the height, velocity, and density of the fluid, respectively. The first term on the left side of the equality in Eq. (A2.45), P, is the static pressure. The second and third terms to the left of the equality, are the dynamic pressure and the hydrostatic pressure, respectively.

Static pressure is the pressure of the free flow, measured by a sensor in motion with the fluid, which is, in practice, difficult to do. The dynamic pressure of the fluid corresponds to its kinetic energy per unit volume and the hydrostatic pressure is the pressure of the fluid at rest, due to the force of gravity. The sum of the static pressure with the dynamic pressure is called the stagnation pressure, corresponding to the pressure exerted when a moving fluid is decelerated by an obstacle, to a zero speed, via a frictionless process. A device used to measure the speed of a fluid flowing in a tube is the Pitot tube. The Pitot tube, inserted into the tube where the fluid flows, allows for the simultaneous measurement of the stagnation pressure, p_o, and the static pressure, defined above, after which the velocity of the fluid, v, can be calculated, using the Bernoulli equation.

It follows that:

$$v = \sqrt{\frac{2(p_o - p)}{\rho}} \tag{A2.46}$$

The measurement of the static pressure of a theoretical fluid without viscosity, moving in a horizontal tube, is based on the principle, which can be demonstrated, that in a flow with horizontal and parallel current lines there is no pressure variation in the direction perpendicular to the flow (e.g. Fox and McDonald 1985). The measurement can then be made through a small hole, inserted in the wall of the tube, with the axis perpendicular to the surface of the tube.

Bernoulli's principle is best illustrated with an example (Asimov 1993). In a column of water flowing over a horizontal tube of constant diameter (Fig. A2.6), water moves at the same rate at all points. Water is under static pressure (otherwise it would not flow) and the pressure is uniform in the pipe. This can be proved using a horizontal tube drilled at several points and with vertical tubes inserted in each outlet. In this condition, the water would rise at the same height in each tube.

The flow of water in a horizontal pipe of variable cross-section with a smaller diameter zone, located in a constricted area of the pipe, and the diameter of the rest of the pipe is similar to that of the pipe in the upper Fig. A2.6. As it is not possible for water to accumulate in any section of the tube, a given volume of water would have to pass through the smaller diameter tube area, in a time interval equal to what would be required to pass through an equal length of a tube of normal diameter. For the volume of water to cross the smaller and normal diameter zones in the same time interval, its velocity must increase as it enters the smaller diameter zone. The result of the increase in speed due to the increase in pressure in the fluid coming from the zone of normal diameter, represents a reduction in pressure in the zone of smaller diameter and, consequently, a smaller rise in the water.

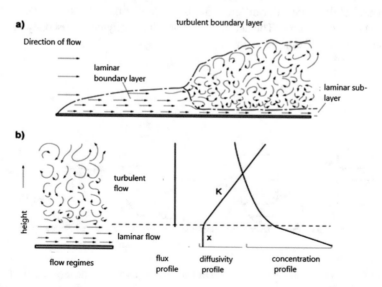

Fig. A2.7 Representative diagram of the development of the transition from laminar flow to turbulent flow in flat plate (**a**), and **b** simultaneous vertical variations of flows and concentrations (adapted from Oke 1992)

The pressure difference induces a force gradient directed towards the smaller diameter zone, inducing an acceleration (increase in velocity) of the volume of water. In turn, when the fluid enters the larger diameter zone, there will be a decrease in the flow velocity and a pressure difference, now oriented in the opposite direction to the displacement of the water, justify the decrease in its velocity and, consequently an increase in pressure.

A2.4.5 Viscosity and Newtonian Fluids

Under real conditions fluids exert friction or viscosity due to resistance of the fluid during laminar flow. Fluids (e.g. oils, water or gases) display viscosity in varying extents. To visualize viscosity, suppose that a thin layer of fluid is placed between two infinite smooth plates, whereby the upper plate moves with a movement, δu, and the lower plate is immobile. The layers of fluid in contact with each of the plates adhere to their surfaces due to forces between the molecules of the fluid and the molecules of the plates. In this way, the upper fluid layer moves with the same velocity of the upper plate and the lower fluid layer remains immobile.

The lower layer of fluid exerts a retarding effect on the top of the adjacent layer of fluid and this process is repeated until the layer contacts the uppermost layer. The velocity of the fluid will vary from 0 to δu, according to a gradient $\delta u/l$, where l is the distance between the plates. A force, F is needed to move the top plate, so that for a given fluid, the required force is proportional to the area of fluid, in contact with each plate, A, at velocity δu and inversely proportional to the separation distance, l of the plates. For different fluids, the higher the viscosity, the greater the plate displacement force. The coefficient of absolute or dynamic viscosity of a fluid μ, can be defined from the equation:

$$F = \mu A \frac{\delta u}{l} \qquad (A2.47)$$

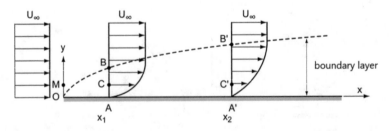

Fig. A2.8 Schematic representation of viscous laminar flow above an infinite plate (after Fox and McDonald 1985)

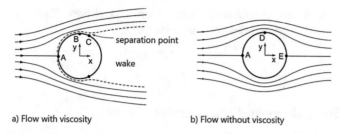

a) Flow with viscosity   b) Flow without viscosity

Fig. A2.9 Representative diagram of the flow around a circular section (adapted from Fox and McDonald 1985)

Viscosity in SI units is expressed by Nsm^{-2} (newton-second per square meter) or Pa.s (pascal-second). Typical values of the absolute viscosity coefficient in Pa.s are 1×10^{-3} for water at 20 °C, 4×10^{-3} for blood at 37 °C 200×10^{-3} for automotive oil at 30 °C, and 1500×10^{-3} for glycerol (Giancoli 2000). Gases, such as air or water vapor, typical have lower viscosity of about 0.0018×10^{-3} or 0.0013×10^{-3}, respectively. The viscosity of liquids decreases with increasing temperature.

The force, F, applied to the upper plate, directly proportional to the viscosity of the fluid, is a tangential force to the fluid layers and, therefore, normal to the perpendicular to the surface of the plates. This induces a tangential deformation of the fluid, characterized, as mentioned, by the tangential movement of some layers of the fluid relative to the others.

This tangential deformation enables fluids to adopt the geometry of the container. A fluid can be defined as a substance that is continuously deformed under tangential force or tension, regardless of its magnitude. The fluid is said to be Newtonian when due to fluid's viscosity, the tangential stress is directly proportional to the velocity gradient (Eq. A2.46). Tangential stresses, perpendicular to normal or pressure forces or stresses, are also referred to as shear stresses.

To the shear stress, F in the present case, a notation of type τ_{yx} can be applied, in which the first subscript y refers to the plane in which the tension τ acts, corresponding to the direction normal to the plane, and the second subscript corresponds to the direction of the tension acting.

Equation (A2.47) can be written in a differential form, per unit area of surface:

$$\tau_{yx} = \mu \frac{du}{dy} \tag{A2.48}$$

The tangential deformation of the fluid is angular. It can be shown that the angular deformation rate, $d\alpha/dt$ is equal to the vertical velocity gradient du/dy (e.g. Fox and McDonald 1985). Concerning flow over smooth plates, Monteith and Unsworth (2013) refer the frictional force per unit surface, τ, as:

$$\tau = 0.66\rho V(Vn/l)^{0.5} \tag{A2.49}$$

with l being the length of surface exposed to airflow. The corresponding resistance to momentum transfer is expressed as:

$$\tau_M = 1.5V^{-1}Re^{0.5} \qquad (A2.50)$$

A2.4.6 Streamlines and the Boundary Layer

The streamlines are the tangential lines, in each moment of time, to the flow direction. Since the current lines are tangential to the velocity vector of each fluid particle, flow perpendicular to the streamlines cannot occur. In steady flow, the velocity at each point remains constant over time, so the streamlines also do not vary between consecutive time instants. This means that particles passing at a given fixed point in space belong to the same streamline, or that they are under stationary flow conditions, e.g. the fluid particle remains in the same streamline (e.g. Fox and McDonald 1985).

The boundary layer concept, in laminar flow, is readily derived from the analysis of tangential tensions in fluids. The concept of the laminar layer limit is a simplification of the real conditions existing in the physical conditions of the environment, under which the turbulent flow, of a more complex nature predominates (Fig. A2.7), which dynamizes the vertical flows of mass and energy.

The transition between laminar and turbulent flow can be analysed in terms of the ratio of the inertial forces associated with the horizontal movement of the fluid to the viscous forces generated by molecular interaction. This ratio (see Chap. 3), the Reynolds number Re, defined as Vd/v, where V is the velocity of the fluid, d is the characteristic dimension of the system, and v is the kinematic viscosity coefficient. This coefficient is defined as the ratio of the dynamic or absolute viscosity to the density of the fluid. Under flow on a flat plate, the characteristic dimension is the distance in abscissa, from the point of fluid considered, to the point of contact between the fluid and the plate. The critical values of Re, for the transition between the laminar and turbulent flow zones, are about 2×10^3.

Suppose, then, a fluid, moving on a flat immobile plate, approaching it with a uniform velocity U_∞. In this case, the fluid layer adjacent to the plate, due to adhesion to its surface, remains immobile, and Eq. (A2.47) remains valid.

Therefore, even though the speed of the fluid on the stationary surface of the plate is zero, there is fluid flow, with velocity gradients and tangential shear stresses. In this case, the slowing effect of sliding fluid layers induces a vertical increase in fluid velocity. At a certain point, located vertically, this retarding effect caused by the surface is no longer present, with the fluid speed being equal to that existing in the previous region, to the fluid contact point with the V_∞ plate.

The fluid velocity, in contact with the immobile plate in the vertical direction $0 \leq y \leq y_B$ varies, therefore, in the range $0 \leq v \leq V_\infty$. The vertical height, influenced by the retarding effect of the plate on the flow of the fluid, increases with the distance to the point of contact between the free fluid and the flat surface (Fig. A2.8).

Fluid flow comprises two distinct regions. One is the boundary layer adjacent to the surface plate, where tangential shear stresses occur. As mentioned before, the height of the boundary layer increases with distance to the point of contact. The outer zone is located above the boundary layer. In this latter zone, the vertical velocity gradient is zero, there are no tangential stresses, and viscosity is not used in the study of laminar flows.

Another important problem concerns stationary, laminar, and incompressible flow around circular sections of solid bodies, such as cylinders, in which both viscous and pressure forces are relevant (Fig. A2.9).

In this case, the streamlines are symmetrical around the circular section while the central line collides with the circular section at point A, divides and bypasses the section. Point A is known as the stagnation point. As in a flat surface flow, a boundary layer is formed in the circular section by the action of the viscosity. The velocity distribution, around the section, can be evaluated by the spacing between the streamlines. If the streamlines are more compacted, because there is no flow between them, the fluid velocity is greater. On the contrary, the fluid velocity will be lower if the current lines are farthest from each other. If the flow is inviscid (fluid without viscosity), then the streamlines are symmetrical relative to the circular section. The speed around the section increases to a point D (Fig. A2.9b) where the streamlines are more compacted, then decreasing as the fluid bypasses and deviates from the section.

According to Bernoulli's principle the speed increase occurs simultaneously with a pressure decrease whereas a decrease leads to an increase in pressure. Thus, in the case of a steady and incompressible inviscid flow, the pressure along the surface section decreases from point A to point D and again increases up to point E. In this ideal flow, a boundary layer is not considered because of the absence of viscosity, and the pressure and fluid velocity fields are symmetrical along section, with no pressure gradient likely to exert a dragging effect on the section. Since the pressure increases again, in the zone posterior section of the section beyond point B, it is natural that the particles in that zone of the boundary layer, experience a balance of pressure forces, in the opposite direction to their movement. From a point C, called the separation point, the fluid inside the boundary layer is brought to rest and separated from the surface. The separation of the boundary layer results in a low-pressure zone at the back of the section, called a wake, with effects of local fluid recirculation.

Under real flow with viscosity (Fig. A2.9a) data suggests that the boundary layer between points A and B is very thin, and it can be assumed that the streamlines above the boundary layer and the consequent distribution of pressures are like that of inviscid flow. As the pressure increases again, in the posterior zone of the section beyond the point B, the particles in that zone of the boundary layer are subject to opposing pressure forces, in the opposite direction to their movement. From the separation point C, called the separation point, the fluid inside the boundary layer is brought to rest and separated from the surface. The separation of the boundary layer results in a low-pressure zone at the back of the section, called as wake, with local fluid recirculation effects.

This dragging effect will be all the greater the larger the wake, so processes such as thinning the geometry of the section of the bodies, tend to delay the separation effect and reduce the formed wake and the consequent dragging force. This effect is also known as form drag because it depends on the shape and orientation of the body. In this way, a pressure gradient oriented in the direction of flow occurs, tending to drag the solid body.

Maximum form drag occurs in surfaces at right angles to the fluid flow, corresponding to bluff bodies matching the maximum pressure that a fluid can exert in contributing to the total form drag on the body. From Eq. (A2.39), the rate of moment transfer, or efflux through a control surface, from the fluid to a unit area of a perpendicular surface of the body is $0.5 \, \rho V^2$. This quantity is the total form drag over the immersed body considering a mean velocity of $V/2$ that exists if there is a decrease of velocity from V, under normal flow, to 0 when flow is stopped at the stagnation point in the body surface. In bluff bodies, the fluid tends to slip around their sides so that the form drag force in the upstream face is lower than $0.5\rho V^2$ (e.g. Monteith and Unsworth 2013).

The real form drag over a unit area will be given by $c_d \, 0.5\rho V^2$, where c_d is the total drag coefficient which is related to the combined form and skin drag. This coefficient ranges between 0.4 and 1.2 for spherical and cylindric bodies under Reynolds numbers between 10^2 and 10^3. For surfaces parallel to the air stream the diffusion of momentum in skin friction is analogous to the diffusion of gases molecules and heat and water vapour. For surfaces perpendicular to airflow, there is no frictional drag in the direction of flow and the similitude will occur only between r_{av} and r_{aH}.

A2.5 Topics on Evaporation Physics

A2.5.1 Fundamental Principles

Atmospheric air usually contains vapor from the evaporation and a key concept in environmental physics is the air relative humidity. It is defined as the ratio of its vapor pressure, e, and the vapor saturation pressure at the same temperature, $e_s(T)$. The relative humidity of the air (in analogy with saturation deficit) is a measure of the drying capacity of the air. In a free water surface, the equilibrium condition of vapor exchanges between that surface and the adjacent air, corresponds to the saturated air with unitary relative humidity. In a situation of contact between air and a porous body, such as soil, wood, or a saline solution, a lower air relative humidity exists under equilibrium of the vapour exchanges (Monteith and Unsworth 1991).

Other concepts used for the characterization of atmospheric humidity (Chap. 4) are the specific humidity, q, defined as the mass of water vapour per unit mass of humid air and the absolute humidity χ, defined as the mass of vapour of water per unit volume of humid air.

To analyse the dynamics of the evaporation process, consider the simple case of an open cup with water, the level of which level falls during the night because of evaporation, due the liquid molecules changing into the vapour phase. This process can be explained in terms of kinetic energy.

The molecules of a liquid move relative to one another in a totally random fashion. The water molecules remain in the liquid state due to attractive forces, which keep them in this state. A molecule in the surface area of the water, with a certain velocity, may momentarily leave the liquid and then eventually be recaptured by attractive forces of other molecules of water if its velocity is not excessive. Otherwise, the molecule will escape from the liquid medium into the gas phase. Only molecules with appropriate velocity and kinetic energy will escape into the gaseous phase, due an increase in the ambient temperature. This confirms the observable reality that evaporation increases with temperature.

As higher velocity molecules escape into the atmosphere, the average velocity, kinetic energy, and temperature of the remaining liquid molecules will decrease. It can be thus anticipated that evaporation is a process of internal cooling of the system. In the case of water, the temperature drop caused by evaporation is due to the release of the latent heat of vaporization, L, for example 2.44 MJkg^{-1} at 25 °C. An example of this is the cooling sensation experienced by a person after intense perspiration followed by a slight breeze, or after a hot shower.

Consider a capped bottle, partially filled with water under vacuum. The water molecules with the highest velocity will migrate to the gas phase. Some of them, due to their random and disorderly movement, collide with the surface of the liquid and are reintegrated into the liquid phase, via condensation process which is the reverse of evaporation. The number of water-vapor molecules thus increases until the number of molecules of water that evaporate is equal to the number of molecules of water-vapour that condense, wherein under these conditions, the vapour pressure is the saturation pressure.

The saturation vapor pressure does not depend on the volume of the container. If the volume above the liquid is suddenly reduced, the density of the molecules in the vapor phase temporarily increases, as well as the intensity of impacts/shocks of these molecules on the surface of the liquid until a new equilibrium state is reached, occurring at the same pressure and temperature. The vapor pressure depends on the ambient temperature. Under higher temperatures, more molecules have enough kinetic energy to separate from the liquid phase into the vapor phase, so that equilibrium is reached at a higher vapor pressure.

In real situations, the evaporation of liquids like water occurs in the atmosphere and not in a vacuum. As in a vacuum, the equilibrium is reached when the number of molecules that evaporate is equal to the number of molecules that condense.

In the case of water vapor, the number of molecules in the condensation and evaporation is not affected by the presence of air, although collisions with air molecules can delay the time required to reach a new equilibrium. So, the equilibriums for evaporation in the atmosphere and in the vacuum will occur at the same vapor pressure. If the water container is large or open, the water may evaporate completely without reaching the saturation of the surrounding air (e.g., Giancoli 2000).

The dryness of the ambient climate depends on the moisture content of the air. In the atmospheric air, there is a mixture of gases, including water vapor, and the total (atmospheric) pressure is the sum of the partial pressures of the constituent gases. The partial pressure of each gas is equivalent to the total pressure that gas would exert if it were isolated in the same control volume. The partial pressure of the water vapor can thus vary from zero to a maximum value equal to the saturation pressure of water vapor at the same temperature. For example, the water vapor saturation pressure is 3.17 kPa at 25 °C.

If the partial pressure of water vapor exceeds the saturation pressure, as it can happen in night cooling conditions, the atmosphere is said to be supersaturated and the excess water vapor condenses in the form of dew (or under peculiar physical conditions such as fog) or precipitation). When a portion of humid air, in isolation conditions, is cooled, the temperature will drop to such an extent that the partial pressure of water vapor is equivalent to the saturation pressure. This temperature value is known as the dew point.

The measurement of the dew point is one of the methods that allow the determination of the relative humidity of the air. For this purpose, for example, a polished metal surface subjected to a cooling process, in contact with the moist air, may be used. If, for example, the ambient temperature is 30 °C and the dew point temperature is 10 °C, the original vapor pressure will be 1.23 kPa (saturation pressure at the dew point temperature), the saturation pressure will be of 4.24 kPa and the relative humidity will be 1.23/4.24 or 29%.

Another process for evaluating the relative humidity of the air, consists of using an aspiration psychrometer, (Appendix A1) whose principle is based on the measurement of two temperatures, dry and wet, by applying appropriate sensors, for example, thermocouples, under normal conditions and soaking in gauze, dipped in

Fig. A2.10 Relationship between dry temperature, wet temperature, equivalent temperature and dew point (after Monteith and Unsworth 2013)

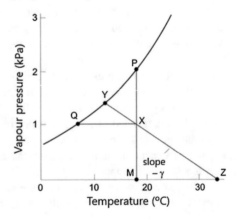

a container of water. If the air is not saturated, part of the gauze water will evaporate, and its cooling, due to evaporation, is recorded continuously by the temperature of the respective thermocouple (wet temperature). The system, which is generally tubular, is isolated from the ambient radiation, for example by a silver foil, and subjected to a steady flow of suction of indoor air by a fan. The difference between the two temperatures is a function of the relative humidity of the air.

It can be demonstrated (Monteith and Unsworth 1990) that the dry temperature, T_{dry} recorded by the thermocouple, is related to the surface temperature of the tubular system exposed to direct radiation, T_s and to the current air temperature T:

$$T_{dry} = \frac{r_H T_s + rT}{r_H + r_r} \qquad (A2.51)$$

where r_H the aerodynamic resistance to heat transfer by convection (sensible heat, Chaps. 2 and 6) and r_r the resistance to heat transfer by longwave radiation, of the order of 300 sm^{-1} at 25° C. By Eq. (A2.51) the measured dry temperature is a weighted average between the current air temperature and the temperature of the thermocouple.

The radiative heat transfer by radiation comes:

$$\tau_r \approx \frac{\rho c_p}{4\sigma T^3} \qquad (A2.52)$$

where σ is the Steffan-Boltzmann constant (Chap. 6).

In practice, the dry and wet temperatures measurement system is optimized by achieving that r_H is much lower than r_r, either by adequate ventilation (speed of the order of 3 ms^{-1}) or using thermocouples, with the smallest possible junction, or even, by using insulation or white paint, on the system surface.

Regarding the wet temperature, it should be noted that the temperature measured by the thermocouple with a wet gauze, approximates the theoretical concept of thermodynamic wet temperature (Monteith and Unsworth 1990).

This theoretical concept can be visualized by considering an isolated system consisting of a closed container with a sample of pure water in a natural air environment. This sample of unsaturated air at an initial temperature T_i, at a vapor pressure e, and a total pressure p, will humify until saturation is reached at saturation partial pressure, at a temperature T_s, lower than T_i (Monteith and Unsworth 1991). In this context it is possible to verify that:

$$e = e_s(T_s) - (c_p p / L\varepsilon)(T_i - T_s) \qquad (A2.53)$$

with the $(c_p p / L\varepsilon)$ term being the psychrometric constant, γ, (Chap. 4) that is about 66 PaK^{-1} at 0 °C or 67 PaK^{-1} at 20 °C.

The rate of increase of $e_s(T)$ with temperature, Δ, is another important parameter in environmental physics, that is given by Eq. (A2.54):

$$\Delta = LM_w e_s(T)/RT^2 \qquad (A2.54)$$

where R, is the perfect gas constant and M_w the molecular weight of water. Eq. (A2.54), is valid up to ambient temperatures of about 40 °C. The rate, Δ, is about 6.5% per °C, between 0 and 30 °C (Monteith and Unsworth 1991).

With an aspiration psychrometer under real conditions, the vapor pressure in the air is related to the wet temperature, T_{wet}, and the dry temperature, T_{dry}, as shown by Eq. (A2.55).

$$e = e_s(T_{wet}) - \Delta(T_{dry} - T_{wet}) \tag{A2.55}$$

where e is the vapour pressure of the air at the dry temperature. The value of e_s (T_{wet}), (or vapor pressure at wet temperature) is obtained from the psychrometric diagram, or by the empiric polynomial Eq. (A2.64) below.

On the other hand, considering the variation of the saturation pressure with the air temperature, Δ, at the average temperature between T_h and T_s, it will come:

$$e(T_{wet}) = e_s(T_{dry}) - \Delta(T_{dry} - T_{wet}) \tag{A2.56}$$

By substituting Eq. (A2.56) in Eq. (2.53) we have:

$$e = e_s(T_{dry}) - \Delta(T_{dry} - T_{wet}) - \gamma(T_{dry} - T_{wet}) = e_s(T_{dry}) - (\Delta + \gamma)(T_{dry} - T_{wet}) \tag{A2.57}$$

Defining the saturation deficit, D, and the temperature difference between the dry and wet thermometer, as B, as expressed in Eqs. (A2.58) and (A2.59):

$$D = e_s(T_{dry}) - e \tag{A2.58}$$

and:

$$B = (T_{dry} - T_{wet}) \tag{A2.59}$$

From Eq. (A2.57) we have that:

$$D \approx (\Delta + \gamma)B \tag{A2.60}$$

In a psychrometric chart (Fig. A2.10) wherein the vapour pressure (kPa) is shown in ordinates and the temperature in abscissa (°C), the line QYP represent the relationship between the vapour pressure of saturated air and temperature, in a given temperature range. The line joining a point X, with coordinates (T, e) representing T_{dry} and respective vapour pressure, with a point Y with coordinates $(T_{wet}, e_s(T_{wet}))$ will be given by Eq. (A2.61):

$$e - e_s(T_{wet}) = -\gamma(T_{dry} - T_{wet}) \tag{A2.61}$$

So, the wet temperature of an air sample can be obtained from the dry temperature, at the interception point of the saturation pressure curve with a line with slope $-\gamma$, passing through a coordinate point (T, e). The intercept of the straight line with the abscissa (point Z in Fig. A2.10), is the equivalent temperature T_e. The coordinates of this point are $(T_e, 0)$. In this figure, the line XQ, corresponds to the dew point temperature at point Q and the line XP, corresponds to the saturation temperature at point P.

The XY line in Fig. A2.10, represents the adiabatic evaporation curve, corresponding to the temperature and vapor pressure changes of a given air sample under adiabatic conditions, i.e. without any heat exchanges with the outside environment.

The equation relating the temperature of a given air sample at a lower vapor pressure than the saturation pressure with the equivalent temperature T_s in line segment XZ, is expressed by:

$$T_e = T + (e/\gamma) \tag{A2.62}$$

The corresponding equation, relating a point of the saturation curve with the corresponding equivalent temperature in line segment YZ is expressed by:

$$T_e = T_{wet} + e_s(T_{wet})/y \tag{A2.63}$$

A2.5.2 Empirical Approach

An alternative approach to calculating the relative humidity of air from dry and wet temperatures using the aspiration psychrometer is based on the application of successive empirical polynomial equations.

The saturation water vapor pressure in the air, at wet temperature, can be calculated using the Eq. (A2.64) below:

$$e_s(T_{wet}) = \left(6.209 * 10^{-5}\right)T_{wet}^3 + \left(2.188 * 10^{-4}\right)T_{wet}^2 + \left(6.319 * 10^{-2}\right)T_{wet} + 0.524 \tag{A2.64}$$

As for the relationship between the absolute humidity of the air χ, for the wet T_{wet} and dry T_{dry} temperatures, we have:

$$\chi_{dry} = \chi_{sTwet} - \rho c_p \left(\frac{\alpha}{D_{if}}\right)^{\frac{2}{3}} \left(T_{dry} - T_{wet}\right)L \tag{A2.65}$$

where D_{if} is the air mass diffusivity, α the thermal diffusivity, and χ_{sTwet} is the absolute saturation humidity, at the wet temperature.

The χ_{sTwet} can be obtained from the following expression:

$$\chi_{STwet} = \left(3.688 * 10^{-7}\right)T_{wet}^3 + \left(3.779 * 10^{-6}\right)T_{wet}^2 + \left(4.221 * 10^{-4}\right)T_{wet} + \left(4.42 * 10^{-3}\right) \tag{A2.66}$$

All other air properties can be obtained from the average temperature T_{wet} and T_{dry} (equals to $(T_{wet} + T_{dry})/2$) using the Eqs. (A2.67)–(A2.71), described below.

The air density (kgm^{-3}) as a function of temperature, can be obtained by the equation:

$$\rho(T) = (2.667 * 10^{-6})T^3 - (1.6 * 10^{-4})T^2 - (1.067 * 10^{-3})T + 1.268 \quad (A2.67)$$

The thermal diffusivity of the air (m^2s^{-1}) may be calculated by:

$$\alpha(T) = -(1.333 * 10^{-10})T^3 + (8 * 10^{-9})T^2 - (1.667 * 10^{-8})T + 1.97 * 10^{-5}$$
$$(A2.68)$$

The mass diffusivity of the air (m^2/s^{-1}), as a function of mean temperature, is calculated by:

$$D(T) = (2.667 * 10^{-10})T^3 - (1.8 * 10^{-9})T^2 + (5.433 * 10^{-7})T + 1.84 * 10^{-5}$$
$$(A2.69)$$

The specific heat at constant air pressure $(kJkg^{-1})$, as a function of mean temperature, may be obtained by:

$$c_p(T) = (5.8 * 10^{-7})T^2 + (5.854 * 10^{-6})T + 1.00512 \quad (A2.70)$$

and finally, the latent heat of vaporization of the air $(kJkg^{-1})$, also as a function of mean temperature, may be evaluated by:

$$L(T) = -2.361T + 2501.55 \quad (A2.71)$$

Relative humidity is then given by the ratio between $\chi_{(Tdry)}$ and $\chi_{s(Thum)}$ calculate respectively from Eqs. (A2.65) and (A2.66).

References

Asimov, I. (1993). *Understanding physics*. Barnes & Noble, 768 pp.

Campbell, G. S. (1997). *Biophysical measurements and instrumentation. A Laboratory Manual for Environmental Biophysics*. International Workshop on Biophysical and Physiological Measurements in Agriculture, Forestry and Environmental Sciences, IPB, Bragança.

Connor, F. R. (1978). *Sinais*. Interciência Editora Lda., 110 pp.

Foken, T. (2008). *Micrometeorology*. Springer, Berlin, 306 pp.

Foken, T. (2017). *Micrometeorology*, 2nd ed., Springer, Berlin, 362 pp.

Fox, R. W., & McDonald A. T. (1985). *Introduction to fluid mechanics*. Wiley, 742 pp.

Giancoli, C. D. (2000). *Physics for scientists and engineers with modern physics*, Prentice Hall, 1172 pp.

Monteith, J. L., & Unsworth, M. H. (1991). *Principles of environmental physics*, 2nd Ed., Edward Arnold, 291 pp.

Monteith, J. L., & Unsworth, M. H. (2013). *Principles of environmental physics*, 4th Ed., Academic Press, Oxford, 403 pp.

Oke, T. R. (1992). *Boundary layer climates*, 2nd ed., Routledge, 435 pp.

Ohmura, A., Duton, E., Forgen, B., Greuell, W., Fröhlich, C., Gilgen, H., Hegner, H., Heimo, A., König-Langlo, G., McArthur, B., Müller, G., Philipona, R., Pinker, R., Whitlock, CH., Dehne, K., & Wilde, M. (1998). Baseline Surface Network (BSRN/WCRP): New precision radiometry for climate research. *Bulletin of American Meteorological Society* 79: 2115-2136.

Paw U. K. T. (1995). Instrumentation II (Temperature, Humidity and Radiation Sensors), Lecture 14. In *Advanced Short Course on Biometeorology and Micrometeorology, Università di Sassari*, Italia, 1995.

Bibliography

Amiro, B. D. (1990). Drag coefficients and turbulence spectra within three boreal forest canopies. *Boundary Layer Meteorology, 52,* 227–246.

Anthoni, P. M., Beverly, E. L., Unsworth, M. H., & Vong, R. J. (2000). Variation of Net radiation over heterogeneous surfaces: Measurements and simulation in a Juniper-Sagebrush ecosystem. *Agricultural and Forest Meteorology, 102,* 275–286.

Aston, A. R. (1985). Heat storage in a young Eucalypt forest. *Agricultural and Forest Meteorology, 35,* 281–297.

Baldocchi, D. D., & Hutchinson, B. A. (1987). Turbulence in an almond orchard: Vertical variations in turbulent statistics. *Boundary Layer Meteorology, 40,* 127–146.

Baldocchi, D. D., Luxmoore, R. L., & Hatfieldf, J. L. (1991). Discerning the Forest from the Trees: An Essay on Scaling Canopy Stomatal Conductance. *Agricultural and Forest Meteorology, 54,* 197–226.

Baldocchi, D. D., UT. Paw, K., Shaw, R. H., & Snyder, R. L. (1995). *Advanced short course on biometeorology and micrometeorology. CNR, CIHEAM, EU.* Università di Sassari, Italia.

Brown, R. A. (1991). Fluid mechanics in the atmosphere. *International Geophysics Series,* Vol. 47, Academic Press, 489 pp.

Brunet, Y. (1999). *Turbulence et Transport.* Oeiras: Seminário sobre Transferências Hídricas em Cobertos Vegetais Descontínuos.

Climate Change, Wikipedia: https://en.wikipedia.org/wiki/Climate_change.Accessed in April 2019.

Climate Change Adaptation, Wikipedia, https://en.wikipedia.org/wiki/Climate_ change_adaptation, (accessed in April 2019).

Donmerge, F. (1997). Environmental change and the mediterranean. (UNU Lectures, 16, 17).

Duffie, J. A., & Beckman, W. A. (1991). *Solar engineering of thermal processes* (2nd ed., p. 910). New York: Wiley.

Grunderbeeck, P., & Tourre, Y., *Mediterranean Basin: Climate Change and Impacts During the 21st Century.* Part 1, Chapter 1, pp. 1.1–1.64. In *Climate Change and Energy in the Mediterranean.* Plan Bleu, Regional Activity Center, Sophia Antipolis. https://www.eib.org/attachments/country/climate_change_energy_mediterranean_en.pdf. Accessed April 2019.

Hinze, J. O. (1959). Turbulence. *An Introduction to Its Mechanism and Theory.* McGraw-Hill Book Company.

Högström, U. (1988). Non-dimensional wind and temperature profiles in the atmospheric surface layer: A re-evaluation. *Boundary Layer Meteorology, 42,* 58–78.

IPCC, 2007 (AR4): Climate Change. (2007). *The Physical Science Basis. Contribution of Working Group I to the Fourth Assessment Report of the Intergovernmental Panel on Climate Change.* In S. Solomon, D. Qin, M. Manning, Z. Chen, M. Marquis, K. B. Averyt, M. Tignor, & H. L. Miller (Eds.). Cambridge University Press, 996 pp.

The Editor(s) (if applicable) and The Author(s), under exclusive license to Springer Nature Switzerland AG 2021
A. Rodrigues et al., *Fundamental Principles of Environmental Physics,*
https://doi.org/10.1007/978-3-030-69025-0

IPCC, 2014 (AR5): Climate Change. (2014). *Synthesis Report. Contribution of Working Groups I, II and III to the Fifth Assessment Report of the Intergovernmental Panel on Climate Change.* In R. K. Pachauri, & L. A. Meyer (eds.). IPCC, Geneva, Switzerland, 151 pp.

Jarvis, P. G., James, G. B., & Landsberg, J. J. (1976). *Coniferous Forest*, pp. 171–240. In J. L. Monteith (Ed.). *Vegetation and Atmosphere*, Vol. II. Academic Press.

Kaimal, J. C. (1991). Time series tapering for short data samples. *Boundary Layer Meteorology, 57,* 187–194.

Kelliher, F. M., Hollinger, D. Y., Schiltze, E. D., Vygodskaya, N. N., Byers, J. N., Byers, J. N., et al. (1997). Evaporation from an eastern siberian larch forest. *Agricultural and Forest Meteorology, 85,* 135–147.

Liu, X., Tsukamoto, O., Oikawa, T., & Ohtaki, E. (1998). A study of correlations of scalar quantities in the atmospheric surface layer. *Boundary Layer Meteorology, 87,* 499–508.

Lynn, P. A. (1985). *An introduction to analysis and processing of signals.* Macmillan Publishers Ltd., 277 pp.

Moore, C. J., & Fisch, G. (1986). Estimating heat storage in amazonian tropical forest. *Agricultural and Forest Meteorology, 38,* 147–169.

Raabe, A. (1983). On the relation between the drag coefficient and the fetch above the sea in the case of the off-shore wind in the near shore zone. *Journal of Meteorolgy, 41,* 251–261.

Rannik, Ü., & Vesala, T. (1999). Autoregressive filtering versus linear detrending in estimation of fluxes by the eddy covariance method. *Boundary Layer Meteorology, 91,* 259–280.

Rannik, Ü., Aubinet, M., Kurbanmuradov, O., Sabelfeld, K. K., Markkanen, T., & Vesala, T. (2000). Footprint analysis for measurements over a heterogeneous forest. *Boundary LayerMeteorology, 97,* 137–166.

Raupach, M. R. (1989). A practical lagrangian method for relating scalar concentrations in vegetation canopies. *Quarterly Journal of the Royal Meteorological Society, 115,* 609–632.

Raupach, M. R., Thom, A. S., & Edwards, I. (1980). A wind tunnel study of turbulent flow close to regularly arrayed plant canopies. *Boundary Layer Meteorology, 18,* 373–397.

Shaw, R. (1985). *On diffusive and dispersive fluxes in forest canopies,* pp. 407-419. In B.A. Hutchinson, & B. B. Hichs (ed.). *The Forest-Atmosphere Interaction.* Reidel Publishing Company.

Shaw, R. (1995f). *Canopy layer micrometeorology.* Lecture 18. In *Advanced Short Course on Biometeorology and Micrometeorology.* Università di Sassari, Italia.

Shaw, R. (1995g). *Mathematical Models of Surface Layer Processes I.* Lecture 19, in: *Advanced Short Course on Biometeorology and Micrometeorology.* Universidade de Sassari, Itália.

Shaw, R. (1995h). *Mathematical Models of Surface Layer Processes II.* Lecture 20, in: *Advanced Short Course on Biometeorology and Micrometeorology,* Università di Sassari, Italia.

Štochlová, P., Novotná, K., Costa, M., & Rodrigues, A. (2019). Biomass production of poplar shot rotation coppice over five and six rotations and its aptitude as a fuel. *Biomass and Bioenergy, 122,* 183–192.

Tanner, B. D., Swiatek, E., & Greene, J. P. (1993). Density fluctuations and use of the Krypton Hygrometer in surface flux measurements, pp. 21–23. In *Management of Irrigation and Drainage Systems.* (Workshop of the Irrigation and Drainage Div./ASCE, July 21–23, Park City, Utah.

Thom, A. S., Stewart, J. B., Oliver, H. R., & Gash, J. H. C. (1975). Comparison of aerodynamic and energy budget estimates of fluxes over a pine forest. *Quarterly Journal of the Royal Meteorological Society, 101,* 93–105.

Verma, B. V., Baldocchi, D. D., Anderson, D. A., Matt, D. R., & Clement, R. J. (1986). Eddy Fluxes of CO2, Eddy fluxes of CO_2, water vapor and sensible heat over a deciduous forest. *Boundary Layer Meteorology, 36,* 71–91.

Weber, K., & Quicker, P. (2018). Properties of biochar. *Fuel, 217,* 240–261.

Wieringa, J. (1980). A revaluation of the Kansas Mast influence on measurements of stress and cup anemometer over speeding. *Boundary Layer Meteorology, 18,* 411–430.

Woo, H. G. C., Peterka, J. A., & Cermak, J. E. (1977). *Wind-tunnel Measurements in the Wakes of Structures.* NASA Contract. Rep. No. 2806, Colorado State University. 226 pp.

Index

A

Acceleration/speedup factor, 150, 152
Advection, 39, 46, 49, 52–54, 56, 59, 101
Aerodynamic drag, 21, 22
Aerodynamic method, 27, 30
Aerodynamic resistance, 21, 23, 24
Aerosols, 271, 272, 281, 283, 290, 294, 296, 297
Anthropogenic contribution, 268, 269, 273, 279, 281, 282, 287, 288, 290, 293, 303, 304
Apparent equivalent temperature, 205
Ascending long wavelength radiation, 196
Atmospheric convergence and/or subsidence, 3, 8
Atmospheric motion scales, 1, 3
Atmospheric turbulence, 34–37, 39, 55, 60
Atmospheric turbulence (general characterization), 2, 3, 9, 10
Atmospheric waves, 1, 2, 10
Autocorrelation, 71, 72

B

Beer's Law, 187, 191, 202
Bending moment, 109
Big-leaf approach, 118
Biochar, 300–304
Boussinesq approximation, 44
Bowen ratio, 106, 115, 117, 123
Brownian diffusion, 216, 217
Brunt-Väisälä frequency, 111, 147
Buoyancy, 168, 179, 213, 225, 227

C

Canopy resistance, 118, 120–124
Carbon dioxide, 268, 271, 273, 275, 276, 287, 292, 298, 301–304, 342

Carbon fluxes partition, 80, 100
Carbon leakage, 301
Carbon sequestration in forest stands, 124, 125, 127, 128
Climate change, 267–273, 278, 281, 282, 291–293, 295, 297, 301
Climate risks, 268, 282, 287, 289, 290
Closed path sensors, 83, 84, 90, 91
Closure of turbulent equations, 34, 60, 61, 63
Coherent structures, 106, 112
Convection coefficient, 173, 174
Convection in cylindrical bodies, 171, 176, 213
Convection in flat surfaces, 168–172, 174–176
Convection in spherical bodies, 176
Coordinate rotation, 80, 86, 87, 93, 100
Cork oak stand, 117, 118, 120
Cospectral analysis, 72, 77, 91
Cross correlation, 70, 71
Crosstalk effect, 92

D

Decoupling coefficient, 118–120, 128
Descending long wavelength radiation, 196
Diffuse radiation, 188, 190, 192–194, 197, 200, 203
Diffusivity coefficient, 16, 20, 25
Dimensional and temporal scales of atmospheric turbulence, 34, 37, 49
Direct radiation, 183, 188, 190, 192, 194
Drag force, 18, 22
Drag/form drag, 107–110, 114
Dry adiabatic gradient, 4, 5
Dust storms, 217
Dynamic similarity, 80, 94, 100

E

Economic blocks, 276

The Editor(s) (if applicable) and The Author(s), under exclusive license
to Springer Nature Switzerland AG 2021
A. Rodrigues et al., *Fundamental Principles of Environmental Physics*,
https://doi.org/10.1007/978-3-030-69025-0

Effective temperature, 205–207
Ejection, 112, 122
Emissive power, 182, 184, 185
Emissivity, 167, 180, 182, 184, 196, 197
Emittance, 183
Empirical functions for atmospheric
    turbulence, 58, 60, 76
Equilibrium evapotranspiration, 119, 121–123
Equivalent temperature, 117
Eucalypts stand, 120, 126–129
Euler formulas, 64, 70
Eulerian and Kolmogorov length scales, 74
Exponential zone in airflow profile, 107, 110
Extreme events, 268, 270, 282, 283, 286, 288,
    289, 297, 298

F
Fast Fourier Transform (FFT), 70
Fetch, 137
Fifth IPCC Report (AR5) assessment, 269,
    270, 273, 278–282, 291, 298, 300
Flow over buildings, 139
Flow over hills, 144, 148, 150, 155
Fohen and Bora winds, 154, 155
Footprint or fetch, 80, 89, 97, 98
Forced convection, 159, 168, 172–179, 209,
    210
Forest coppice, 126–128
Fourier's Law, 160, 165
Fourier analysis, 63
Fourth IPCC Report (AR4) assessment, 269,
    270, 278, 281, 292, 294, 296, 299
Free convection, 177–179, 210
Frictional drag, 1, 3
Friction delay, 15
Froude number, 233

G
Gap-filling procedure, 80, 101
Gas analyser, 83
Global energy budget, 269, 271
Global warming, 268, 269, 271, 272, 278–280,
    283–286, 291, 295, 297
GPP, 125–128
Grashof number, 178, 179, 210
Greenhouse radiative properties, 201

H
Hardwood stand, 111, 114
Harmonic change, 163, 206
Heatwave, 280, 282–290, 292, 293, 298

High frequency peaks, 80, 89, 90, 100
Holm oak stand, 126, 128
Hydraulic jump, 147, 155, 156

I
Impaction, 215–217
Imposed evapotranspiration, 119, 120, 126
Inertial sublayer, 15
Inertial subrange, 73–75, 77
Internal boundary layer, 133–136
Internal climate variability, 286
Irradiance, 183, 187, 190, 191, 193, 202, 203

K
Katabatic winds, 153–155
Kirchhoff's Law, 182, 188

L
Laminar boundary layer, 15, 16, 21
Land use, 279, 295
Lewis number, 210
Logarithmic zone in airflow profile, 107
Low frequency corrections, 89–91

M
Mean free path of gas molecules, 212
Mediterranean basin, 285, 292–297, 299
Mixed layer, 6, 8–10
Monin-Obhukov length, 26, 29

N
Navier-Stokes equations, 40, 44
NEE, 125–128
Net radiation, 187, 203–207
Night-time carbon storage, 80, 95, 99, 100
Numerical simulation for particle entrainment,
    223
Nusselt number, 172–176, 179, 209, 210

O
Open path sensors, 81, 84–86

P
Particle creeping, 218–220
Particle mass transfer, 211, 213
Particle rebound, 217, 223
Particle Reynolds number, 212, 214
Penman-Monteith equation, 106, 118–120
Phase diagram, 65
Pine stand, 108, 121–123, 127, 128
Potential evapotranspiration, 289, 297

Potential temperature, 5, 8, 9
Potential temperature profile, 111, 119
Prandtl number, 170, 174, 210
Precipitation, 267, 270, 280–283, 286, 287, 289, 290, 292, 293, 295–299

R
Radiation intensity, 167, 184, 186, 202
Radiative absorption, 188, 191, 195
Radiative Forcing (RF), 272
Radiative reflectivity, 181, 183, 198
Radiative shape factor, 183, 186
Radiative transmissivity, 181, 190, 192, 194, 198
Radiative window, 192
Ramp change, 204, 206–208
Recalcitrant biochar, 302
Regional tendencies, 281, 289, 293–297
Relative motion of particles, 211, 212
Representative Concentration Pathway (RCP), 279–281, 284, 288, 291, 300
Residual layer, 9
Return period, 285, 290
Reynolds number, 34, 169, 172, 174, 175, 209, 213, 228
Richardson number, 25, 26
Roughness length, 17, 19, 31
Roughness sublayer, 14, 105, 106, 110, 114

S
Saltation, 217–221
Saturation deficit, 122, 123
Schmidt number, 209
Schotanus corrections, 80, 83, 100
Sedimentation velocity, 213–215
Sediment transport, 218, 224–226, 228, 233
Sherwood number, 209, 210
Shields parameter, 225, 228, 229, 231, 232
Skin friction, 107, 108
Softwood stands, 112, 124
Soil loading capacity, 303, 304
Solar declination, 189
Solar height, 189, 200
Solid angle, 184
Sonic anemometry, 80–83, 86, 89, 93
Spectral analysis, 35, 63, 65, 66, 72, 73, 77, 82
Spectral power, 72, 75–78
Specular reflection, 183, 198, 200
Stability functions, 24, 25, 27
Stationary conduction, 160, 162
Stefan-Boltzmann's Law, 167, 182
Step change, 204, 206, 207

Stokes law, 212, 213
Stokes number, 215
Stopping distance, 215
Stream power, 224–226, 231, 233
Supercritical flow, 233
Surface layer, 13–17, 21
Surface roughness, 134–138, 152
Synthesis and analysis equations, 65

T
Taylor hypothesis, 38, 39, 73, 93
Thermal conduction, 160, 171, 204
Thermal conductivity coefficient, 160, 162, 165, 171–173
Thermal damping diffusivity, 162, 167
Thermal damping layer, 165
Thermal internal boundary layer, 136
Thermal stability, 19, 24–27
Thermal stability influence on atmospheric flow, 134, 136, 138, 146–148, 150, 153, 154
Thermal storage, 167, 206
Time constant, 204, 206–208, 214
Time constant for evapotranspiration regime, 120
Top atmosphere boundary layer, 4, 8, 9
Top inversion, 6, 9
Total Ecosystem Respiration (TER), 125, 126
Transient conduction, 160–162
Turbulent dissipation, 34, 37, 48, 49, 51, 54, 55, 57, 59, 62, 74, 75
Turbulent flow, 34, 40, 44–47, 50, 60, 62, 72
Turbulent intensity, 35, 36, 39, 46, 54
Turbulent isotropy, 48, 50, 51, 53, 55, 62, 74, 75
Turbulent Kinetic Energy (TKE), 34, 35, 37, 48, 49, 54–56, 58–60, 62, 73, 74, 114, 115
Turning moment, 109

U
Understory strata, 122, 123, 125
Urban boundary layer, 137, 138

V
Vertical profiles in the internal boundary layer, 138, 144

W
Wake turbulence, 109, 114, 115
Weather forecast, 271, 279–281, 284, 285, 291, 292, 294, 296, 297, 299

Wet deposition, 216
Wien's Law, 181
Wind tunnel experimentation, 107–109, 113
WPL correction, 62, 80, 84–86, 100

Z
Zenith angle, 183, 184, 189, 190, 192, 194,
    197, 200, 203